AF393624

Advanced Topics
in Materials Science
and Engineering

Advanced Topics in Materials Science and Engineering

Edited by

J. L. Morán-López

Universidad Autonoma de San Luis Potosí
San Luis Potosí, S.L.P., Mexico

and

J. M. Sanchez

The University of Texas at Austin
Austin, Texas

Springer Science+Business Media, LLC

Library of Congress Cataloging-in-Publication Data

Advanced topics in materials science and engineering / edited by J.L.
Morán-López and J. M. Sanchez.
       p.   cm.
    "Proceedings of the First Mexico-U.S.A. Symposium on Materials
Science and Engineering, held September 24-27, 1991, in Ixtapa,
Guerrero, Mexico"--CIP t.p. verso.
    Includes bibliographical references and index.
    ISBN 978-1-4613-6230-2   ISBN 978-1-4615-2842-5 (eBook)
    DOI 10.1007/978-1-4615-2842-5
    1. Materials science--Congresses.   I. Morán-López, J. L. (José
L.), 1950- II. Sanchez, Juan M. III. Mexico-U.S.A. Symposium on
Materials Science and Engineering (1st : 1991 : Ixtapa, Mexico)
TA401.3.A38 1993
620.1'1--dc20   93-12843
                                                              CIP

Proceedings of the First Mexico – U.S.A. Symposium on Materials Science and Engineering , held
September 24 – 27, 1991, in Ixtapa, Guerrero, Mexico

ISBN 978-1-4613-6230-2

©1993 Springer Science+Business Media New York
Originally published by Plenum Press in 1993
Softcover reprint of the hardcover 1st edition 1993

To the memory of Francisco Mejía-Lira and John K. Tien

Preface

This volume contains the papers presented at the First México-U.S.A. Symposium on Materials Sciences and Engineering held in Ixtapa, Guerrero, México, during September 24–27, 1991. The conference was conceived with the primary objective of increasing the close ties between scientists and engineers in both México and the U.S. with an interest in materials. The conference itself would have not taken place without the drive, determination and technical knowledge of John K. Tien of the University of Texas at Austin and of Francisco Mejía Lira of the Universidad de San Luis Potosí. This book is dedicated to their memory.

The event brought together materials scientists and engineers with interests in a broad range of subjects in the processing, characterization and properties of advanced materials. Several papers were dedicated to structural materials ranging from ferrous alloys to intemetallics, ceramics and composites. The presentation covered properties, processing, and factors that control their use, such as fatigue and corrosion. Other materials and properties were also explored by U.S. and Mexican participants. Several papers dealt with the characterization and properties of magnetics, optical and superconductor materials, nanostructured materials, as well as with computational and theoretical aspects likely to impact future materials research and development.

In addition to the technical presentations, the Conference provided a forum for numerous discussions and on-going materials research programs carried out in each country individually and cooperatively. These discussions were suplemented by oral presentations and by two plenary panel discussions on Cooperative Programs and on Materials Science and Engineering Education. The Symposium was significantly enriched by participation of different Agencies and Organizations from México and the U.S., which included representatives from Consejo Nacional de Ciencia y Tecnología, Secretaría de Educación Pública, National Science Foundation, Air Force Office of Scientific Research, National Institute of Standards and Technology, The Metallurgical Society of AIME, National Materials Advisory Board (National Research Council), ASM International, and the Materials Research Society (México).

We thank the participants for their interest in all aspects of the Symposium. We would also like to acknowledge the sponsorship of Consejo Nacional de Ciencia y Tecnología, Secretaría de Educación Pública, National Science Foundation, and the Organization of American States.

Finally, we wish to thank Dr. J.M. Montejano-Carrizales for his role in the technical review, organization and editing of this book.

J.L. Morán-López
San Luis Potosí, S.L.P. México

J.M. Sanchez
Austin, Texas

October, 1992

Contents

Advances in Superalloys and High Temperature Intermetallics

J.K. Tien, E.P. Barth, A.M. Gyurko, M.W. Kopp,
A.B. Rodriguez, and G.E. Vignoul

Strategic Materials R & D Laboratory
The University of Texas at Austin
Austin, Texas, 78712
U.S.A.

I. Introduction

High temperature structural materials have been a major determining factor for economic and military strength as a result of their critical nature in aircraft and aerospace systems. In order to meet the demands imposed by both military and commercial end users, the development and commercialization of advanced high temperature structural materials must anticipate the future requirements of advanced propulsion technology. To date, nickel and iron-base superalloys have been the materials of choice in critical high temperature structural applications. The evolution of these materials, as measured in terms of increased effective use temperature, has largely kept pace with the demands of the jet turbine industry. However, as a consequence of the conservative nature of the jet turbine industry, no truly new class of high temperature structural materials have found acceptance within the jet turbine engine since its appearance.

While significant effort and resources have been devoted to the development of a new generation of high temperature structural materials, these endeavors have largely been focused on improving the performance of existing technologies. The classic example of this trend has been the extensive work devoted to the research and development of monolithic Ni_3Al as a replacement for the structural superalloys, which already contain upwards of 80% of this phase as strengthening precipitates. However, it has long been known that this particular intermetallic compound, either by itself or alloyed with ternary or quaternary elements, offers little advantage over that of the superalloys. Indeed, the beneficial effects of boron and hafnium alloying additions have long been recognized in the superalloy literature.[1-3] Further, even with alloying additions that improve ductility, the level of ductility produced is generally considered insuffi-

Advanced Topics in Materials Science and Engineering, Edited by
J.L. Morán-López and J.M. Sanchez, Plenum Press, New York, 1993

cient. While this effort on Ni_3Al has resulted in a body of scientific knowledge, its future as a monolithic intermetallic is not in the aerospace sector. However, it should be noted that some of the titanium, iron and nickel aluminides, do offer certain near term gains for special turbine applications.[4]

Marginal improvements in current high temperature material performance may be sufficient to carry the turbine industry through the next decade, but the driving force imposed by the desire for and needs of hypersonic flight, for example, implicitly demand a quantum leap in philosophy and technology. In essence, the future of high temperature structural materials is encompassed by intermetallics whose melting points exceed those of any being actively studied today and composite materials that are metallic-based, intermetallic-based, ceramic-based and carbon-based. While the brute-force temperature capabilities of carbon/carbon and other ceramic matrix composites make them the ultimate solution in terms of future high temperature applications, the problems imposed by service environment and manufacturing/engineering concerns will delay their use until they are satisfactorily addressed. Consequently, this paper will discuss the present and future of modern superalloys, high temperature intermetallics, and metallic and intermetallic matrix composites.

II. Superalloys

The continued evolution of superalloys has enabled these materials to fulfill the requirements of advancing jet turbine technology. While many of the improvements to date have involved the development of optimal microstructures, such as the advent of directionally solidified and single-crystal structures, and oxide dispersion strengthening, all of which have been addressed in Japanese national projects, it seems clear that further significant performance advances are most likely to come as a consequence of improved processing techniques.

The cleanliness of nickel-base superalloys has been shown to be highly critical due to the strong correlation of high temperature fatigue and fracture properties to inclusions introduced during primary melt processing. This strong defect sensitivity is pronounced for both cast and wrought materials. Dynamic fractures in superalloys are almost always initiated at inclusion sites, particularly when these sites are in proximity to free surfaces, and they can be affected by oxidizing atmospheres. It is well known that the fatigue endurance limits for superalloys are directly controlled by the quantity, size, and distribution of inclusions.

In the past decade, several improved refining processes have gained commercial acceptance for improving superalloy cleanliness. These have included the introduction of ceramic foam filters in the vacuum induction melting (VIM) process and the reintroduction of electroslag remelting (ESR). Perhaps the most promising of the emerging clean metal processes is electron beam, cold hearth refining (EBCHR). The remarkable success of EBCHR can be directly linked to avoiding melt contact with ceramic components and by removing both high and low density inclusions from the melt.

Recent studies have examined the effectiveness of EBCHR on Inconel 718.[5] In these studies Inconel 718 feedstock was remelted by means of EBCHR and the inclusion content and low-cycle fatigue (LCF) behavior of both the VIM feedstock and EBCHR material were examined. The cleaning effect of EBCHR was shown to be quite remarkable in that more that 95% of low density inclusions were removed, and all particles larger than 50 μm were eliminated. As a result, the interinclusion distance was increased by about an order of magnitude.

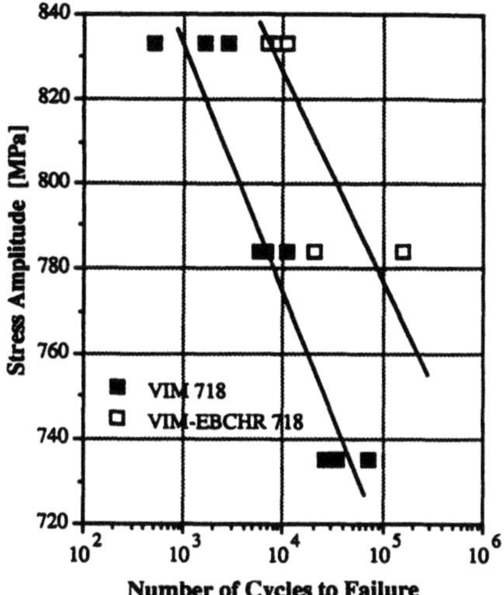

Figure 1. S-N curves obtained for low cycle fatigue of VIM Inconel 718 and VIM/EBCHR Inconel 718 (R= −1, 538°C).

Stress controlled LCF tests were performed at 538°C using a sinusoidal waveform (n=1 sec^{-1}, R= −1). The results of these tests show a dramatic increase in the fatigue life of the VIM/EBCHR material relative to the VIM feedstock material. S-N curves for these two materials are shown in Fig. 1. Clearly there is a dramatic statistical difference in the LCF life of clean and dirty Inconel 718. Specifically, the EBCHR material exhibited a life at a stress amplitude of 784 MPa that was about twice that of the VIM feedstock, and at 833 MPa this increase was about a factor of three when considering the worst EBCHR test material to the best VIM test material. On average, the LCF life of EBCHR material was six times that of the VIM material.

III. Intermetallics

As has been recognized for some time, ordered intermetallic compounds have a number of properties that make them intrinsically more appealing than other metallic or ceramic systems for high temperature use. The primary requirements for high temperature structural intermetallics, as with any structural material used at elevated temperature, are: 1) they have a high melting point, 2) they possess some degree of resistance to environmental degradation, 3) they maintain structural and chemical stability, and 4) they retain high specific mechanical properties. In addition to meeting these criteria, intermetallics are manufacturable by a variety of techniques and have better thermal conductivities than ceramics.

Melting point is a useful first approximation of the high temperature performance of a material, as various high temperature mechanical properties (*i.e.* strength, creep resistance) are limited by thermally assisted or diffusional processes and thus tend to scale with the melting point of the material. Therefore, the intermetallics can be crudely ranked in terms of their melting points to indicate their future applicability as high temperature structural materials. As may be seen in Fig. 2, metallic materi-

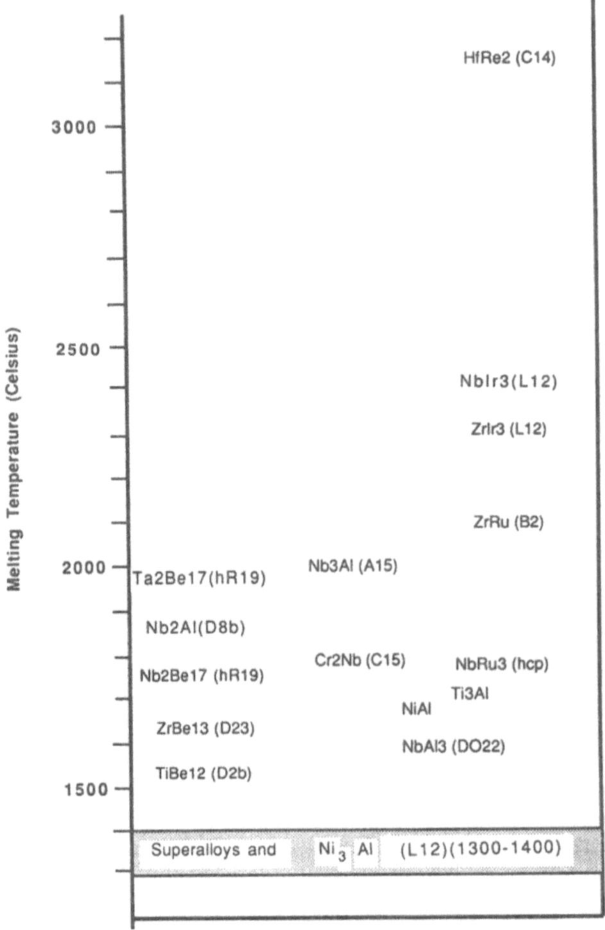

Figure 2. Melting points of selected intermetallics relative to superalloys.

als (intermetallics or otherwise) which are currently in use or being studied melt at temperatures much lower than 1650°C. The intermetallics in Fig. 2 may be roughly divided into two groups; those that fall in the temperature range up to 1650°C and those whose melting points extend to much higher temperatures.

To date the vast majority of research on developing structural intermetallics has been focussed on materials that fall into this lower temperature class, in particular iron, nickel, and titanium aluminides. Although many of the perennial complaints (low ductility, poor creep resistance, environmental problems) have been largely mitigated, the net benefit of implementing these materials is low when compared to the possibilities presented by other intermetallics that will require similar engineering considerations.

One class of these intermetallics that has garnered renewed recent interest in the 1650°C range is the niobium aluminides.[6] Indeed, there are significant current programs both in the U.S. and Japan investigating these intermetallics. Fig. 3 shows hardness *vs.* temperature data for two niobium aluminides compared to the much

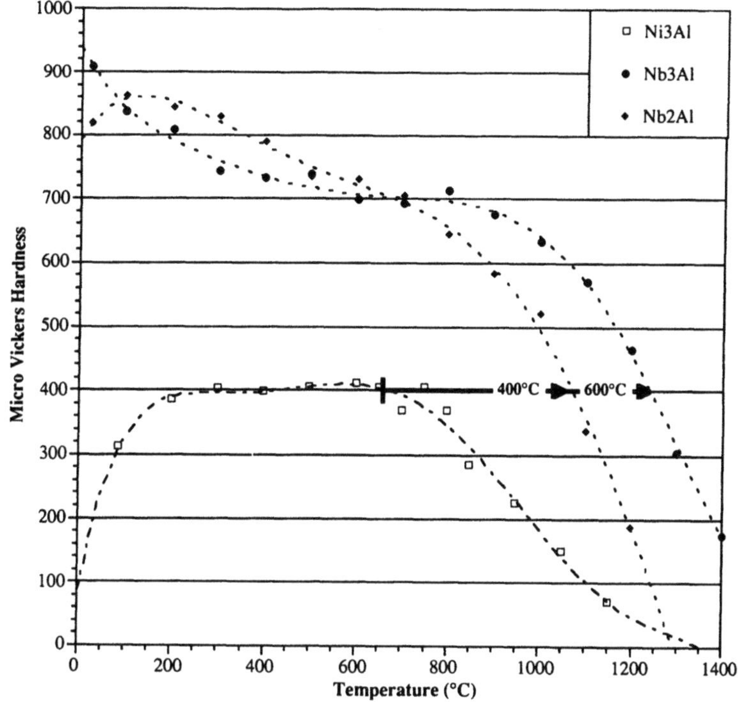

Figure 3. Comparison of hardness of Nb_2Al, Nb_3Al, and Ni_3Al *vs.* temperature.

studied Ni_3Al. Although these materials are clearly stronger in Ni_3Al's temperature regime, perhaps the most meaningful way to view this data is to compare use temperatures at a constant strength. As is illustrated, Nb_2Al yields an approximate 400°C increase and Nb_3Al yields about a 600°C benefit.

Many of the intermetallic compounds whose melting points exceed 1650°C belong to a group of intermetallics which are predicted on the basis of the Engel-Brewer phase stability theory. These compounds are predicted to be very stable.[7-9] For example, ZrRu has a free energy of formation of −21.5 kcal/g-atom, which is about three times the −7.4 kcal/g-atom value reported for Ni_3Al.[10] It should be noted that large negative free energies of formation may translate not only into very high melting points, but also into potentially high inherent oxidation resistance because the constituent compound atoms may prefer one-another more than they prefer oxygen. Of course, kinetics will be the determining factor with respect to this issue.

The Engel-Brewer compounds of particular interest should be the high symmetry $L1_2$ systems. It might be expected that these materials might combine the low temperature ductility necessary for fabrication with the anomalous strengthening behavior found in the favorite $L1_2$ Ni_3Al compound, but with a significant boost in high temperature capability. However, as illustrated in Fig. 4, this simplistic approach has not as yet been successful. This can be explained as follows. Sigli and Sanchez have established a relationship for the ordering energy of an *fcc*-based lattice using the cluster variation method (CVM).[11,12] Using this model, the ordering energy of

Ni$_3$Al was determined to be 0.075 eV. By comparison, the ordering energy of ZrIr$_3$ compares at 0.114 eV and that of NbIr$_3$ at 0.121 eV. The low ordering energy of Ni$_3$Al corresponds to a low antiphase boundary (APB) energy, which allows for the thermally activated dislocation locks necessary for anomalous yield behavior.[13] While it has yet to be conclusively established that high ordering energy corresponds to a high APB energy, the mechanical behavior exhibited by ZrIr$_3$ and NbIr$_3$ is consistent with a model that predicts that systems with a high APB energy will have superpartial dislocation cores that are planar and therefore glissile, resulting in little resistance to deformation.[14] However, while the high ordering energy of ZrIr$_3$ is a detriment to its yield strength, it is the source of its remarkable creep resistance. Microindentation creep studies have found a stress exponent of 18.3 and an activation energy of 467 kJ/mol for this intermetallic.[15] Thus, although this material is not particularly strong at elevated temperatures, it is very resistant to thermally assisted processes.

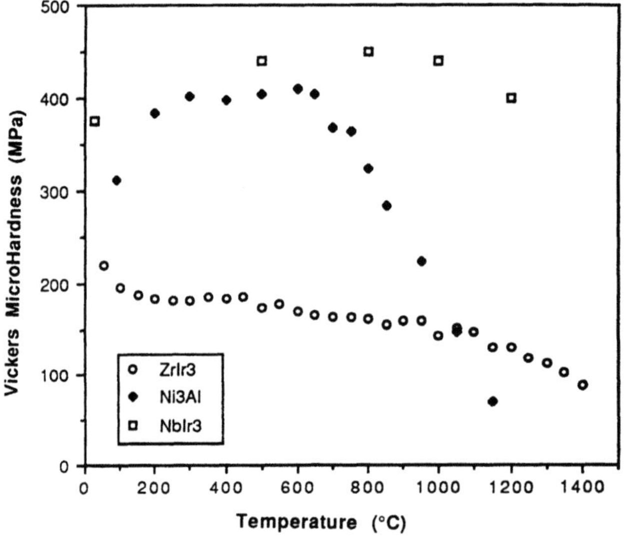

Figure 4. Comparison of hardness of ZrIr$_3$, NbIr$_3$, and Ni$_3$Al *vs.* temperature. 0.5

Of recent intense interest in the intermediate temperature regime are intermetallic beryllide materials. Several of the transition metal beryllides such as Nb$_2$Be$_{17}$ Ta$_2$Be$_{17}$, and ZrBe$_{13}$, for example, have properties that make them ideal for certain applications. Perhaps the most significant are their extremely low densities and high coefficients of thermal expansion. Although these materials have high s/E ratios they are rather brittle so their eventual incorporation into composite structures is the goal of the majority of current research. However, before realistic attempts for this can be attempted, basic monolithic study is required and is underway. As illustrated in Fig. 5, the high temperature strength shows the clear advantage that the beryllides

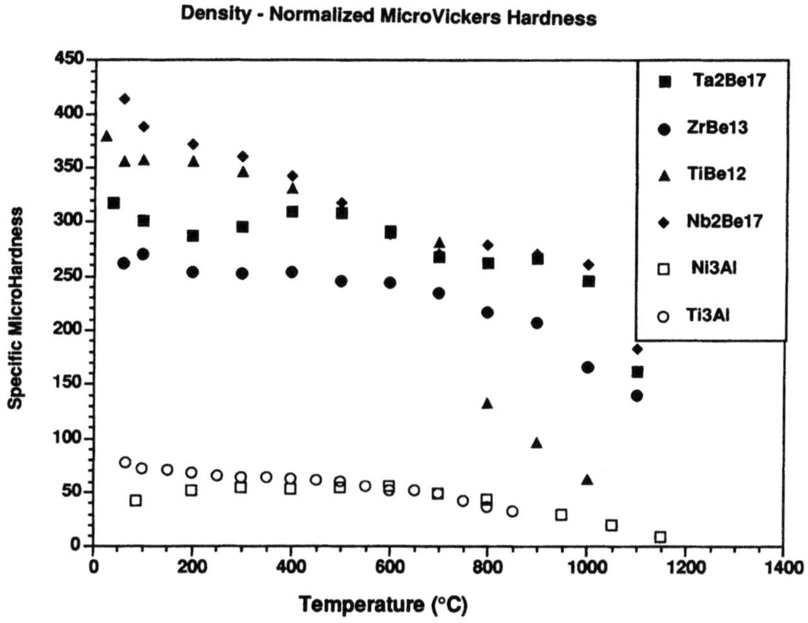

Density - Normalized MicroVickers Hardness

Figure 5. Hardness *vs.* temperature for selected beryllides and Ni$_3$Al.

can have over the more conventional intermetallics.[16] However, the cause of this benefit also is the cause of its greatest liability. Both their high strength and concomitant lack of ductility can be attributed to the very low symmetry of their crystal structures.

Within the group of intermetallics with melting points greater than 1650°C, but not so high as the typical Engel-Brewer intermetallic, recent studies have indicated that there are several materials with low symmetry akin to the beryllides that have similarly attractive high temperature mechanical properties. For example, the Cr$_2$Nb binary system has been examined in some depth.[17] As may be seen in Fig. 6, the results of this study have shown that this intermetallic system possesses strength that is up to three times greater than that found for Ni$_3$Al at comparable temperatures. In fact, the strength of the Cr$_2$Nb system at 1300°C is comparable to or greater than the peak strength exhibited by a Ni$_3$Al(B, Hf) compound.

Further, the microindention creep behavior of the Cr$_2$Nb system was studied. Analysis of the data showed a stress exponent of 24 and and activation energy of 478 kJ/mole. These unusually high values, particularly for the stress exponent, are indicative of the existence of an effective resisting stress against creep.[18] This is somewhat surprising, given that resisting stresses against creep are normally associated with such complex multi-phase systems as high-volume fraction γ' superalloys and ODS alloys. When the data were fit against a microindention creep deformation law which was modified to incorporate an effective resisting stress term, it was determined that the stress exponent was 4.5, the activation energy was 357 kJ/mole, and there was a resisting stress term of 300 MPa. While the actual mechanistics for this apparent creep resistance have yet to be fully explored, it is believed that the resistance to creep is, at least in part, due to the lack of active glide planes in the C15 crystal structure and the resultant complex dislocation-dislocation interactions that occur during deformation.

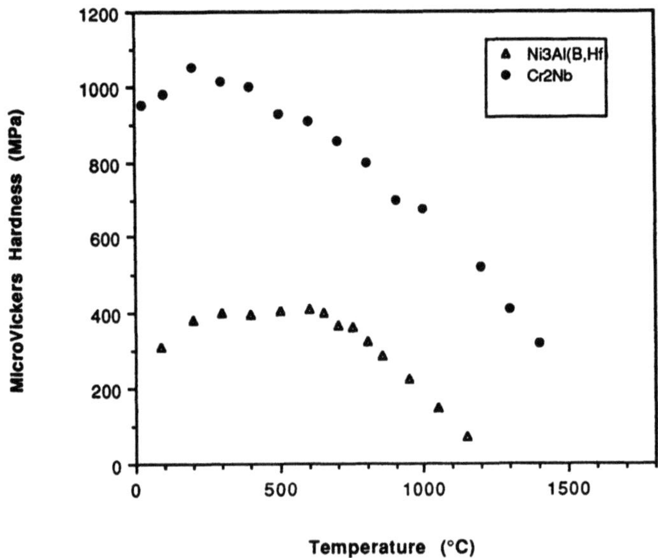

Figure 6. Hardness *vs.* temperature for Cr_2Nb and Ni_3Al.

IV. Metal and Intermetallics Matrix Composites

The predominant factor affecting the implementation of metal matrix composites (MMCs) for elevated temperature applications are the degree of chemical interaction between the fiber and matrix components and thermal-mechanical stability issues due to mismatch of the coefficients of thermal expansion (CTE) for fiber and matrix. Interdiffusional phenomena can take the form of fiber dissolution, fiber properties being "poisoned" by matrix element influx, fiber/matrix reactions, and fiber coarsening. CTE mismatch can result in thermal fatigue and fiber/matrix load transfer problems due to debonding at the interface.

A great deal of work has been focussed of late on developing methodologies to assess the effect of these phenomena. Efforts in recent years to address these problems directly have focussed on developing methodologies for assessing these phenomena.

Previous work on very simple W/Nb single phase composites (*i.e.* complete solid solution) have resulted in a methodology that allows long term prediction of interdiffusional behavior.[19] The first priority was determining composition dependent interdiffusion coefficients. This was easily accomplished by Boltzmann-Matano analysis of planar interface diffusion couples. Having determined the interdiffusion coefficients for the temperatures of interest, composition profiles were calculated using numerical solutions to Fick's second law. This finite difference computer code, which was adapted from the program of Tenney and Unnam,[20] calculates diffusion profiles for diffusion couples with planar, cylindrical, or spherical geometry with finite boundary conditions. Using this method, forecasts of the level of interdiffusion for W/Nb composites for very long term exposures are possible. For this effort, the ability of the numerical solutions utilized to handle finite boundary conditions (*i.e.*, overlapping diffusion fields) was crucial. Fig. 7 illustrates radial diffusion profiles for a 40 volume percent 200 micron diameter fiber reinforced composite annealed from 1 month to 10

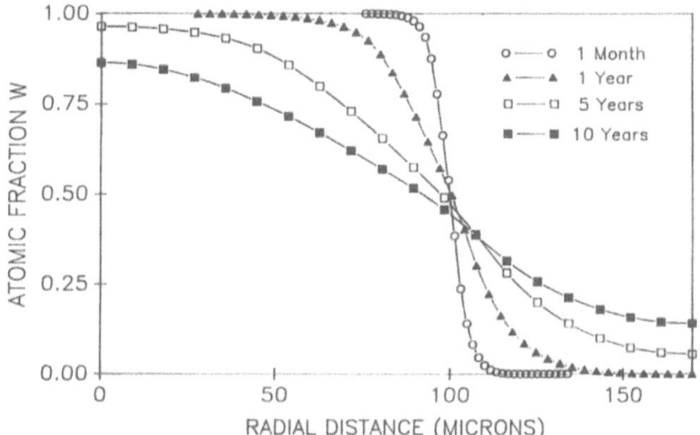

Figure 7. Long term predictions of interdiffusion in a 40 v/o fiber W/Nb composite at 1500°K.

years. Clearly, significant degradation of fiber properties can be expected at the very long times. It must be noted that a great deal of care is necessary in determining diffusion coefficients. Normal experimental uncertainty can lead to large systematic errors when predicting composition profiles for very long times.

Tungsten fiber reinforced superalloys (TFRS) have long been candidates for high temperature composites allowing for significant increases in operating temperatures through increased creep resistance and strength. TFRS composites present several concerns including diffusion induced recrystallization of the tungsten fibers and reaction zone formation at the fiber/matrix interface. Recrystallization has a pronounced effect on the strength and creep resistance of tungsten fibers. Fiber/matrix diffusional reactions in TFRS have been shown to produce brittle intermetallic phases that, as they continue to grow, may adversely affect mechanical strength and toughness. Thus, this system presents a more difficult set of problems than the simple W/Nb system.

Reaction zone growth has been determined to be rate controlled by interdiffusion across the reaction zone phase.[21] The interdiffusion coefficient for these systems using a pseudobinary approximation can be expressed as

$$D(T) = AK_{rz}(T)^{1/2}K_f(T)^{1/2}, \qquad (1)$$

where A is a constant of proportionality dependent on the interfacial chemistries at the fiber/reaction zone and reaction zone/matrix interfaces, $K_{rz}(T)$ and $K_f(T)$ are the parabolic rate constants for the growth of the reaction zone and the growth of the portion of the reaction zone that displaces the fiber. Several matrix alloys have been ranked according to the product of the roots of these rate constants. The ranking of these matrix alloys by reaction zone growth kinetics is shown in Table I. Alloys 89 and 90 are experimental alloys developed in an effort to minimize reaction zone growth kinetics. As can be seen, reductions of iron and cobalt decrease interdiffusion across the reaction zone and thereby decrease the kinetics of reaction zone formation.

Recrystallization of the tungsten fibers reinforcement is also a primary concern for TFRS composites. Recrystallization of ThO_2 doped tungsten wires normally occurs at about 2000°C. A number of studies have shown that a number of elements, most notably nickel, cause this recrystallization temperature to drop dramatically when they are in contact with the fibers. In fact conventional wisdom to date has been

Table I

Reaction zone growth and recrystallization kinetics
for selected TFRS composites annealed at 1093°C & 1100°C*.

Matrix	Fe	Fe+Co	Ni	$K_{rz}^{1/2} K_f^{1/2}$ ($\times 10^{-12}$ cm^2/sec)	Recrystallization** Penetration (mm)
FeCrAlY	71	71	0	3.5	0
SS316	70	70	12	2.9	–
Incoloy 907	57	70	25	1.7	–
Incoloy 903*	42	57	38	0.8	53.3
Hastalloy X	17	19	46	–	30.9
Waspaloy	0	13	56	0.3	–
Alloy 90	0	0	64	–	10.0
Alloy 89	0	0	66	0.05	11.8

* Annealed at 1100°C
** All annealed at 1100°C and data interpolated/extrapolated to 60 hour exposure.

that any increases in nickel content of the matrix alloy in TFRS composites would
be accompanied by increased recrystallization kinetics of the fibers in that composite
system. Since very pure tungsten wires recrystallize at about the same temperature
as the poisoned ThO$_2$ doped wires, it appears that the infusion of the poisoning
elements affects the recrystallization inhibiting nature of the dopant. A complication
has been some uncertainty regarding the nature and distribution of dopant particles.
Perhaps the most notable theory to explain the diffusion promoted recrystallization
of doped tungsten wires is that the poisoning species, which diffuses primarily along
grain boundaries through short circuit paths, lowers the interfacial energy of the grain
and subgrain boundaries, thereby overcoming the effect of the pinning dispersoids.
However, the effect of the poisoning species on the dispersoid/bulk tungsten energy
may also be playing a significant role.

As stated above, the conventional wisdom, with regard to matrix nickel content,
has been that increases in nickel content should promote accelerated tungsten fiber
recrystallization. The purpose of this investigation pivoting about alloys 89 and 90 is
to attempt to elucidate the previously reported matrix chemistry effect.

Unexpected result were found when considering the level of fiber recrystallization
for these various matrix materials.[22] The FeCrAlY matrix composites did not exhibit
significant recrystallization at the temperatures and exposure times studied. Table I
contains recrystallization penetration data for selected TFRS, including some of the
data of L.O.K. Larsson.[23] The earlier composites of Larsson show more accelerated
recrystallization than the higher and lower nickel containing alloys of the present
study. Although considerably more nickel is available in the matrix of alloys 89 and
90 to source diffusion induced recrystallization, these alloys were specifically designed
to minimize diffusion across the anticipated reaction zone. Thus, it appears some level
of competition exists between total nickel availability and interdiffusional kinetics that
provide nickel to the fiber surface.

As discussed above, beryllide intermetallic have been of great interest for composite application. This has lead to attempts to employ them as the reinforcing element in

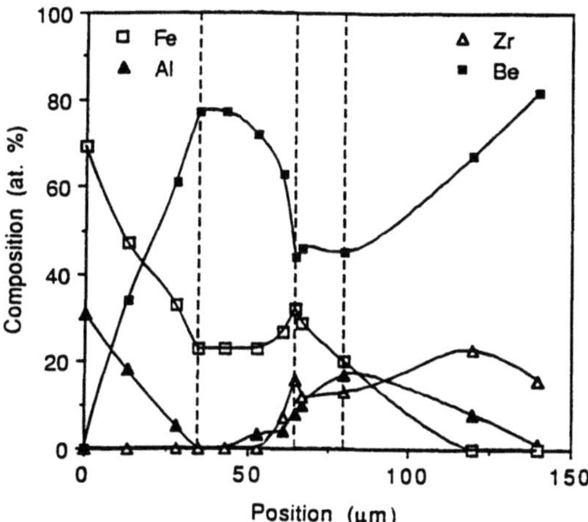

Figure 8. Composition-position profile of the $ZrBe_{13}$ / Fe-40% Al-system annealed for 100 hours at 1220°C.

intermetallic matrix composites. One of the prime matrix candidates has been FeAl due to its mechanical and environmental properties. Four beryllide reinforced Fe-40% Al matrix systems have been studied.[24] The beryllides studied included $TiBe_{12}$, $ZrBe_{13}$, Nb_2Be_{17}, and Ta_2Be_{17}. A similar dual phase reaction zone evolution was observed for each system. This growth behavior took the form of a decreasing rate of growth from the initially observed parabolic rate law. This deviation from parabolic behavior implies that some mechanism is operating that increasingly retards diffusion across the reaction zone as annealing time increases.

Examination of the composition-position plots for these systems has revealed some very interesting phenomena that are believed to account for this decrease in reaction zone growth kinetics. As a typical example, the plot for the $ZrBe_{13}$ / Fe-40% Al system is shown in Fig. 8. The most striking features of these plots are the concentrations of beryllium and aluminium from the terminal phases through the reaction zone phases. For both these elements, concentration drops in the immediately adjacent reaction zone phase from the element rich terminal phase, climbs in the second and more distant phase, then drops to zero in the opposite terminal phase. In the case of the aluminum profile, the aluminum concentration in the reaction zone phase adjacent to the aluminide drops below detection limits for both EDS and SAM. This phase has been identified as $FeBe_5$. The other phase for these systems have yet to be identified.

Based on these diffusion profiles that were obtained, an *in situ* diffusion barrier mechanism is believed to have been formed in these Fe-40% Al systems. Although no ternary data exists for the Fe-Be-Al system, apparently $FeBe_5$ has no, or at most trace, solubility for aluminum, and thus is a barrier to its further diffusion. The aluminum present on the beryllide side of the $FeBe_5$ is believed to have diffused there early on in the fabrication process, prior to the phases formation. It is similarly possible that the longer term, slow growth may be a result of the reaction zone phases picking up their needed elements from the relatively small amounts that diffused into the terminal phases adjacent to them early on in the fabrication process. Thus, the $FeBe_5$ may be getting the necessary beryllium from the Fe-40% Al, for example.

V. Concluding Remarks

Clearly, the challenge of developing high temperature structural materials to exceed current capabilities is being approached from a wide variety of different points of view. Although the relative maturities of the basic technologies vary for these different materials, none have yet made any real inroads into real world application. To be sure, EBCHR superalloys cannot be far from military and commercial engines, but due to the high ambitions for hypersonic flight, advanced intermetallics and composites cannot, and must not, be too far behind.

Acknowledgements

The assistance and financial support for this work has been provided by: Nippon Mining Company, the U.S. Air Force Office of Scientific Research, US Office of Naval Research, and the NASA-Lewis Research Center.

References

1. R.F. Decker and J.W. Freeman, *Trans. AIME,* **218**, 277 (1960).
2. C. Lund, J. Hockin, and M.J. Woulds, *U.S. Patent 3,677,747,* (1972).
3. J.E. Doherty, B.H. Kear, and A.F. Giamei, *J. Metall.,* **11**, 59 (1971).
4. F.H. Froes, *J. Metall.,* **9**, 6 (1989).
5. P.L. Bretz, J.K Tien, T. Denda, S. Himeno, F. Shimizu, and N. Mori, in: *High Temperature Materials for Power Engineering 1990, Part II*, edited by E. Bachelet, R. Brunetaud, D. Coutsouradis, P. Esslinger, J. Ewald, I. Kvernes, Y. Lindblom, D.B. Meadowcroft, V. Regis, R.B. Scarlin, K. Schneider, and R. Singer,(Kluwer, Deventer, 1990), p. 1675.
6. Barth reference
7. N. Engel, *Powder Metall. Bull.,* **7**, 8 (1954).
8. L. Brewer, in: *Electronic Structure and Alloy Chemistry of Transition Elements,* edited by P. A. Beck, (Wiley-Interscience, New York, 1963), p. 221.
9. L. Brewer, in: *High Strength Materials*, edited by V.F. Zackay, (Wiley, New York, 1965), p. 12.
10. J.K. Gibson, L. Brewer, and K.A. Gingerich, *Metall. Trans. A,* **15**, 2075 (1984).
11. C. Sigli and J.M. Sanchez, *Acta Metall.,* **33**, 1097 (1985).
12. D. de Fontaine, in: *Solid State Physics, Vol. 34*, edited by H. Ehrenreich, F. Seitz, and D. Turnbull, (Academic Press, New York, 1973).
13. M. Yamaguchi, V. Paidar, D.P. Pope, and V. Vitek, *Phil. Mag. A ,***45**, 867 (1982).
14. V. Paidar, M. Yamaguchi, D.P. Pope, and V. Vitek, *Phil. Mag. A,* **45**, 883 (1982).
15. A.M. Gyurko, *Masters Thesis,* (The University of Texas, Austin TX, 1991).
16. A.B. Rodriguez and J.K. Tien, unpublished research.
17. G.E. Vignoul, J.M. Sanchez, and J.K. Tien, in: *High Temperature Ordered Intermetallic Alloys IV, Materials Research Society Symp. Proc. Vol. 213*, edited by L.A. Johnson, D.P. Pope, and J.O. Stiegler, (Materials Research Society, Pittsburgh, PA, 1991), p. 739.
18. V.C. Nardone, D.E. Matejczyk, and J.K. Tien, *Acta Metall.,* **32**, 1509 (1984).

19. M.W. Kopp and J.K. Tien, in: *Proc. 9th Int. Riso Symp. on Metall. and Mater. Sci.*, edited by S.I. Anderson, H. Lilholt, and O.B. Pederson, (Riso National Laboratory, Roskilde, Denmark, 1988), p. 247.

20. D.R. Tenney and J. Unnam, *NASA-TM-78636*, (National Aeronautics and Space Administration, Washington DC, 1978).

21. T. Caulfield and J.K. Tien, *Metall. Trans. A*, **20**, 255 (1989).

22. M.W. Kopp, K.E. Bagnoli, and J.K. Tien, in: *Recrystallization '90*, edited by T. Chandra, (TMS, Warrendale PA, 1990), p. 261.

23. L.O.K. Larsson, *Ph.D. Dissertation*, (Chalmers University of Technology, Goteborg, Sweeden, 1981).

24. M. W. Kopp, A.J. Carbone, and J.K. Tien, *Mater. Sci. Eng. A*, in press.

Influence of the Superlastic Metals
in the Future
of the Metal Forming Industry

G. Torres-Villaseñor

Instituto de Investigaciones en Materiales
Universidad Nacional Autónoma de México
Apartado Postal 70-360
México, D.F.
MEXICO

Abstract

The phenomenon of superplasticity is present in ceramics and alloys when theirs grain size is lower than 10 μm, and consists of a strong reduction in the mechanical strength at a temperature close to 0.5 T_f (K) . At this temperature the deformation of such materials can reach more than 2000% in a tension test without the observation of necking. The transformation of the phenomenon into a manufacturing technique has been pushed ahead over the last two decades, mainly by the aerospace industries in the developed countries.

Mexico is a world producer of metals that can be transformed into superplastic alloys. This give us the opportunity to start the use of this metals in the metal forming industry, taking advantage of the low forming energy required for this process.

In this work this possibility is analyzed on the basis of the research results obtained in the IIM-UNAM, and the advances produced in developed countries.

I. Introduction

The great interest displayed in the metallurgical phenomenon of superplasticity is understandable because now new perspectives are opened for metallic materials concerning ductility. Superplasticity is the capability of certain polycrystalline (fine grained)

Advanced Topics in Materials Science and Engineering, Edited by
J.L. Morán-López and J.M. Sanchez, Plenum Press, New York, 1993

materials to undergo extensive tensile plastic deformation at a given temperature. Extensive plastic deformation means 3000 to 10,000% in a tensile test. The temperature at which the superplastic phenomenon appears is around 0.5 of the melting point in K. Apparently the term "superplasticity" appeared for the first time in a 1959 paper by Lozinsky an Simeonova.[1] The first scientifically documented report on superplasticity was done by Bengough[2] in 1912, and the first published photograph of a superplastically deformed sample was by Jenkins[3] in 1928.

Although occasional papers appeared on the subject after this date, the mayor increase in interest came with Underwood's[4] review article in 1962 on work in the USSR. In 1964, Backofen at M.I.T. showed the spectacular formability of a superplastic Zn-Al eutectoid alloy, starting the rapid growth in the field of superplasticity that took place after the publication of his paper in 1964.[5] The increase in commercial interest reflects the fact that superplastic materials exhibit low resistance to plastic flow (in specific temperature and strain-rate regimes), as well as high plasticity. This combination of properties is ideal since it is possible to form a complex shape with a minimum expenditure of energy, and the fine structure can give better service properties in the finished product. The feasibility of the commercial application of superplasticity has recently been reviewed for alloys based on titanium, nickel, aluminum and iron.[6]

The eutectoid Zn-Al has been the typical alloy for superplastic research. Langdon and coworkers[7] found in this material the classical behavior of a superplastic metal when tested at different strain rates. Fig. 1 shows this behavior for a copper modified Zn-Al eutectoid alloy called Zinalco tested at room temperature.

The relationship between the flow stress, σ, and the strain rate, when the data are plotted logarithmically, shows a sigmoidal relationship. The slope of this curve is the strain rate sensitivity m which is a parameter that describes the ductility of the superplastic alloys. According to the value of m we can differentiate three regions. Region I, with the lowest value of m, is weakly grain-size dependent; region II has an m value of typically 0.3 to 0.5 and is strongly grain size dependent; and region III, where the material behaves the strongest, is almost grain size independent and has a low value of m.

Large tensile elongations require a high value of m and therefore, the maximum elongation is achieved in region II, as we can see from the upper portion of Fig. 1. The high rate sensitivity of superplastic alloys has been used almost as a definition of superplasticity. The fact that there is a strong correlation between ductility and the value of m is now well established. For industrial applications, alloys are required which can be deformed at as high a strain rate as possible in order to get maximum productivity.

The main deformation mechanism of a superplastic metal is grain sliding, suggesting that the interfaces between the grains are an important component of a superplastic metal. The lower the coherence of the grains the better the superplastic behavior. This is because the slip of one grain on the other will require lower energies and, at the same time, this process will produce internal cavitation and premature failure.

It is possible to obtain a general overview of the deformation process occurring in a superplastic metal by direct observation of the deformation in an scanning electron microscope equipped with a tensile stage. Fig. 2 shows a sequence of deformation steps in a superplastic Zn-Cd alloy. It was observed following some features on the surface that grain sliding is a main mechanism of deformation at room temperature of this alloy. The shape of the grains did not change even at high deformations ($\epsilon = 140\%$). At the beginning of the deformation ($\epsilon = 10\%$) some fissures are observed inside of a conglomerate of zinc grains and they become apparent when deformation reaches 35%, Fig. 2b. These fissures do not propagate by the specimen to produce fracture. As the deformation progresses, they get wider without increasing their lengths by the migration of the two parts of the fractured grain in opposite directions parallel to the tensile axis. After 60% to 80% deformation, Fig. 2c, several small grains of less than 2 μm coming from the bottom of the crack start to fill up the region in between the two parts of the grain. Higher deformation (up to 140%) exposes a bigger portion

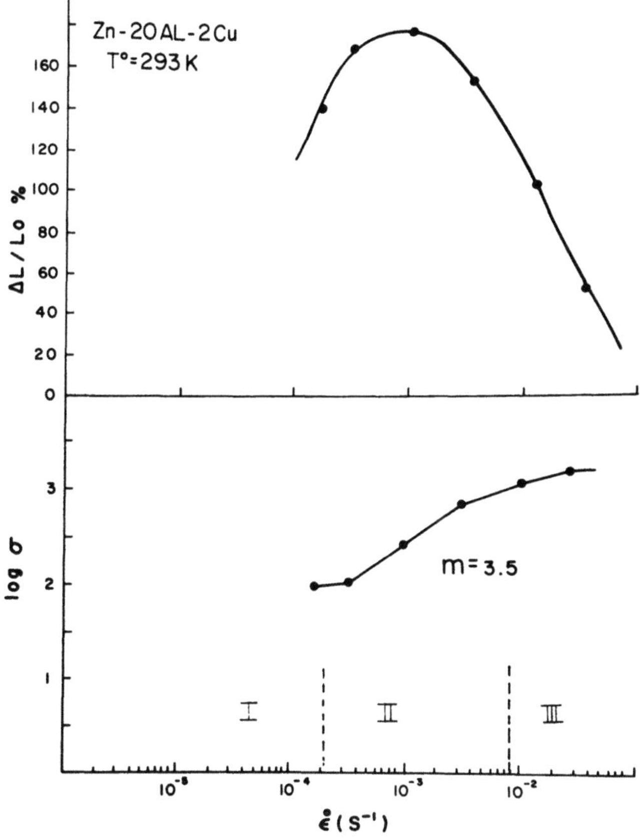

Figure 1. Elongation to failure (upper) and flow stress (lower) *vs.* initial strain rate for Zinalco alloy with 2 μm grain size.

of the boundary facet and more grains are observed to move to the surface (A and B in Fig. 2d). In addition to these grains that emerge, the lateral contraction of the specimen pushes some grains from the rim of the fissure, contributing to restore the surface. The observed movement of material from the lower layers to the surface maintains the correct surface area of the specimen that, as is well known, has to be increased when the specimen is deformed.

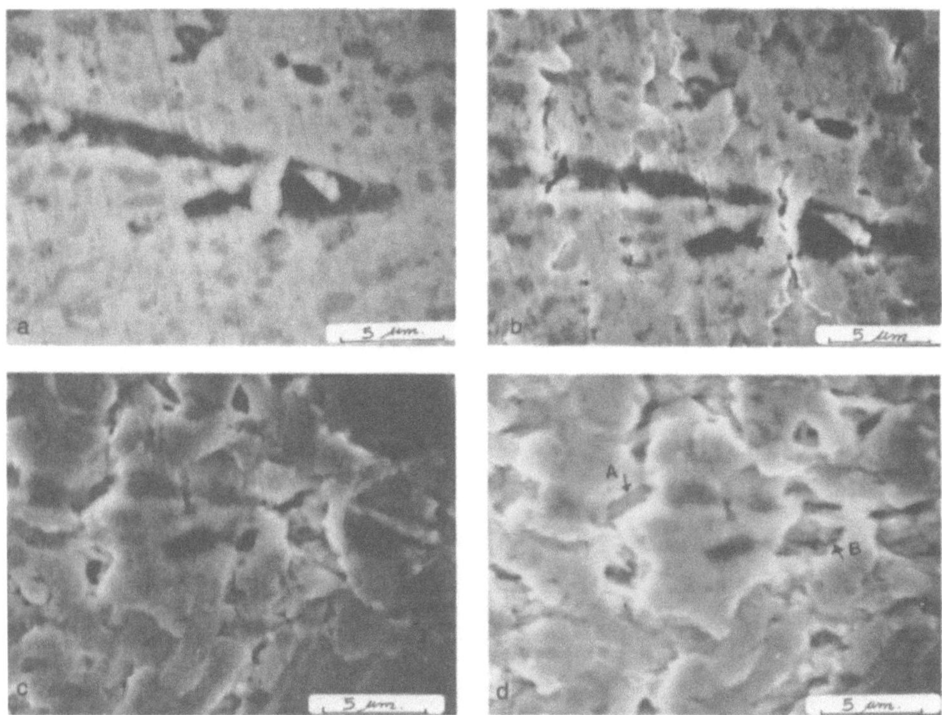

Figure 2. *In situ* observation of the deformation stages. a) Initial structure $\epsilon = 0\%$. b) The grain separation is apparent $\epsilon = 35\%$. c) Grain sliding produce cavitation on the surface, $\epsilon = 60\%$. d) Several small grains emerge to fill the gap between grains, restoring the surface $\epsilon = 140\%$.

II. Superplastic Forming Design Concepts

Superplastic sheet forming allows the realization of complex structures with high local deformations in one forming operation. A comparable, conventionally produced part must either be realized in different steps with intermediate annealing, or has to be built of many different parts bonded together afterwards with time consuming and expensive joining techniques.

Those materials that shows superplastic behavior have the following characteristics;

    a. Grain size should be fine ($< 10\mu$m) and equiaxed.
    b. Dual phase material.
    c. The temperature of superplastic behavior is generally in the range 0.4–0.7 T_m
    d. The strain rates for superplastic flow are generally in the order of 10^{-6} to 10^{-3} per second.
    e. The grain boundaries should no be prone to ready tensile separation.

At present time it is possible to produce superplastic effects in most of the engineering materials such as stainless steels, Al, Ni and zinc alloys, including ceramics such as yttria stabilized zirconia, or the high temperature superconductor ceramics.[8] It is interesting to point out that many of the metals produced in Mexico such as Bi,

Cd, Zn and Pb, can be used in the confection of superplastic alloys; this is a good reason to conduct research in Mexico on the superplastic behavior of alloys based on this metals.

III. Economic Aspects

The main disadvantage of superplasticity and its related forming processes is the low strain rate of about $\sim 10^{-4}$ s^{-1}. The strain rates are low in comparison to all other forming processes, and the possibilities of increasing forming speed are very limited. This excludes in most cases mass production of materials which are easy to form as, for instance, the production of steel parts for body parts of mass produced cars. New designs are required to use the superplasticity adequately. The new body car, designed to be produced with superplastic technologies, has to be made of no more than 2 or 3 pieces, avoiding welding and the use of a chassis. Currently only the military aerospace industry applies superplastic forming in the production of fighter airplanes, helicopters, missiles, etc. The main superplastic material is titanium. In the case of Mexico where no airplane industry exists, the use of superplasticity looks uneconomical. This is true if we try to reproduce with superplastic techniques the items that are manufactured with normal forming processes. A new proper design, for example of the beverage cans, can rise the production rate of superplastic forming and this type of industry is well applicable in Mexico for many kinds of containers. Large cost and weight savings (through redesign) and the high cost of the conventional forming machines will provide the driving force in commercial manufacturing for change from conventional to superplastic forming. A mayor factor in cost savings arises from the elimination of expensive machining and joining operations.

IV. Materials

The high elongations of up to 1000 per cent and more realized under superplastic conditions depend on a high strain rate sensitivity of these materials. Metallurgical prerequisites are fine grain size, typically less than 10 μm, high stability *versus* grain growth, and high resistance to cavitation.

A countless number of metallic alloys between Al alloys and Zr alloys with more or less superplastic properties have been developed on the laboratory scale; the number of materials of industrial importance today is rather low. The conversion of superplasticity to an industrial forming process is confined to some titanium alloys and aluminum alloys, and, with reservations, to some steels. A further decisive advantage with titanium alloys (Ti_6Al_4V) is the relative simple simultaneous application of superplastic forming and diffusion bonding (SPF/DB). This technique allows the forming and welding in one step. Aluminum alloys do not exhibit superplastic properties following a conventional production technique. Additional alloying and/or thermomechanical treatments are necessary to obtain a fine and stable microstructure, which, however, leads to an increase in materials price. The high-strength aluminum alloy 7475, on the other hand is a qualified aerospace alloy which obtains superplastic properties through thermomechanical treatment, including static recrystallization.

The aerospace industry is showing increasing interest in the Al-Li alloys due to about 10% reduction in density and about 10% increase in elastic modulus. The distinctly higher price of these Al-Li alloys particularly justifies the use of superplastic forming for producing complicated structures.

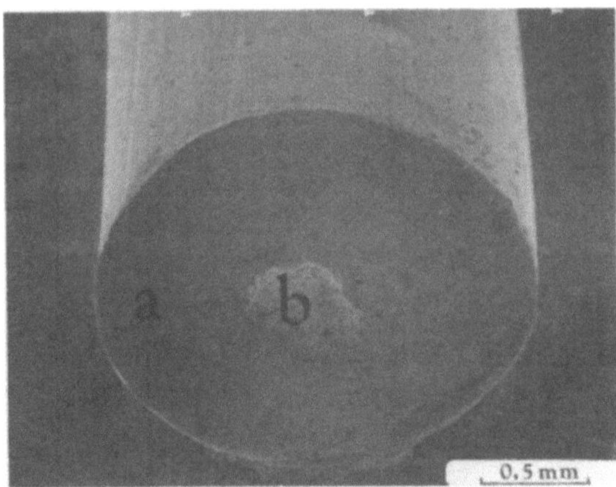

Figure 3. Composite wire obtained by extrusion. a) Superplastic metal, b) Superconductor ceramic.

All these alloys described before are not appropriate for the mexican industry because we have not aerospace industry. The mexican industry requires food or beverage containers, new materials for the automotive industry and architectural applications. The superplastic forming technique has good future in the national industry because many of the metals produced in our country show superplastic behavior when alloyed. The eutectoid Zn-Al alloy reinforced with Cu (>2%) is an alloy developed and intensively studied during the last 10 years at the IIM-UNAM. It is now available at commercially under the name "zinalco". Its mechanical properties are similar to a medium carbon steel[9] and can behave superplastic after conventional fabrication by extrusion or rolling, so that no additional material costs are involved. Because its high strength, it can be applied for the manufacture of body cars with low energy; it is 35% lighter than steel and has high corrosion resistance. In the field of communications, the zinalco alloy can be used to fabricate parabolic antennas, because the reflectivity of electromagnetic waves in zinalco is almost the same as in the aluminum. The Cd-Zn alloys have the consistence of rubber at room temperature and can be applied instead of plastic in many applications including credit cards. In all cases the superplastic alloys are fully recyclable.

Recently it was found an important application of superplastic metals, in forming brittle ceramics,[10] as the superconductor based on YBaCuO. This superconductor ceramic (scc), can sustain strong plastic deformation in compression, when it is imbedded in a superplastic metal matrix (spm). Deformations as high as 300% true strain can be obtained in compression. The composite scc-spm can be extruded to obtain fine wires of scc (0.8 mm) surrounded by a metal envelope Fig. 3.

The diameter of the ceramic fibre after extrusion is not uniform along the length of the wire; it shows variations of ± 0.3 nm. The wire showed a great flexibility; it was possible to bend it up to 180° without disrupting the continuity of the ceramic. Fig. 4 shows the bent ceramic wire after removing its metal envelope. It is possible to observe how the scc has enough consolidation to keep its shape without breaking.

Alloys based on the Ag-Cu system preserve their two-phase structure up to the melting point, so they are suitable to be used for high temperature superplastic forming. The alloy Ag- 28 wt. % Cu prepared from 99.999 per cent pure materials, shows extensive plastic deformation (up to 600%) between 250 and 600°C; the equiaxed structure was obtained by cold rolling the as cast alloy. The amount of cold rolling

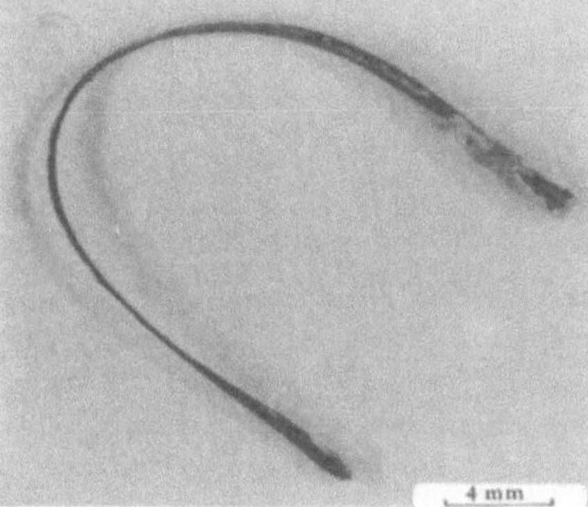

Figure 4. Superconductor wire obtained by extrusion of a cylindrical superconductor ceramic bar, inside of a superplastic metal matrix. The metal envelope was removed after the composite wire was bent 180°. No fractures were developed in the ceramic during the bending process.

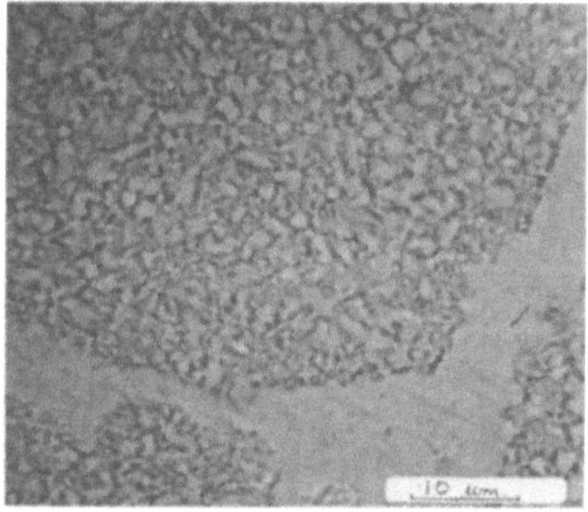

Figure 5. Structure of the silver-copper eutectic, after 60% cold rolling.

given was 60% high reduction. Scanning electron microscopy was used to resolve the details of the microstructure. The fine wrought structure is shown in Fig. 5.

This research is still in progress. The current results shows that an alloy of this composition can be made superplastic using cold rolling to obtain equiaxed grains. The resistance to neck formation (m-value) found is sufficiently high and the flow stress sufficiently low in the strain-rate range useful for commercial forming operations ($> 2 \times 10^{-4}$ s^{-1}), to suggest that this processing route would be capable of producing commercially useful superlastic sheets from this alloy. The applications can be in the production of silver appliances, in jewelry or in production of superconductors wires, using the method described before. In the latter case, it is expected to be able to oxygenate the superconductor ceramic without removing the metal cover, because the well known properties of the silver to be permeable to oxygen.

The applications of the superplastic properties of several metals as those described here can open a wide field of applications, not expected for these metals, producing benefits to the producer countries.

V. Summary

Superplasticity is a flourishing field of study that is very broad in the range of materials it covers, and the span of interest extends from basic research to engineering applications. The immediate future for the field is very exciting because of the discovery of superplasticity in ceramic based materials as well as the opportunities that are available for research in metallic and intermetallic systems and in composite systems.

References

1. M.G. Lozinsky and I.S. Simeonova, *Acta Metall.*, **7**, 709 (1959).
2. G.D. Bengough, *J. Inst. Met.*, **7**, 123 (1912)
3. C.H.M. Jenkins, *J. Inst. Met.*, **40**, 21 (1928).
4. E.E. Underwood, *Jour. Met.*, **14**, 914 (1962).
5. W.A. Backofen, I.R.Turner, and D.H. Avery, *Trans ASM* **57**, 980 (1964).
6. J. Wadsworth, T.G. Nieh, and O.D. Sherby, in *Superplasticity in Advanced Materials. ICSAM-91*, edited by N. Furushiro, (Japan, 1991).
7. F.A. Mohamed, M.M. Ahmed, and T.G. Langdon, *Met. Trans.* **A 8A**, 933 (1977).
8. J.E. Moreno y G. Torres-Villaseñor, in *Superplasticity in Metals, Ceramics, and Intermetallics*, edited by M.J. Mayo, M. Kobayashi, and J. Wadsworth. (Materials Research Society, **196**, 1990), p. 337.
9. G. Torres-Villaseñor, *Ciencia* **39**, 103 (1988).
10. J.E. Moreno, N. Floriano, and G. Torres-Villaseñor, in *Superplasticity in Advanced Materials*, edited by S. Hori and N. Furushiro, *The Japan Soc. for Research on Superplasticity, 1991*, p. 281.

Advanced Aerospace Materials: Titanium Aluminide Intermetallic Compounds and Metal Matrix Composites

F.H. Froes*, C. Suryanarayana* and I.S. Polkin#

*Institute for Materials and Advanced Processes (IMAP)
College of Mines
University of Idaho
Moscow, ID 83843-4195
U.S.A.

All Union Institute for Light Alloys (VILS)
Moscow
U.S.S.R.

Abstract

The aerospace systems of the twenty-first century will have mission requirements considerable beyond present day vehicles; including operation in a cost-effective manner. This in turn will require materials with enhanced mechanical property characteristics compared to current state-of-the-art materials. The mission scenarios of the future air and space systems will be briefly reviewed followed by a general consideration of the materials of construction. Two specific types of advanced materials will then be discussed - intermetallic compounds and metal matrix composites. It will be pointed out that while both generic types of materials offer increased strength, stiffness and temperature capability, both also suffer from low ambient temperature "forgiveness" (ductility, fracture toughness, fatigue crack growth rate, etc.) and high cost. The advances which have been made in resolving these concerns, while maintaining other characteristics at attractive levels, will be discussed. A detailed presentation of monolithic and composite titanium aluminides (Ti Al, where $x = 1$ or 3) will be made.

Advanced Topics in Materials Science and Engineering, Edited by
J.L. Morán-López and J.M. Sanchez, Plenum Press, New York, 1993

I. Introduction

The three most important industries in driving technological change, national security considerations, and economic advances into the next century are information/communications systems (computers), biotechnology, and advanced materials and syntheses/processes.[1-15] Further, of these three, advanced materials and syntheses/processes are the most critical and considered vital to advancements in the other two fields, hence the major emphasis being given to advanced materials and the synthesis and processing of these materials in many parts of the world.

Advanced materials may be defined as materials which have enhanced mechanical and physical characteristics compared to traditional materials, such as aluminum and steel, currently manufactured in large-volume assembly line type facilities. The characteristics either allow for very significant improvements in product or device performance or, of even greater significance, allow for new technologies that are not achievable using conventional materials. The advanced synthesis and/or processing techniques are those methods used to produce these advanced materials.

The present paper will discuss developing aerospace systems and their missions, and the material characteristics needed to meet performance demand. Particular emphasis will then be given to two specific types of advanced materials which could see considerable use in aerospace applications - intermetallic compounds and metal matrix composites. While performance of these materials will be emphasized, another major concern with advanced materials will also be addressed - that of affordabilty (cost).

II. Developing Aerospace Systems

The systems of the twenty-first century will be more maneuverable, including very short take-off and landing (V/STOL) concepts to reduce dependence on large and fixed operating bases, and will also have increased speed capability. They will include aircraft with greatly increased range and reduced fuel consumption over current systems. Pay load will be increased while building-in varying degrees of stealth, instantly reactive knowledge-based defensive and offensive systems, and increased confidence (decreased inspection, increased reliability and decreased maintainability). Access to space in an affordable and predictable manner will be required, with greatly decreased turn-around times and much larger payloads.

Among the systems under various stages of development are supersonic V/STOL concepts, air superiority fighters featuring agility with thrust vectoring, high Mach number interceptors and bombers, advanced helicopters, including tilt-rotor concepts, light-weight fighters and high speed civil transports (Mach 2-5) which combine low noise and emission, with fuel efficiency (Fig. 1). There is likely to also be trend towards unmanned autonomous systems.

In the space arena major emphasis is on development of single stage to orbit hypersonic flight vehicles. These include programs in the U.S., U.S.S.R., Britain, France, Germany, and Japan. In the US the National Aerospace Plane (NASP) program has been in full swing for about 5 years with a first flight scheduled for the mid to late 1990's. This experimental manned vehicle will require low density-high temperature materials for construction, with contending materials including monolithic or composite titanium aluminides, reinforced conventional titanium alloys, carbon/carbon composites, and other advanced metal matrix composite concepts. The British con-

Figure 1. Artist's conception of the high speed civil transport (courtesy Boeing Company).

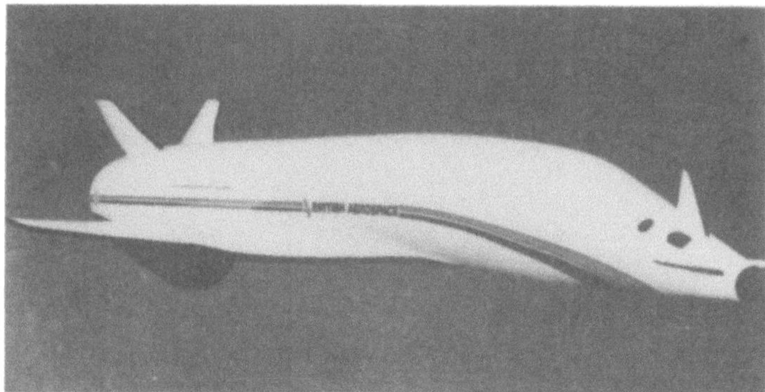

Figure 2. The unmanned horizontal take-off and landing (HOTOL) transatmospheric vehicle.

tender, the horizontal takeoff and land vehicle (HOTOL), is planned to be unmanned with somewhat less severe material requirements (Fig. 2). For the future, these experimental vehicles will evolve into a raft of transatmospheric systems, such as the single-stage-to-orbit (SSTO) and generic USAF vehicle (Figs. 3 and 4).

The power units for the advanced systems will include low bypass turbofan engines at speeds up to Mach 2, and turbojet engines with afterburners to speeds above Mach 3 (Fig. 5).[16] Hypersonic aircraft will require multiple-mode engines for conventional take-off, acceleration to hypersonic or transatmospheric cruise, and conventional landing. This could include take-off and landing with conventional turbine engines, acceleration to Mach 6-12 with a scramjet (supersonic combustion ramjet), and final

Figure 3. Single-Stage-To-Orbit (SSTO) concept that would take-off vertically from a launch pad and land on a runway (Courtesy Rockwell International).

Figure 4. Artist's concept of a United States Air Force transatmospheric vehicle of the next century.

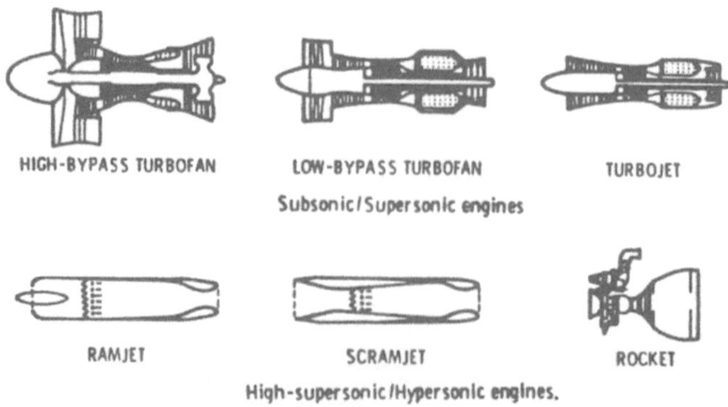

Figure 5. Types of aircraft engines used for different flight envelopes.[16]

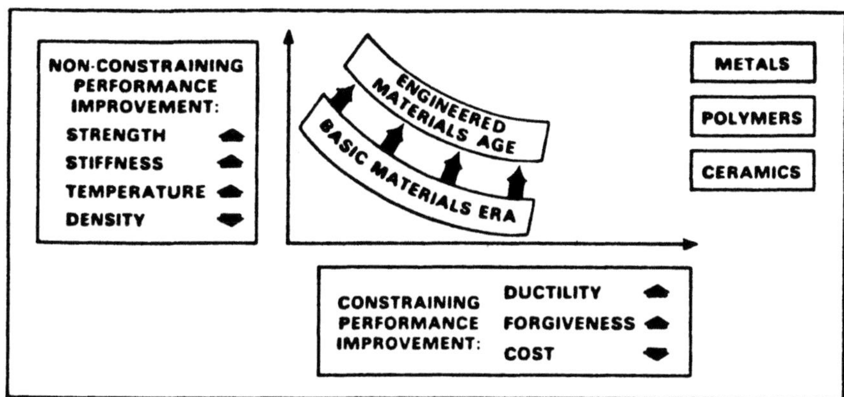

Figure 6. Trend bands of conventional materials of today and advanced tailored materials of tomorrow.[10]

attainment of orbital velocity with a rocket engine. Even more advanced concepts include use of antiproton, directed energy, and nuclear propulsion systems.

In the USA the Integrated High Performance Turbine Engine Thrust (IHPTET) program is attempting to double the present thrust-to-weight ratio of the engine for fighter aircraft of the early next century, while decreasing fuel consumption by 50%.[17] Basically, this means hotter running engines, more highly loaded components, and a decreased part count. The material challenge involved is demonstrated by considering the limited capability of nickel-based superalloys to continue to meet increasing turbine inlet temperatures (TIT), particularly under conditions of reduced internal cooling air. Presently as much as 20% of the compressor delivery air is used for cooling turbine components, elimination of which could require a 500°C enhancement in material capabilities.[18]

III. Advanced Aerospace Metals

Structural metals with improved mechanical characteristics compared to present day metals will be necessary to enhance the performance characteristics of aerospace systems of the next century. These systems will require new materials* which are "stronger, stiffer, hotter and lighter" than traditional materials of construction. Additionally, the materials may be "tailored" to have the properties required for a given application by use of composite concepts.[1-15]

In using engineering structural materials, it is necessary to balance the so-called "unconstrained" characteristics against the "constraining" characteristics. It is necessary to move from the present day trend band to an enhanced trend band by innovative chemistries, synthesis/processing, and microstructures (Fig. 6). To achieve these latter characteristics we can no longer stay with the "basic" materials of today and must instead move on to the "tailored" or "engineered" materials of tomorrow.

---

* "materials" is used here to recognize that advances are also being made in materials other than metals.

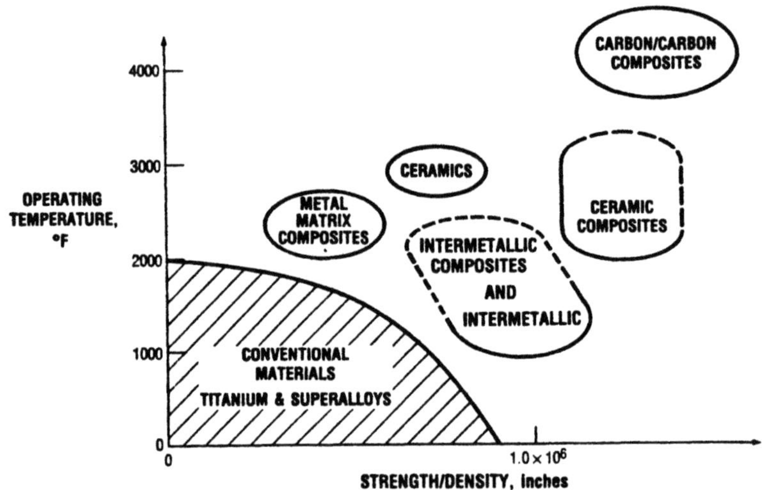

Figure 7. Comparison of maximum operating temperatures and strength/density ratios of conventional materials with developing advanced elevated temperature materials.[19]

A number of advanced materials offer the tantalizing attraction of high temperature operating capability in conjunction with relatively high specific strength (strength divided by density). These materials include metal matrix composites, intermetallics and intermetallic composites, ceramics and ceramic composites, and carbon/carbon composites (Fig. 7).[19] However, while all of these materials offer high specific strength characteristics, their reliability is generally significantly below that of conventional metals. Here reliability generically refers to features such as environmental resistance, ductility, toughness, etc. The intermetallics and intermetallic composites offer the best near term (less than 15 years) opportunity for significantly increased operating temperatures combined with a realistic chance of achieving acceptable reliability levels.

IV. System Costs

The aerospace market is a "niche" market in which advanced structural materials tend to be technology driven, with the major emphasis on performance rather than cost, in contrast to other industries (Fig. 8), and cost strongly dictates sales volume (Fig. 9).[10] From the producer's perspective, volume sales require use in industries such as the automotive arena, but this will require low-cost fabrication techniques to be developed, contrasting with the low-volume aerospace industry.

Despite the emphasis on performance, cost of advanced aerospace systems will be a major concern, and only an integrated design, manufacturing, and use approach can lead to cost effective application compared to conventional materials use. These new materials must be considered as structures rather than in the same way as traditional materials such as metals, or initial acquisition costs could negate use, and prevent cost savings during operation due to the improved behavior.

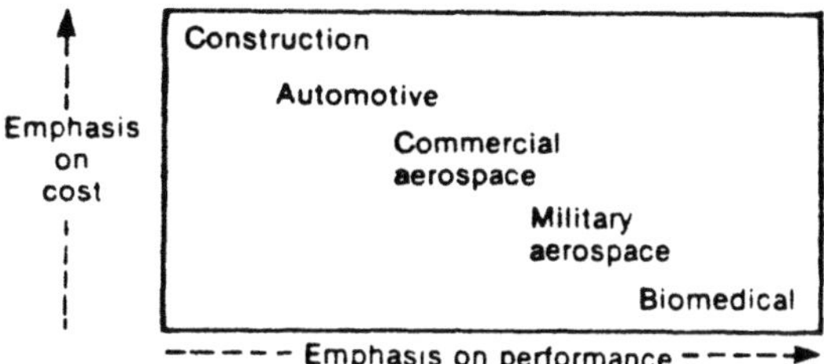

Figure 8. Cost performance emphasis in various industries.

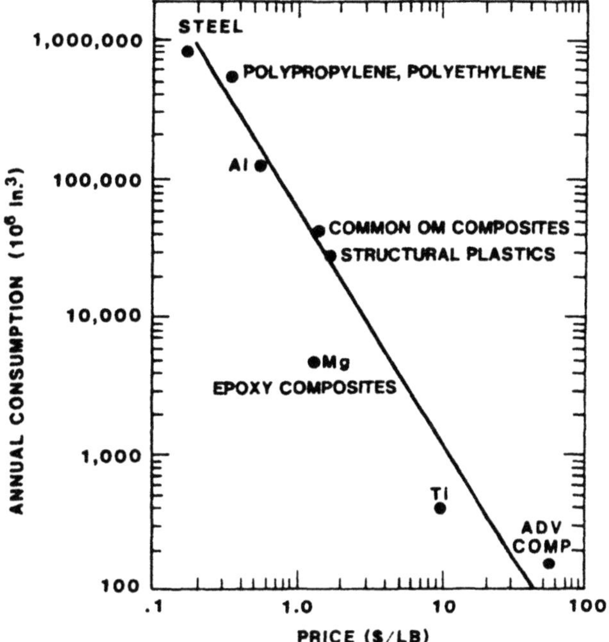

Figure 9. Effect of cost on sales volume of various materials.

A further cost concern from the US point of view is the impact that increasingly expensive, high-technology aircraft coupled with restrictive US export control policies may have on the ability for the US to compete for foreign sales with European manufacturers in particular.[20] Even when the F-117A stealth fighter-bomber, the F-22 advanced tactical fighter and the planned AX follow-on to the A-6 become available for export, their high costs and sophisticated maintainability are likely to prove prohibitive to prospective overseas customers (Fig. 10). As just one example of price differential the F-22 will cost the USAF $104M per copy (with a 648 aircraft buy) while the Light Combat Aircraft being developed in India will carry a $15M price tag.

Figure 10. Maturity of various advanced aircraft.[20]

V. Intermetallic Compounds

V.1 Science

Intermetallic compounds are phases which occur in the central parts of the phase diagram between two or more metals with a characteristic crystal structure, and may have a very specific composition or a range of compositions.[21-23] Because of the strong attraction between the unlike atoms involved there is a strong preference in the selection of nearest neighbors which in turn leads to an ordered structure, a high resistance to deformation (movement of lattice defects), and a high melting point. The high melting point in combination with the difficulty of movement of lattice defects leads to high strength and retention of strength to elevated temperatures. However, these same features lead to the "achilles heel" of the intermetallic, that of very low ductility (brittleness or lack of forgiveness)* at ambient temperatures. This low ductility can be attributed to three major reasons (a) flow stress can be higher than the fracture (cleavage) stress, up to a temperature termed the ductile-brittle transition temperature, below which brittle failure occurs (Fig. 11),[24] (b) a limited number of available deformation modes (slip systems etc.), a particular concern with large complex crystal structures of low symmetry, and (c) early crack initiation at grain boundary regions, particularly when impurities have segregated to these regions (such as S, P, Sn or Sb)

---

   * The reliability of components used in demanding applications is generally determined using a fracture mechanics approach in which a pre-existing crack is assumed to be present in the materials. Component life is then calculated by the time required for this crack to reach a critical length corresponding to the fracture toughness of the material at which point catastrophic crack growth occurs. However, designers also demand some minimum level of ductility (generally 4–5% minimum elongation) as a further "comfort" factor.

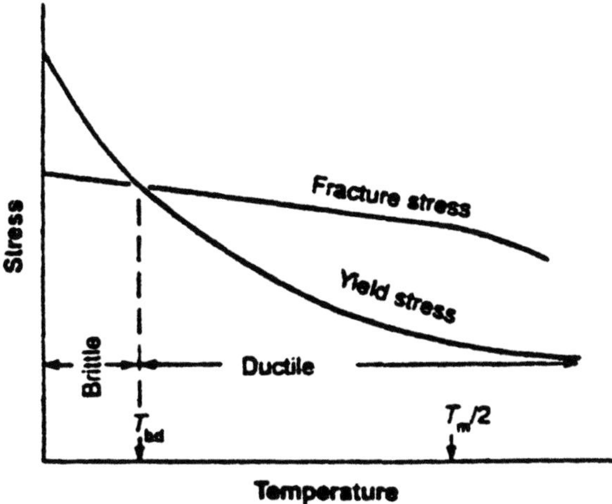

Figure 11. Schematic plot of variation of fracture stress and flow stress with temperature. Below the cross-over point (T_{bd}) the material is brittle; above this point it is ductile.[24]

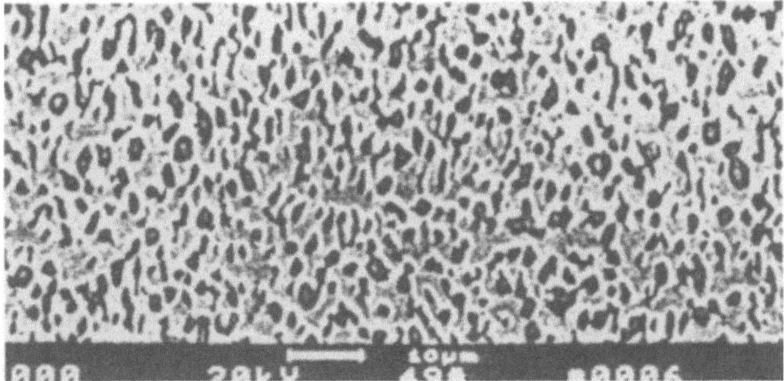

Figure 12. Ti$_3$Al-Nb alloy processed to contain fine equiaxed α_2 grains and a room temperature ductility of 26% elongation.[25]

causing reduced cohesive forces and intergranular fracture. Thus the major thrust in the development of useable intermetallics has been to "build-in" forgiveness (where forgiveness is a general term used for ductility, fracture, toughness etc.) while maintaining attractive elevated temperature properties. This has been achieved to an extent using chemistry/processing/microstructure control and by use of composite concepts in which a reinforcing phase is incorporated into the intermetallic matrix to increase strength/stiffness and to deflect crack propagation (increase toughness).

Since weight saving is critical, particularly in aerospace applications, the lower density intermetallics are particularly attractive, a number of which are listed in Table I. The present paper will be restricted to a discussion of the titanium aluminides [Ti$_x$Al, where $x = 1$ or 3: Ti$_3$Al (α_2) and TiAl (γ)] because of their low density and relatively high level of maturity.

Table I

Characteristics of selected intermetallics

Intermetallic Compound	Crystal Structure (Ordered)	Melting Point °C (°F)	Density g/cm^3 (lbs/in^3)	Young's Modulus GPa (10^6 psi)
Ni$_3$Al	face centered cubic	1390 (2530)	7.50 (0.274)	178.5 (25.9)
NiAl	body centered cubic	1640 (2980)	5.86 (0.214)	294.2 (42.7)
Fe$_3$Al	body centered cubic	1540 (2800)	6.72 (0.245)	140.0 (20.4)
FeAl	body centered cubic	1250 (2.280)	5.56 (0.203)	260.4 (37.8)
Ti$_3$Al	hexagonal close packed	1600 (2910)	4.20 (0.153)	144.7 (21.0)
TiAl	tetragonal	1460 (2660)	3.91 (0.143)	175.6 (25.5)
Al$_3$Ti	tetragonal	1350 (2460)	3.40 (0.124)	
Nb$_2$Be$_{17}$	rhombohedral	1705 (3100)	3.28 (0.120)	296.3 (43.0)
Al$_3$Nb	tetragonal	1600 (2910)	4.54 (0.166)	

Because of their low density in combination with attractive elevated temperature properties the titanium aluminides have received considerable attention for aerospace applications particularly in the USA and USSR. The titanium aluminides have improved resistance to elevated temperature oxidation, better creep resistance and higher modulus than conventional titanium alloys, however as with the other intermetallics they suffer from extreme brittleness at ambient temperatures. Thus the strategy for both classes of alloys has been to enhance room temperature ductility while maintaining the attractive elevated temperature behavior.

The major success to date with the Ti$_3$Al composition has been attained by additions of alloying elements which stabilize the beta phase, particularly Nb (at about the 10 at. % level). The increased amount of the beta phase promotes ductility. A fertile area for further ductility enhancement is by microstructural control; particularly by refinement and spheridoization of the microstructure (Fig. 12).[25]

There also appears to be good potential for development of useful Ti$_3$Al alloys at even higher levels of Nb (about 25 at. %) where an orthorhombic phase forms; however these alloys require further maturation. Investigation of the potential of enhanced behavior using rapid solidification processing of the Ti$_3$Al type alloys has shown some promise, but a lower than optimum volume percentage of second phase particles (generally rare-earth oxides) and relatively rapid coarsening rates are disappointing. However, the enhanced elevated temperature mechanical properties and attractive oxidation resistance continue to make the Nb-modified Ti$_3$Al-type of alloys a potential material for aerospace use.

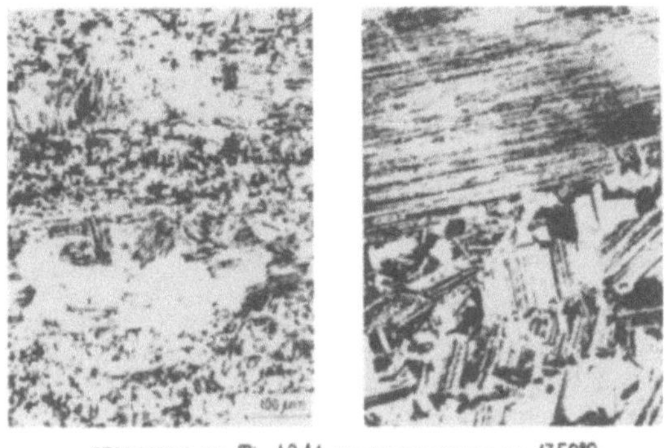

Figure 13. Bands of recrystallized and lameller structure in TiAl deformed at 1300°C.

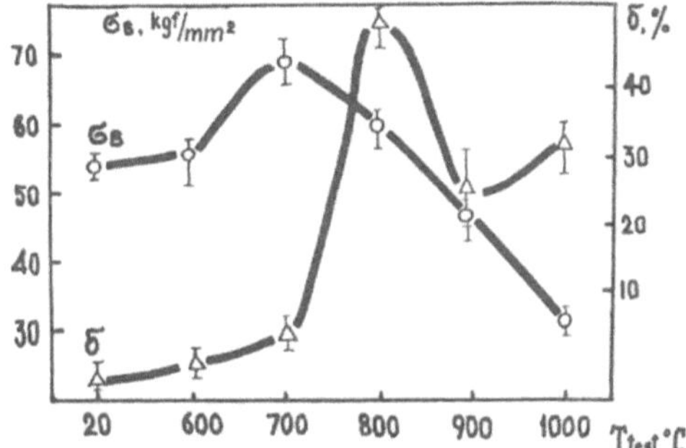

Figure 14. Tensile properties of Ti-48 at. % Al alloy with microstructure shown in Fig. 13.

Figure 15. Vane produced from Ti-48 at. % Al using a powder metallurgy/isothermal forging approach.

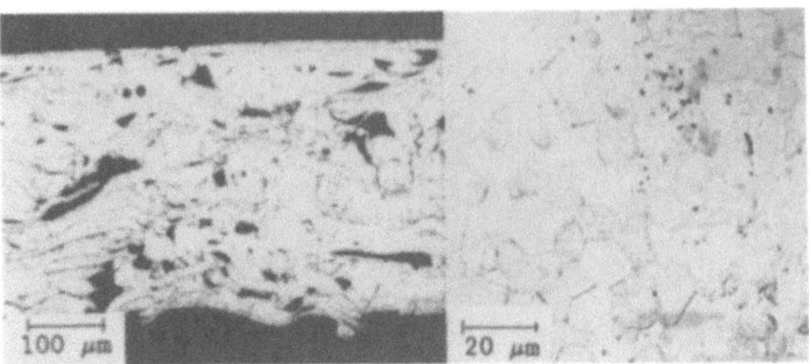

Figure 16. Microstructure of As-sprayed and consolidated Ti-14Al-21Nb (at. %) α_2 alloy processed by induction plasma deposition.[26]

Figure 17. Artist's rendition of a conceptual National Aerospace Plane, NASP (Courtesy Rockwell-International).

The TiAl composition has largely resisted all attempts to achieve useable ductility levels. However, the significantly better elevated temperature behavior compared to the Ti_3Al family of alloys, at a reduced density, has continued to create great interest. Alloying to the lean-side of stoichiometry (equiatomic TiAl) into the two-phase TiAl + Ti_3Al region combined with careful control of the microstucture, to produce a banded structure of fine recrystallized and non-recrystallized regions, has improved ductility, but only to a maximum of 2–3% at room temperature (Fig. 13). However, this ductility increases to beyond 50% at 800°C (Fig. 14) allowing fabrication of complex shapes such as the vane shown in Fig. 15 using a powder metallurgy/isothermal forging technique.

V.2 Technology

The vast funding put into the titanium aluminide has resulted in development of the technology to produce all normal mill products from Ti_3Al compositions, including 4550 kg (10,000 lbs) ingots, forgings, flat products (including foil), powder products, and castings; albeit with processing more difficult than with conventional titanium

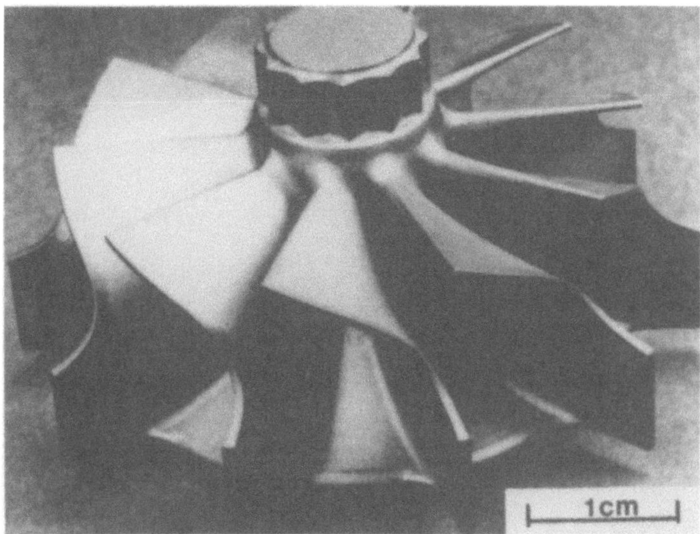

Figure 18. Cast turbocharger rotor produced from TiAl.[27]

Figure 19. High pressure stator ring fabricated from Ti_3Al (Courtesy GE).

alloys. The increased difficulty in fabrication of the TiAl class of alloys has to date restricted ingots to 200 kg (440 lbs) in the USA. Subsequent conversion is considerably more difficult than Ti_3Al compositions, with near net shape processes such as castings, powder metallurgy and isothermal forging being favored. Generally the normal producers of titanium alloy products are the suppliers of titanium aluminide material, with specialized processes such as induction plasma spraying (Fig. 16) being developed by engine manufacturers such as GE.[26]

In the USA the major potential use of the titanium aluminides is in advanced gas turbine engines (generically the Integrated High Performance Turbine Engine Technology [IHPTET] program), and more recently for use in the transatmospheric

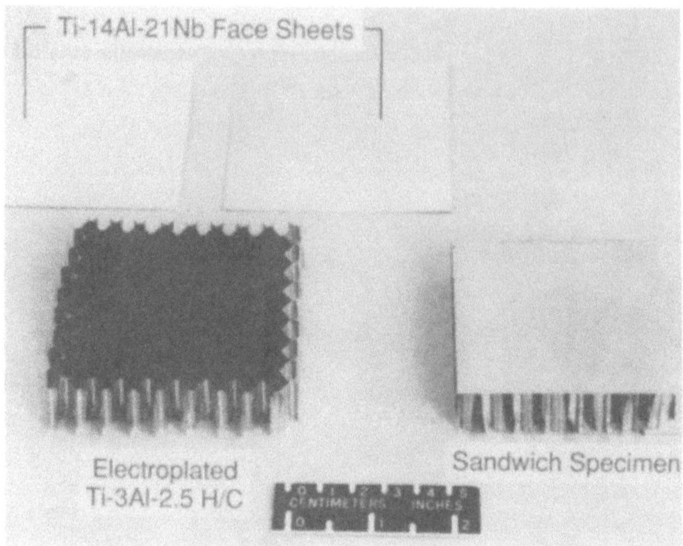

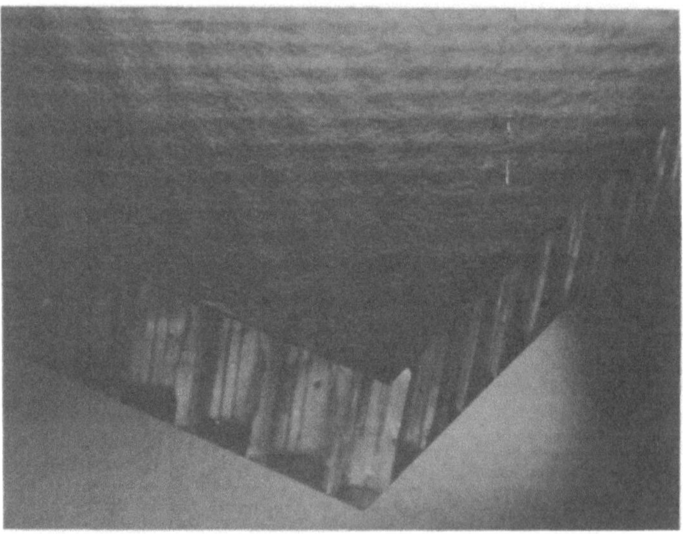

Figure 20. Honeycomb structure produced using diffusion bonding from Ti_3Al face sheets.[28]

National Aerospace Plane (NASP) (Fig. 17). However, it is quite likely that the first application for the titanium aluminides will be a cast automotive turbocharger in Japan (Fig. 18).[27]

Other demonstration aerospace parts produced from the titanium aluminides include a Ti_3Al high pressure stator ring which can result in a 43% weight savings compared to a superalloy part (Fig. 19), and the turbine blade made from TiAl using a powder metallurgy and hot forging approach discussed earlier (Fig. 15). An example of a honeycomb structure fabricated using a diffusion bonding method, for potential application in the NASP, is shown in Fig. 20.[28]

Table II

Thoughening Mechanism Provided by
Reinforcing Continuous Fibers[29]

1. Plastic deformation of the matrix
2. Fiber pull-out
3. Presence of weak interfaces/fiber separation, and deflection of the crack.

Table III

Desirable Characteristics of a Reinforcing Fiber[23]

1. Compatible coefficient of thermal expansion (CTE) with the matrix
2. Chemical compatibility with the matrix even after extended elevated temperature exposures
3. Low density
4. High modulus of elasticity
5. High strength at elevated temperature
6. Good oxidation resistance
7. 50–250 μm diameter continuous monofilament
8. Mass production feasibility (including affordable cost)

VI. Metal Matrix Composites

VI.1 Science

In a composite concept a second phase is introduced into the matrix to enhance physical and mechanical behavior. Geometrically, the second phase can have either a particulate, short fiber, continuous fiber, or lamellar type configuration.[29] A number of metallic matrices have been studied but only those based on the titanium aluminides will be discussed here.

Generally, the type of reinforcement which has been used with the intermetallic titanium aluminides has been continuous fibers (SiC and to a lesser extent Al_2O_3), although attention has also been given to an *in-situ* concept, which will also be discussed, and to so called "designer" microstructures. The major attributes of the continuous fiber approach are that elevated temperature strength and stiffness are enhanced, and at the same time the fibers can provide increased toughness by at least three mechanisms (Table II).[29]

With the move to "damage tolerant" designs in which parameters such as fracture toughness and fatigue crack growth rate are of more importance than ductility, this latter attribute of the composite approach could play a major role in application of titanium aluminides. This same trend has occurred with superalloys where some monolithic cast alloys in flying applications have ductility levels of less than 2% at room temperature but acceptable "damage tolerance".[23]

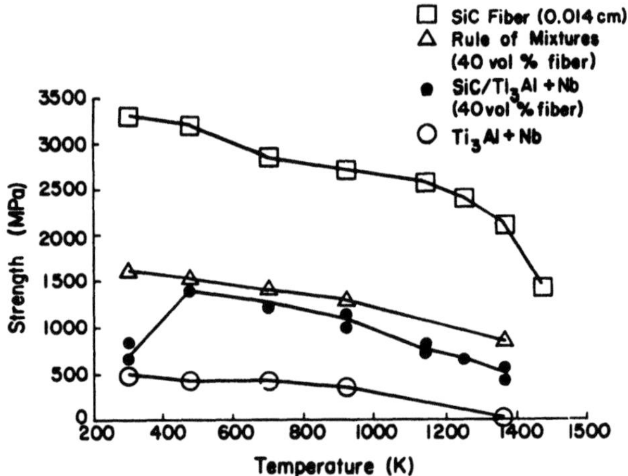

Figure 21. Tensile strength *versus* temperature for SiC fiber, a Ti_3Al monolithic material, a Ti_3Al/SiC composite, and rule-of-mixture predictions.[30]

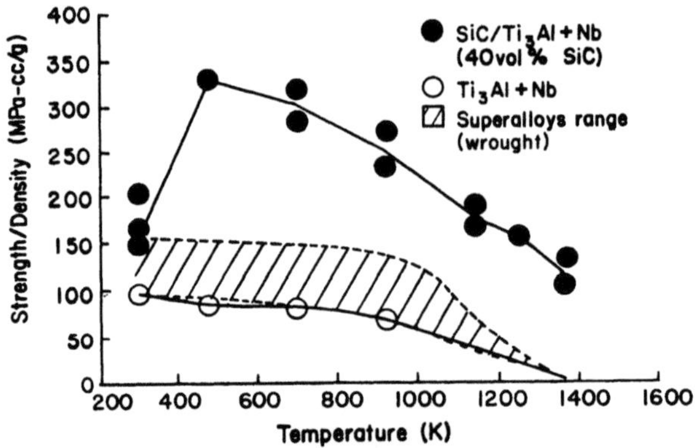

Figure 22. Comparison of density normalized strength for monolithic and composite Ti_3Al, and wrought superalloys for a range of temperatures.[30]

VI.2 Continuous Fibers

The enhancement in longitudinal tensile strength which is theoretically predicted from rule-of-mixtures (ROM) calculations, and has actually been achieved over a substantial temperature regime, is shown in Fig. 21[30] with the actual values averaging about 77% of the ROM predicted values.

A density normalized plot of the SiC/ Ti_3Al+Nb data *versus* wrought superalloy data (Rene 41 and Hastelloy X) demonstrates the attraction of the intermetallic system (Fig. 22).[30]

The desirable characteristics of a reinforcing fiber are summarized in Table III.[23] The latter six of these characteristics are attributes of the fiber itself, whereas the first two items depend on the behavior of both the fiber and the matrix. Silicon carbide

Figure 23. Chemical vapor deposition process used to produce SiC fibers (Courtesy Textron Specialty Materials).

Table IV

Typical SiC Fiber Properties[31]

Fiber Type[a]	UTS (MPa)	E (GPa)	Diameter (μm)	CTE[b] (per °C)	Coating[c]
SCS-2	3400	400	140	1.5×10^{-6}	1μm C-rich
SCS-6	3400	400	140	1.5×10^{-6}	3μm C-rich[d]
SCS-8	3400	400	140	1.5×10^{-6}	1μm C-rich[e]
SM-1040	3500	400	100	1.5×10^{-6}	none
SM-1240	3500	400	100	1.5×10^{-6}	SCTB[g]
SM-1030[f]	3800	440	75	1.5×10^{-6}	none

a. SCS Textron Specialty Materials.
   SM BP Metal Composites.
b. Coefficient of Thermal Expansion.
c. Actual coating more complex, see ref. 31 for further details.
d. Primarily developed for use in titanium based materials.
e. Improved version of SCS-2.
f. Limited production experience.
g. Structural carbon titanium boride.

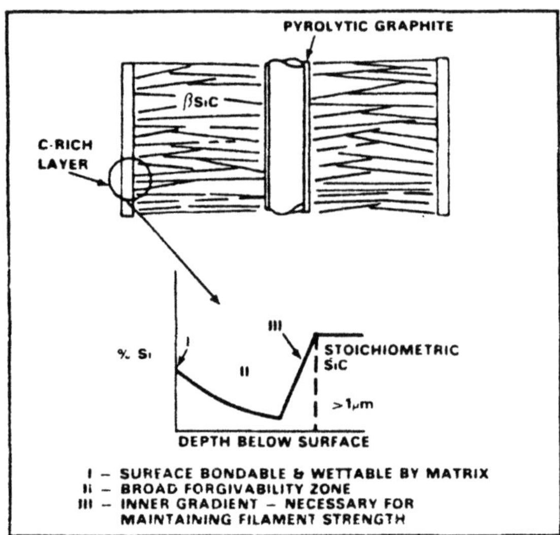

Figure 24. Schematic of SCS-6 (SiC) fiber, including details of carbon rich coating (Courtesy of Textron Specialty Materials).

fibers with the required last six characteristics, are currently available and a number of other fibers (such as Al_2O_3) are in various stages of development.

The SiC fibers are manufactured using a chemical vapor deposition (CVD) process which has been extensively refined over the past few years (Fig. 23).[23,31]

Because of the extreme reactivity of the titanium matrix with SiC at elevated temperatures ($\geq$ 1600°F) a coating is placed on the SiC fiber, which also aids in reducing the crack-sensitivity of the fiber. A schematic of one such coating is shown in Fig. 24.

However, while this coating provides considerably more protection than uncoated SiC, reaction still takes places and can lead to mechanical property degradation. Typical SiC fiber properties are shown in Table IV.[31]

There are a number of techniques which have been used for composite fabrication using either foil or powder matrix material in combination with a continuous fiber. These include hot isostatic pressing, vacuum hot pressing, arc-spraying, plasma spraying (Fig. 16), powder cloth technique, electron beam vapor deposition and a woven fiber mat.[32]

The thermal expansion and chemical compatibility between the fiber and the matrix (items 1 and 2, Table III) present major challenges to the fabrication of high performance composites which must be overcome.[23,31,32] The difference in coefficient of thermal expansion (CTE) between the aluminide matrix and the fiber can lead to cracking in the inherently low-ductility matrix during cool-down after fabrication and particularly under thermal cycling even without an external load.

The chemical reaction between the fiber and the matrix has also been suggested to be a major reason for less than rule-of-mixtures (ROM) for many of the mechanical properties in titanium matrix composites, with increasing thickness of the complex reaction zone being related to a general degradation in mechanical properties.[23,31,33] Studies of titanium matrix fiber composites indicate that while an extensive amount

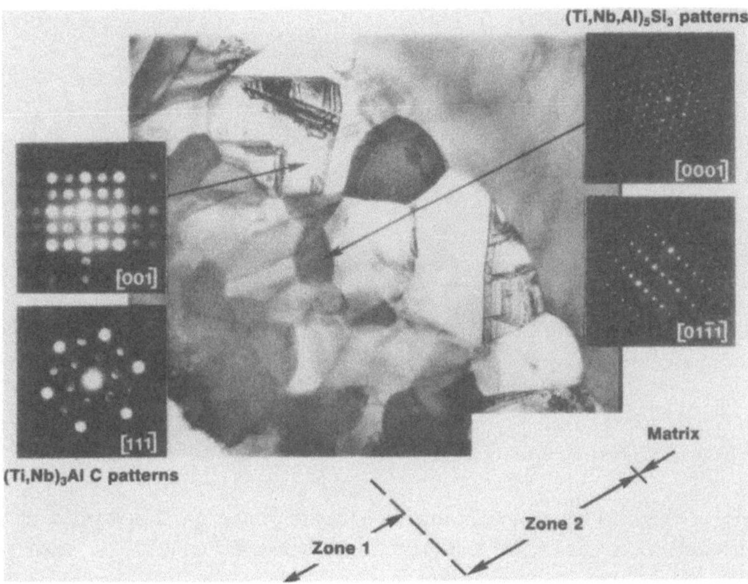

Figure 25. Extensive complex reaction zone occurring between a SCS-6-fiber and Ti$_3$Al-Nb matrix.[34]

of reaction can occur between a Ti$_3$Al matrix and a SiC-based fiber (Fig. 25[34]), this reaction is less than that observed in Ti-6Al-4V/SiC composites,[33] and that with increasing Nb addition to the Ti$_3$Al the reaction zone width decreases.[23,31,32]

Thus with the inherently brittle nature of the titanium aluminide matrices not only is a diffusion barrier necessary to control interfacial chemical reactions, but also a compliant layer to accommodate CTE differences.

The SCS-6/(Ti-24Al-11Nb) composite exhibits excellent elevated temperature strength and stiffness and improvements in fracture and growth crack resistance compared to monolithic material.[35] Longitudinal strength is close to ROM values with the behavior being controlled by the fibers, but the transverse properties are significantly lower with the strength being controlled by the matrix capability. Creep mirrors tensile behavior with the rupture stress in the longitudinal direction being an order of magnitude better than at right angles to the fiber direction. Fatigue crack initiation behavior appears to be controlled by the matrix strain-life characteristics; with notches degrading life in the longitudinal direction, but not in the (much lower) transverse direction. This suggests that under transverse loading the fibers act as potent notches. Under cycling loading, cracks propagated five orders of magnitude faster in the fiber direction than orthogonal to the fibers. The stress intensity at fracture in the transverse direction ranges from 14–19 MPa$\sqrt{m}$, close to the fracture toughness of monolithic material. However, in the longitudinal direction a substantial toughening occurs, presumably due to crack bridging, with toughness values ranging from 110–150 MPa$\sqrt{m}$ at room temperature. As discussed previously thermal fatigue is a great concern with the work of Russ[36] demonstrating that cycling in air can reduce the initial strength by almost 90%, attributed to heavily oxidized surface cracks.

Thus while the composite α_2 material does demonstrate enhanced mechanical property behavior in a number of areas compared to monolithic material, several concerns remain including transverse properties, environmental resistance and designing with a material of limited ductility. It is also unclear exactly what the roles of CTE differences and chemical reaction play in determining mechanical properties, particularly under complex loading conditions such as fatigue.

Higher ductility α_2 alloys (Fig. 12) could be optimized as a matrix for "forgiveness" and environmental resistance, letting the fibers bear the stresses - the concept used in polymeric matrix composites.[29] Further enhancement[23,31,32] in fiber quality also appears possible.

The extremely low inherent "forgiveness" of continuous fiber reinforced γ compositions has precluded the development of meaningful mechanical property data to date.

VI.3 Exothermic Dispersion (XD^{TM})

A dispersion of thermodynamically stable second phase particles produced using an exothermic reaction in which the reinforcement is formed within the matrix at molten metal temperatures has been evaluated with a γ-matrix.[37] This process results in an *in-situ* reinforcement such as TiB_2, with an interface free of reaction products or externally developed contamination such as an oxide layer. Easy decohesion of the reinforcing phase, and early failure, should be prevented by this "clean" reinforcement.

The microstructure of the XD^{TM} γ containing the TiB_2 phase can be controlled both in terms of matrix grain size and morphology (lamellar or equiaxed), and the shape of the reinforcing particles. The XD^{TM} material results in material with enhanced room temperature and elevated temperature strength. The best room temperature ductility is obtained from an equiaxed matrix morphology in combination with particulate reinforcement ($\sim 1\%$ elongation), while a lamellar/short fiber combination optimizes strength, creep resistance and fracture toughness (17 $MPa\sqrt{m}$). To date XD^{TM} material has been developed in various mill product forms including forgings, extrusions, investment casting, powder metallurgy, and rolled sheet.[37] Secondary fabrication techniques such as superplastic forming, welding, and brazing/diffusion bonding have also been demonstrated.

VI.4 Designer Microstructures

The extremely low "forgiveness" of ceramic materials has led to studies of a dispersion of a ductile phase in the brittle ceramic matrix in order to enhance toughness levels. However, the characteristics of the ductile phase necessary to generate optimum toughness have not yet been fully defined. Factors of importance include ductile phases which exhibit extensive plastic stretching after being intersected by the crack, large mean strain for hole initiation, and appreciable residual compressive strain in the matrix, arising from thermal expansion mismatch.[23]

This concept has recently been applied to the titanium aluminides, particularly the γ-alloy which behaves in some ways in a similar manner to a brittle ceramic material, where it was found that second phase ductile particles arrested cracks propagating through the gamma matrix. It appears that for best results no brittle reaction zone should be present between the reinforcing phase and the matrix. This is a potential fruitful area for further exploration with developments in the ceramics area serving as useful guidelines to follow.

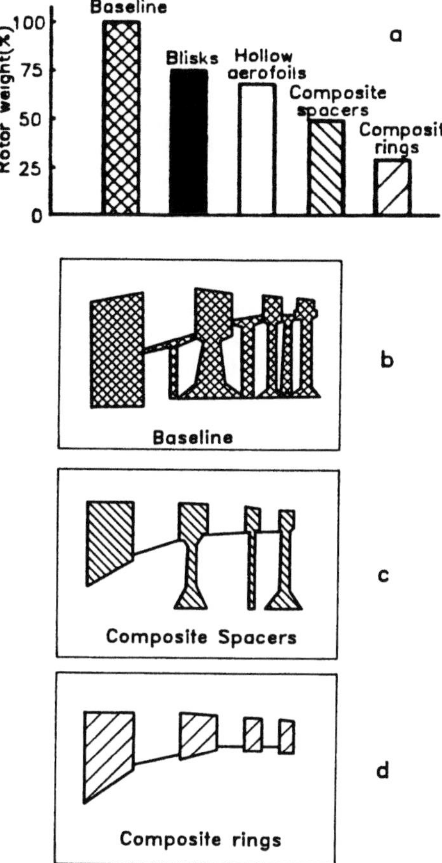

Figure 26. Weight savings achieved by replacing a conventional disc and spacer assembly with a titanium aluminide composite ring.

VI.5 Technology

The Ti_3Al-based continuous fiber reinforced composites demonstrate clear advantages over monolithic titanium aluminides with excellent elevated temperature strength and stiffness and a significant enhancement in fracture and growth resistance. Longitudinal properties are very attractive; however, those in the transverse direction leave much to be desired. Thermal fatigue in air can have a very large negative effect, suggesting that in addition to the necessity for innovative design concepts with these "brittle" anisotropic materials, environmental effects will require resolution with the α_2 composites. Further tailoring of the matrix for composite assemblies, rather than taking "off-the-shelf" monolithic alloys should be pursued including both chemistry and processing/microstructure in this optimization.

Wide-spread use of the titanium aluminides will also require a cost-affordable product. The reinforced Ti_3Al materials offer great potential for weight savings.[38,39] This is the situation for components such as complex shaped blades, but even more so in axi-symmetrical parts such as rings, cases, and discs where the filament direction can be oriented so that the largely hoop stresses are accommodated by these filaments. The dramatic weight saving which can be achieved (up to 75%) by replacing a conventional disc and spacer assembly with a titanium aluminide reinforced ring is shown in Fig. 26.[39]

The composite α_2 titanium aluminides are being seriously considered for use on transatmospheric vehicles such as the NASP.[40,41] Here applications could include skin surfaces, internal structural parts, rocket nozzles, and engine parts.[40] An example of a (presently unreinforced) honeycomb face construction was shown in Fig. 20.[28]

The TiAl-based continuous fiber reinforced material has yet to be developed as a viable structural material. However, the discontinuously reinforced XD^{TM} and "designer microstructure" material offer promising properties; with the XD^{TM} material already being available in various mill product forms. This latter material is likely to receive serious consideration for terrestrial and aerospace applications, initially in non-critical components.

VII. Conclusions

As a result of extensive development programs both in the USA and USSR the intermetallic titanium aluminides, both in monolithic and composite configurations, offer a number of attractive mechanical properties, in combination with low density, which could lead to use in aerospace applications. Lack of forgiveness and high cost are challenges which remain to be resolved before extensive use will occur. However, compared with other potential elevated temperature materials, such as monolithic or composite ceramics, use of the intermetallics as "bill-of-material" in aerospace applications could occur in a relatively short time-frame (by the year 2005).

Acknowledgements

The assistance of T. Bales, A. Begg, B. I. Bondarev, P. K. Brindley, L. Christodoulou, D. Driver, D. Eliezer, R. MacKay, M. Mittnick, A.Notkin, and S. Russ in writing this article is acknowledged. The authors also greatly appreciate the contributions of Mrs. Kandy Nelson, Mrs. Leslie Cossairt, and Mrs. Tracy Gronbeck in formatting and typing the manuscript.

References

1. Congress of the US, Office of Technology Assessment, *Advanced Materials by Design*, (June 1988).
2. A.R.C. Westwood, *Met. Trans. B* **19B**, 155 (1988).
3. F.H. Froes, in: *P/M in Aerospace and Defense*, edited by F.H. Froes, (MPIF, Princeton, NJ, 1990), p. 23.
4. F.H. Froes, in: *1991 P/M in Aerospace and Defense Technologies*, edited by F.H. Froes, (MPIF, Princeton, NJ, 1991), p. 5.
5. *Scientific American*, **255, No. 4**, Oct. 1986.
6. F.H. Froes, in: *Fourth Israel Materials Engineering Conference*, edited by D. Itzhak and D. Eliezer, (Weizmann Press, Jerusalem, Israel, 1991), p. 1.
7. F.H. Froes, *Material and Design*, **X**, No. 3, 110 (May/June 1989).

8. F.H. Froes, *Swiss Materials*, **2**, 23 (1990).

9. F.H. Froes, *Materials Edge*, 19 (May 1988).

10. F.H. Froes, in: *US BOM Conference on Advanced Materials-Outlook and Information Requirements*, edited by Louis J. Sousa and Charles A. Sorrell, (US BOM, Washington, D.C., 1990), p. 41.

11. F.H. Froes, *Materials Edge*, 17 (May/June 1989).

12. F.H. Froes and C. Suryanarayana, *Workshop on Advanced Materials*, Minsk, USSR, May 29-June 2, 1989, *Rapid Solidification and Mechanical Alloying: State-of-the-Art and Technological Implications*, and to be published in the proceedings.

13. F.H. Froes, in: *Proceedings of Second Int. SAMPE Metals and Metals Proc. Conference on Space Age Metals Technology*, edited by F.H. Froes and R.A. Cull, (Dayton, OH, 1988), p. 1.

14. *Materials Science and Engineering for the 1990's*, NRC National Academy Press, NW, Washington, D. C., (1989).

15. M.F. Ashby, *Phil. Trans. Roy. Soc., London*, **A322**, 393 (1987).

16. D.L. McDaniels, T.T. Serafini, and J.A. DiCarlo, *J. of Energy Sys.*, **8**, 80 (1986).

17. Robert A. Sprague, *Ad. Mats. and Proc.*, 67 (Jan. 1988).

18. G.W. Meetham, in: *High Temperature Alloys for Gas Turbine and Other Applications*, (Reidel Publishing Co., 1986).

19. David L. McDaniels, and Joseph R. Stephens, NASA Tech. Memo. #100844, (June 1988), Lewis Research Center, Cleveland, OH.

20. *Defense News*, **6, No. 24**, 1 (June 17, 1991).

21. H.A. Lipsitt, in: *High Temperature Ordered Intermetallic Alloys*, edited by C.C. Koch, C.T. Liu, and N.S. Stoloff, (MRS, Pittsburgh, PA, **39**, 1985), p. 352.

22. C.T. Liu, F.H. Froes, and J. Stiegler, *Metals Handbook*, Tenth Edition, *ASM Int., Materials Park, OH* **2**, 913 (1990).

23. F.H. Froes, C. Suryanarayana, and D. Eliezer, *Iron and Steel Inst. of Japan Int.*, **31**, 1235 (1991).

24. A.I. Taub and R.L. Fleischer, *Science*, **243**, 616 (1989).

25. Zou Dunxu, Private Comm, 16 Jan. 1991, Central Iron and Steel Research Inst. Beijing, PRC.

26. J.J. Jackson and D.A. Siemens, reported by John H. Moll, Charles F. Yolton, and Brian J. McTiernan, *Int. J. of P/M*, **26**, 149 (1990).

27. Y. Nishiyama, T. Miyashita, S. Isobe, and T. Noda, in: *High Temperature Aluminides and Intermetallics*, edited by S.H. Whang, C.C. Liu, D.P. Pope, and J.O. Stiegler, (TMS, Warrendale, PA, 1990), p. 557.

28. E.K. Hoffman, R.K. Bird, and T.T. Bales, in: *Light-Weight Alloys for Aerospace Applications*, edited by E.H. Chia and N.J. Kim, (TMS-AIME, Warrendale, PA, 1989), p. 481.

29. K.K. Chawla, *Composite Materials-Science and Technology*, (Springer-Verlag, New York, 1987).

30. Pamela K. Brindley, in: *High Temperature Ordered Intermetallic Alloys II*, edited by N.S. Stoloff, C.C. Koch, C.T. Liu, and O. Izumi, (MRS, Pittsburgh, PA, **81**, 1987), p. 419.

31. M. Mittnick, F.H. Froes, and R. MacKay, in: *Metal Matrix Composites*, edited by A.R. Begg, (Edward Arnold, Sevenoaks, Kent, UK, 1993).

32. R. MacKay, P.K. Brindley, and F.H. Froes, *JOM*, **43**, 23 (May 1991).

33. P.R. Smith and F.H. Froes, *JOM* **36**, 19 (March 1984).

34. J.I. Eldridge and P.K. Brindley, *J. of Mat. Sci. Letts.* **8**, 1451 (1989).

35. James M. Larsen, Katherine A. Williams, Stephen J. Balsone, and Monica A. Stucke, in: *High Temperature Aluminides and Intermetallics*, edited by S.H. Whang, C. T. Liu, D. P. Pope, and J. O. Stiegler, (TMS, Warrendale, PA, 1990), p. 521.
36. S.M. Russ, *Met. Trans.*, **21A**, 1595 (June 1990).
37. L. Christodoulou and J.M. Brupbacher, *Materials Edge*, 29 (Nov/Dec 1990).
38. D. Dix, *Defense News*, 7 (July 16, 1990).
39. David Driver, in: *High Temperature Materials for Power Engineering*, edited by E. Bachelet *et al.*, (Kluwer Acad. Pubs. Dordrecht, The Netherlands, part 11, 1990), p. 883.
40. T.M.F. Ronald, *Ad. Mats. and Proc.*, **135, No. 5**, (1989).
41. Vicki P. McConnell, *Advanced Composites*, 37 (Nov/Dec 1990).

Microalloyed Steels:
New Alternatives for the Steel
Industry of Mexico

Lorenzo Martínez

Instituto de Física
Universidad Nacional Autónoma de México
Apartado Postal 139B
62191 Cuernavaca, Morelos
MEXICO

Abstract

A review of the investigation in the field of microalloyed steel, high strength, weldable reinforcing bars is presented. The world market for reinforcing bars is now evolving toward higher strength grades which will eventually substitute the 415 MPa, nonweldable rebars now produced in North America. Since 1991, the Euronorm 18 has established a 500 MPa weldable grade for all the European Community, and there are initiatives to set a 590 MPa weldable grade for the reinforcing bars to be used in the Pacific Rim. Experiences in fabricating 590 MPa weldable grades have been performed in the steel mills of SICARTSA employing niobium and vanadium as microalloying elements of steel. The research to be described has been supported by transmission electron microscopy, microanalysis and mechanical testing. The results point toward the modification of the hot rolling facilities, to introduce controlled cooling, to reduce carbon and carbon equivalent for better weldability and to minimize the ammount of microalloying elements to reduce cost.

I. Introduction

The steel industry fabricates a wide variety of products for the many structural components that are made of this material. One of the items which accounts for a big fraction of the total is the rebar steel. In Fig. 1 the share of the steel rebar fabrication is shown in terms of the total steel production of Mexico and the United States of America. In Mexico the development of new infrastructure is intense and the con-

Advanced Topics in Materials Science and Engineering, Edited by
J.L. Morán-López and J.M. Sanchez, Plenum Press, New York, 1993

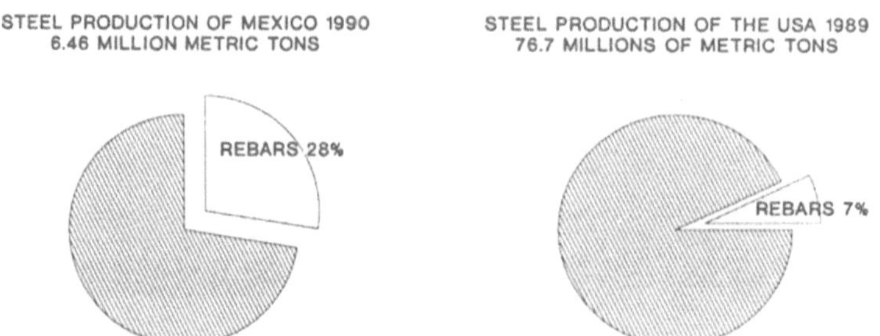

Figure 1. Steel production in Mexico and the USA. The fraction of steel dedicated to rebar production is depicted.

struction work represents a vital sector of the economy. Also part of the rebar steel production is sold to other countries, therefore the share of the rebar fabrication is big. This is not the case in the USA where the economy favours other uses of steel and imports of rebar steel are important. However, as a single item, the 7% (Fig. 1) is still a significant fraction of the total steel production of the USA.

The technology for fabricating rebar steels in America as a continent has remained dramatically static for decades.[1-5] The last big transition of the technology of fabricating reinforcing steel bars in Mexico and the USA occurred about 25–30 years ago when the old "high strength grade" rebars where the elastic regime limit was set above 415 MPa (60 ksi). The economic advantages associated to the significant increase of the steel strength and consequent reduction in the volume of steel required for construction were considerable, since the cost of fabricating the new rebars was almost the same. The rebar steels traditionally have been fabricated employing plain carbon steels which are alloys of Fe, C, Mn and Si (many other elements may also be present in the alloy but as impurities such as S and P). The main change introduced to the steel fabrication process during the transition to the high strength grade was the element carbon wich increased from 0.2–0.25 wt. % to the range of 0.45–0.50 wt. %. Manganese was also increased in some steel mills but generally was maintained within the range of 0.9–1.4 wt. %.

Soon after the transition to the high strength grade it was recognized that the weldability of the steel was severely compromised, because the high levels of carbon and carbon equivalent of the steel. The carbon equivalent represents the contribution of carbon and the other elements to the formation of structures susceptible of hydrogen embrittlement during welding and is calculated by

$$C.E. = C + Mn/6 + (Cr + Mo + V)/5 + (Ni + Cu)/15. \tag{1}$$

where as usual the elements represent weigth percent. In the carbon-carbon equivalent diagram of Fig. 2,[1] the transformation is schematically represented. The rebar steels moved from a region of optimum-medium weldability (C) to a region of high risk weldability (A). The specifications for the high strength rebars such as the ASTM A615 (USA) or the NOM B6(Mexico) reflected this problem and explicity excluded the weldability.

The specification ASTM A706[7] was issued in order to cover the rebars of high strength low alloy (HSLA) steel which would be weldable. However, in practice, this kind of rebars have been essentially non-existent in the American market.

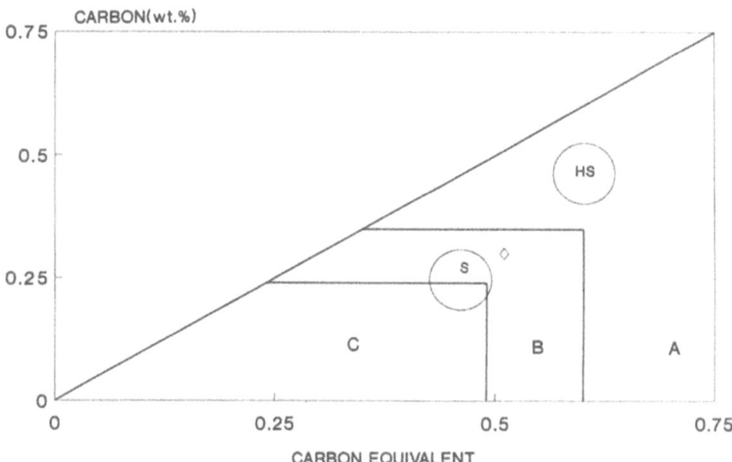

Figure 2. Carbon-carbon equivalent diagram indicating the areas of optimum weldability (C), regular (B) and of high risk of cold crack formation (A). The circle S indicates approximately the levels of carbon and carbon equivalent of the alloys used for fabricating the rebars of the old structural grades. The circle HS correspond to the medium carbon high strength grades. The diamond represents the 590 MPa minimum yield strength rebar to be discussed in this paper.

Other interesting properties of the carbon steels were also compromised when the rebars began to be fabricated employing medium carbon steels. Among these was the type of elastic-plastic transition. Most theories of concrete reinforced structures design consider that the mechanical behavior of the steel is represented by only two parameters: The elastic modulus E, and the yield stress σ_y. Typically it is expected that the steel will respond to eventual deformations of the structure first in a linear mode during the elastic regime and then it will yield plastically at constant stress up to a deformation of about 1.5%. The elastic-perfect plastic behavior of the steel is assumed by almost every model or code for earthquake resistant construction. The main reasons being predictability and adequate energy absorption by plastic yielding.[1,8] To produce a steel with a near elastic-perfect plastic regime means in metallurgical terms to have a stress strain curve with a wide Luders deformation after yielding. This is a characteristic property of hot rolled plain low carbon steels that have a carbon content below 0.3 wt. %. The rebars produced after the transition to high strength do not yield at constant stress but rather show a "round house" type of stress strain curve which cannot be easily introduced to the current models of earthquaque resistant construction.

The technological progress in the field of steel rebars in Europe has been very dynamic during the last decades. The search of high strength combined with high ductility, weldability and toughness has produced innovative methodologies for rebar fabrication. The two main developments followed by the European producers are the Tempcore process and microalloying. Tempcore is a process where the hot rolled deformed bars are quenched at the end of the rolling mill by applying high pressure jets of cold water on the red hot steel surface. This process hardens a crust near the steel surface while the bar core remains with high ductility. The overall behavior of the

Table I

Chemical Composition of the Rebar Steel

Element	C	Mn	Si	P	S	Nb	V	Al	N
wt. %	0.30	1.36	0.32	0.038	0.011	0.036	0.15	0.004	0.0053

steel is improved and a low carbon, high stength, ductile weldable and tough material is obtained. Microalloying as will be described later in this paper, depends on precipitation for grain refining and hardening a low carbon ferrite-pearlite microstructure.

Recently the European community gave a very important step: Since 1991 all the rebars to be fabricated or used in Europe will be alloyed with a carbon content below 0.2 wt. % and a maximum carbon equivalent of 0.49. The steels within this limits are within area C of optimum weldability in the carbon-carbon equivalent diagram shown in Fig. 2. The new European specification sets a minimum yield strength of 500 MPa (about 72 ksi) and 8% minimum uniform elongation.

Many countries in the Pacific Rim are influenced by European standarization. Usually they have been employing British standards for rebars and have followed the general trend of increasing strength within the high ductility and weldable limits.[8] There is now an initiative for this area to set a market trend for weldable rebars of minimum yield strength above 590 MPa (about 85 ksi).

II. Microalloyed Steel Fabrication

The technology of microalloyed steels has provided a variety of alloy systems which can combine the advantages of the high strength to the high ductility, weldability and toughness of the lower carbon and formerly low strength grades.[3,10,11] Niobium, vanadium, titanium, boron and aluminum are the most frequently used microalloying elements of steel.[10,11] The combination of niobium and vanadium in small amounts in steel has been widely studied.[11-14] Generally the niobium rich precipitates are known to form and act during the transformations at the higher temperatures of the austenite field. The vanadium rich precipitates occur at the lower temperatures of austenite and have a significant role during precipitation hardening of ferrite.[15] Some synergetic mechanisms of the combination of niobium and vanadium have been reported. Both elements act as effective recrystallization and grain growth inhibitors during the thermomechanical processing of austenite, leading to a finer grain ferrite and pearlite structure.[6]

The steel used for the present study was produced in an integrated blast furnace-basic oxygen furnace plant. The process also works employing steel scrap which amounts up to 20% of the load of the basic oxygen furnace. The final alloy preparation of the chemical composition, given in Table I, is arranged in a ladle furnace while the melt is agitated with nitrogen and/or argon jets. Standar ferroalloys such as ferrovanadium, ferroniobium and silicomanganese were incorporated to the melt also in the ladle furnace.

The liquid steel was solidified in a continuous casting machine in the form of billets of square section of 150 mm per side, and air cooled to room temperature and stored. Later, the billets were reheated at 1150 C and hot rolled. The final diameter

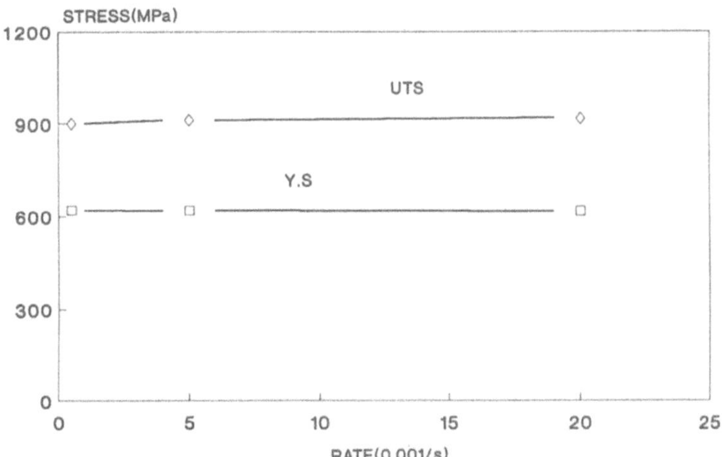

Figure 3. Yield stress and ultimate tensile strength of the steel rebars in a wide interval of deformation rates.

of the deformed bars was 19 mm and left the rolling mill at a temperature of 1050°C. The rolling mill was not equiped with auxiliary cooling systems to allow temperature control.

III. Mechanical and Microstructural Characterization

Samples of the bars were employed to prepare tensile specimens to be tested in a Model 4200 Instron. The samples for optical and scanning electron microscopy were gradually polished with alumina powders of 1, 0.3 and 0.05 μm. The metallographic samples were further etched with Nital. the optical microscope used is an Olympus PMG3 and the scanning electron microscope is a Jeol JSM-200.

The samples for the transmission electron microscope were prepared by cutting 0.1 mm thick and 3 mm diameter disks from the steel bar. The disks were exposed to a twin jet of a solution of 10% perchloric acid in ethylic alcohol and cooled at −50°C. The scanning electron microscope used was a Jeol 100CX.

IV. Results

IV.1 Mechanical Behaviour

The yield stress and the ultimate tensile strength of the material have average values of 650 MPa and 901 MPa, respectively. The measurements were performed at deformation rates between $8.3 \times 10^{-5} \text{s}^{-1}$ and $4.4 \times 10^{-3} \text{s}^{-1}$ as is indicated in Fig. 3. The average elongation to fracture is 12%. A full stress strain curve is presented in Fig. 4. The yield point is clearly defined and the strain hardening regime is extended in strain, but mainly in stress. The average ratio UTS/YS is 1.4.

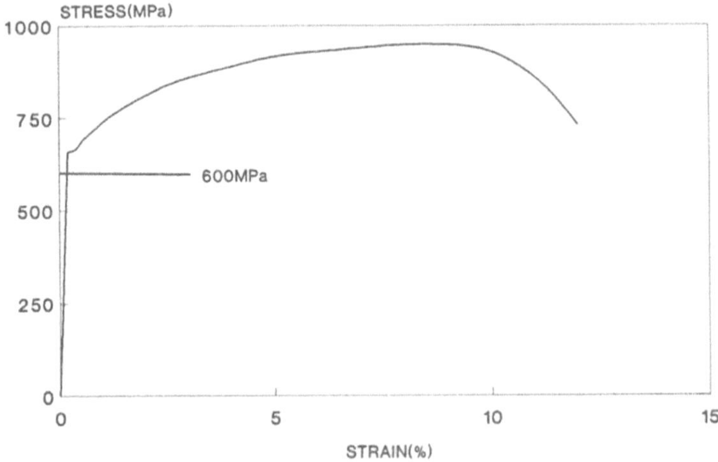

Figure 4. Engineering stress-strain curves of one steel rebar. The elastic limit is clearly defined and the strain hardening regime is extended in stress and in strain.

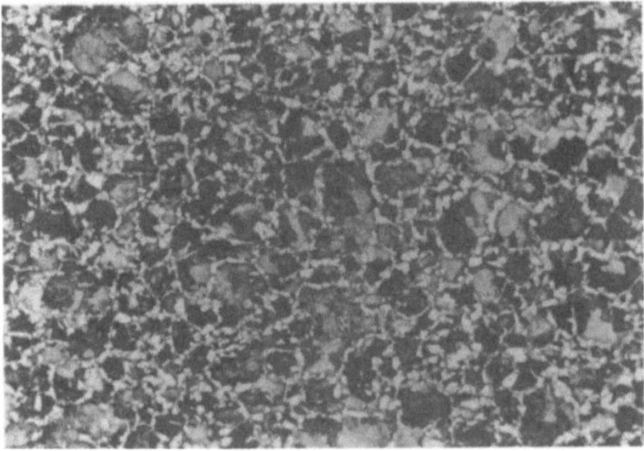

Figure 5. Microstructure of the steel in an optical microscope. The pearlite grains dominate over the microstructure.

IV.2 Microstructure

The steel has the fine microstructure that is shown in Fig. 5, which is essentially clean. The inclusions are mainly manganese sulfides and aluminum oxides. The average grain size is 15 μm. The grains of pearlite are larger and dominate over the microstructure. The ferrite grains are smaller and contour the pearlite grains as shown in Fig. 6. The average width of the lamellae is 8 nm. The particles in the nanometric scale are niobium and vanadium carbonitrides. The faceted particles were found to be rich in niobium. Apparently the finer particles are vanadium carbonitrides, which are shown in the bright field TEM image of the interior of a ferrite grain in Fig. 7.

Figure 6. Ferrite grains are formed along the limits of the pearly grains.

Figure 7. Fine precipitation in a ferrite grain.

V. Discussion

The results indicated that the steel rebars of minimum yield strength above 590 MPa can be fabricated employing the alloy described above. The strain hardening regime is wide, both in stress and strain. The UTS reaches about 1.4 times above the yield strength. Other steel rebars of similar strength have the problem of yielding at a stress very close to the UTS and to fracture. This may be troublesome for structural designers and has been conveniently avoided in the present case.[19] The deformation to fracture is reasonably large and above the minimum of most steel rebar specifications.

The role of the components of the microstructure on the mechanical properties of the steel has been schematically represented as follows[9]

$$\sigma = \sigma_0 + \sigma_{SSS} + \sigma_{ISS} + \sigma_{PPT} + \sigma_{DSL} + \sigma_{SUB} + \sigma_{SPH} + Ky/d^{0.5} \qquad (2)$$

where σ_0 is the basic steel matrix strength. σ_{SSS} is the contribution to strength of the alloy elements in substitutional solid solution which in this case are manganese and silicon. σ_{ISS} is the contribution of interstitial elements in solid solution, σ_{PPT} of the fine precipitation, σ_{DSL} of the dislocations, and σ_{SPH} of second phase. The term $Ky/d^{0.5}$ represents the contribution of grain size. Appart from the σ_0 and the σ_{SSS}, all the other contributions to the strength are very sensitive to the thermomechanical processing of the steel.

Several contributions to the steel strength are known to be dependent of each other. For instance the fine precipitates harden the ferrite matrix but it is widely accepted that the main role of the precipitates in microalloyed steels is to refine the grain structure. In this case, the fine precipitation of the vanadium carbonitrides is clearly playing its role in precipitation hardening, but the relatively large grain size indicates that the role of the precipitates as grain refiners is not working at its full potential.

Vanadium and niobium in microalloyed steels participate in the enhancement of grain refining by retarding the recrystallization of austenite, and inhibiting the grain growth at subsequent stages. By itself niobium is a good retarder of the recrystallization of austenite. It is reported that the austenite recrystallization of a 0.03% Nb microalloyed steel is about three orders of magnitude slower than the austenite recrystallization of a plain carbon steel of the same alloy composition, except for niobium, at a given temperature. The addition of vanadium in the range of 0.2% approximately reduces austenite recrystallization by another order of magnitude.[11-14]

The roles of vanadium and niobium as grain refiners are optimized by controlling the temperature during hot rolling. In the steel bars reported here the temperatures along the rolling mill go from 1150°C at the entrance to 1050°C at the end of the finishing stage. The temperature 1050°C is relatively high in terms of the several transformations involved in the material. The reduction of the finishing temperature from 1050°C to about 950°C is known to introduce at least one order of magnitude of reduction in the recrystallization and further grain growth kinetics.[15] Changes in the rolling mill facilities have been initiated in order to instrument the temperature control of the bar at the finishing stage of rolling.

The weldability of the steel rebars is now demanded by many constructors. According to the chemical composition given in Table I, the location of this steel in a carbon-carbon equivalent diagram is depicted in Fig. 2, with a diamond. The weldability of the steel is not poor, but it is not within the optimum area. The steel can be welded if proper preheat and intercaps temperatures are provided during welding. It is however very desirable to reduce the carbon content of the steel below 0.24% in order to have it in a better condition for welding. Actually this could be done maintaining the composition of the other elements without impairing the steel strength, if the rolling temperature is adjusted to let the microalloying elements have a better action as grain refiners.

VI. Conclusions

We reported the experience provided by the production of rebars of minimum yield strength of 590 MPa employing a heat of steel microalloyed with niobium and vanadium. The mechanical properties indicate that the rebars are suitable for reinforcing concrete construction both in terms of the yield strength that was aimed and of the strain hardening behaviour. The microstructural characterization exhibited the role of the microalloying elements in forming precipitates and allowed us to determine the limitations of the thermomechanical processing. The relatively high temperature in the finishing stage of the rolling mill was associated to the large grain size developed by the steel.

Acknowledgements

This work was developed partially at the industrial facilities of SICARTSA, thanks to the enthusiastic support of Ing. Gabriel Magallon. The valuable support of the TEM of the Instituto de Investigaciones Eléctricas (IIE) is acknowledged. We thank the technical support of J.L. Albarrán, A. Sánchez Ariza, A. González and O. Flores Cedillo at the Instituto de Física UNAM, Alberto Brito at IIE and J.J. González at SICARTSA.

References

1. V. Castaño and L. Martínez, *Journal of Materials Research,* **5**, 658 (1990).
2. L. Martínez, J.I. Albarran, and J. Fuentes, *Welding Journal,* **66**, 23 (1987).
3. A. Hey, H. Weise, and W.G. Wilson, in: *Niobium*, edited by H. Stuart, (TMS, Warrendale, 1984), p. 967.
4. F. Estevez y L. Martínez, *Siderurgia Latinoamericana,* **329**, 42 (1987).
5. J.L. Albarran, B. Campillo, F. Estevez, and L. Martínez, *Scripta Met.,* **23**, 1099 (1989).
6. B. Campillo, J.L. Albarran, F. Estevez, D. López, and L. Martínez, *Scripta Met.,* **23**, 1363 (1989).
7. Specification ASTM A706-76, American Society for Testing and Materials, p. 739 (1978).
8. J.F. McDermott, in: *Proc. of Workshop on Earthquake Resistant Reinforced Conrete Building Construction*, edited by V. Vertero, (Berkeley, Ca., 1978), p. 629.
9. British Standar B-4999, London (1985).
10. P.E. Pepas, in: *Microalloyed HSLS Steels, Procc. Microalloying 88*, (ASM international, Metals Park, 1988), p. 3.
11. A.J. de Ardo, J.M. Gray, and L. Meyer, in: *Niobium*, edited by H. Staurt, (TMS Warrendable, 1984), p. 685.
12. J.R. Michael, J.G. Speer, and S.S. Hansen, *Met. Trans.,* **18A**, 481 (1987).
13. M.J. Crooks, A.J. Garrat, J.B. Vander, and W.S. Owen, *Met. Trans.,* **12A**, 1999 (1981).
14. J.G. Speer, J.R. Michael, and S.S. Hansen, *Met. Trans.,* **18A**, 211 (1987).
15. M.J. Crooks and J.M. Chilton, *Met. Trans.,* **15A**, 1137 (1984).
16. L.J. Cuddy, *Thermomechanical Processing of Microalloyed Austenite*, (TMS-AIME, New York, N.Y., 1982), p. 129.

Future Ferrous Technologies

H.W. Paxton

Carnegie Mellon University
Pittsburgh, Pennsylvania, 15 213
U.S.A.

Abstract

The patterns of changing technology in the world steel industry are discussed from a base in 1970 to help understand the changes which are likely to occur especially in the USA The point is made that the industry is no longer expanding at its historical rate, and thus the emphasis must be on efficient production in an arena where competitive materials, environmental regulations, and changing energy sources set the constraints. While changes are necessarily slow, the industry is still reasonably robust and can probably survive quite well with the introduction of the new technological concepts here or on the way.

I. Introduction

The world steel industry has undergone some very significant changes over the last twenty years or so. Many groups have made unfavorable comments on the viability of the industry, both in the USA and on a worldwide scale. The comments often reflect a feeling that the management within the industry is Neanderthal, and has caused a lot of its own problems by an unwillingness to invest in the necessary technology to remain competitive in the world markets which have grown in the seventies and especially in the eighties. As in most generalizations, there is some truth in the accusations but the picture is much more complex than the simplistic view of "experts" (often self appointed).

This short paper will attempt to outline some of the issues which affect the future of the world steel industry, and will find that, while a new boom is unlikely, the current industry is much more robust than many would expect. I apologize in advance to readers familiar with steel processing for what may seem like unnecessary detail, but others would miss the significance of my theme without this background.

Advanced Topics in Materials Science and Engineering, Edited by
J.L. Morán-López and J.M. Sanchez, Plenum Press, New York, 1993

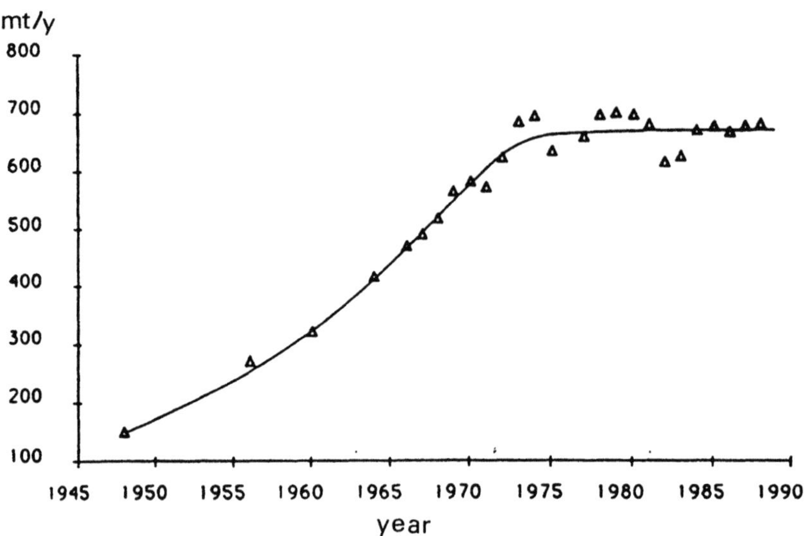

Figure 1. World crude steel production 1948–1988 A .Kelly.

II. Status in 1970

In order to appreciate the changes which have taken place, it is convenient to use the situation in 1970 as a baseline. Discussion of events since offers an idea of the rates of change and can be used to examine possible and plausible projections for the future.

1970 is not an entirely arbitrary choice. At that time, the US was the dominant world steelmaker with a sizeable export business in both products and technology. Japan (using much technology imported from the US and Europe) and most of the European companies had made great strides in replacing and expanding facilities destroyed in World War II. The Comecon countries had considerable capacity installed, although the technology (with the exception of some in the blast furnace) was, and is, not the most modern.

1970 also represents the year in the US when the volume of steel produced in open hearths was equal to that in top blown converters with electric furnaces a poor third at about 15%. "Integrated" plants were the norm with most companies having control over their own resources such as ore and coal mines (sometimes even limestone), railroads and shipping, foundries, and many auxiliary service and sales companies. World trade in steel largely to non-producing countries was about 100 million tons per year in a total production of nearly 700 million tons (Fig. 1).

The standard technology used a feedstock mix of coke from by-product ovens (which generated substantial methane and hydrogen for fuel and heavier gases which were used for aromatic chemicals) and beneficiated ore as either sinter or pellets. The blast furnaces into which these were charged were often small by today's standards although the Japanese were experimenting with what is about the current size. Steelmaking, by whatever method, made a substantial number of heats which did not fully meet chemical specifications and offered marginal control of pouring temperature at the ingot stage. Continuous casting was rare and in any case, was in a rudimentary stage as far as large slabs were concerned; breakouts, slab cracking and nozzle clogging were too frequent for reliable production scheduling.

Electric furnaces were usually three-phase AC with graphite electrodes and varied in size from 5 tons for specialty steels to 100 tons or more for carbon steels. Furnaces have since been built over 300 tons but a more typical size today in regular use is often well under 200 tons. The increasingly important single electrode DC furnace has certain operating advantages and was just coming into use at about 15 tons; current models are much larger.

Ingots needed to be reheated ("soaked") before further mechanical working. Some adventurous efforts to roll the ingot while the core was still molten led to much excitement but not to a practical process! Ingots contain a number of defects both on the surface and internally and can not be reduced directly to finished products without inspection and conditioning (surface defect removal) at an intermediate size. Since this must be done at room temperature, further working of this "semi-finished" state requires yet another heating to make plates, bars, strip, etc. In 1970, few efforts were made to control properties by heat treatment on the mill or by careful cooling afterwards; although the necessary understanding was reasonably well established, the equipment was not yet adequate.

Some of the hot rolled material was pickled (the oxide scale removed in acid) and rolled or drawn further at ambient temperature to sheet, strip, bars or wire. Often these were given a recrystallization anneal to restore ductility (and possibly other properties) in an atmosphere controlled furnace. Because of thermal inertia, this was an inherently slow process, taking days, and with obviously poor productivity.

By this time, even the neophyte to steel will realize that there were many batch processes each with considerable waste ("yield loss"), and often poorly connected to the next stage so the time to produce a specific order was long. While plants could often handle all processes within their own boundaries, the layout was such that much time, effort, and expense was spent moving product from one department to the next. Sometimes it was necessary to move a material in an intermediate stage to a sister plant, often several hundred miles away. By 1970, the first of the new Japanese seacoast plants with a logical product path had been built. US plants had been located in locations to serve fairly local needs from 1870 on, with no thought of world competition. Capital costs were always high and as a result, frequently only replacements for specific obsolete equipment were made on an *ad-hoc* basis. Some new plants were built between 1950 and 1970 but the technology was chosen conservatively.

This should not be taken to imply that the research competence of US industry was second-rate. In fact, into the seventies it was extremely venturesome in large scale process developments, some of which were installed in operating plants and some of which are now being revisited. These included in-line rolling of CC slabs, coal-ore briquettes, formcoke, reduction of liquid iron oxide by carbon, thin slab casting and production of coated sheet by vapor deposition. These activities were gradually curtailed on an individual company basis as the industry was restructuring, but have recently reappeared in part with the formation of industry consortia.

III. External Changes Since 1970

The picture I have tried to draw for 1970 is of a successful and profitable steel industry in the US in spite of, in hindsight, many inefficiencies in siting, installed technology, quality control, productivity, and unconcern about the cost of energy and effects on the environment. As world economics, politics and social judgements changed, the industry responded slowly and parts did in fact disappear or come very close to disappearing before dramatic changes were instituted. Several important issues will now be discussed.

III.1 Overproduction

In the rest of the world, basically new plants had been built in Japan and Europe and were looking for markets wherever they may be. Other countries were developing their economies and as the old joke goes, the first priorities were a national airline and a steel industry. Between 1970 and 1974, the demand for steel was increasing rapidly and dire predictions were being made of a global shortage by people inside and outside the industry. Many plans and purchases were made with this as a hypothesis and caused a great deal of the subsequent turmoil. At a conference which I attended in Vienna in 1975, there was some doubt whether the demand in 1985 would be 1.5 or 2 billion tons. In actual fact, 1974 was the high point in world production for practical purposes, was approximately the same in 1978–80 and has not reached this level since. In 1985, world production was less than 700 million tons (Fig. 1).

Since steel production in integrated plants is a capital intensive process, the break-even point is uncomfortably high. As a result, there has been a long history in some cases of running the facilities as close to 100% as possible and looking for markets at discounted prices to dispose of the product. For industries which are partially or totally government owned, this also serves to keep employment high. Privately owned companies in the US can not operate with this philosophy, although for reasons which will not be discussed here in detail, employment costs were very high on a total hourly cost basis relative to world costs in the seventies (the disparity is now much less).

As new facilities came on around the world, often funded by international agencies on something other than plausible economic plans, the logical outlet for them was the US which had by far the most open market. Eventually the import levels into the US rose above 20% of the market, and while certainly not all of this was "dumped" at unfair levels, enough was so that political actions were generated, causing in turn considerable resentment.

III.2 Energy costs.

In 1974, the first of the OPEC oil price increases quintupled the cost of crude oil with enormous repercussions on world commerce. The effect was not so much directly on the energy costs of the steel industry, since the great majority of these were from coal and the companies by and large owned their own mines. However, the general inflationary climate which developed made the costs of other goods and services rise sharply. One particularly significant change was in the already high cost of labor (and management whose salaries went up in lockstep with the unionized labor). In 1972, as part of the Experimental Negotiating Agreement which eliminated strikes as a mechanism to avoid the triennial disruption of hedge-buying before contract discussions, a cost-of-living adjustment was introduced. On a historical base in the sixties, this COLA was entirely reasonable but in a time when the prime rate approached 20%, the COLA adjustment became a killer. Hourly total costs were $25/h and with far too many man-hours per ton, trouble was certain.

Prices could be and were adjusted upward to a point but eventually ran into consumer resistance. The days when USS could tell its customers what the price would be, and be followed by the other companies, were no more. The comfortable ROI's which had been expected disappeared. With the second "oilshock" in 1980, a whole new attitude arose and the total plan on how to run a steel business was changed.

Energy conservation projects were started but had rather modest effects because of necessity they nibbled at small pieces of the operation. The large energy consumers are ironmaking, (which is often more than 75% of total energy), reheating for hot working operations and the basic yield loss which occurs inherently with ingot casting because of the need to crop the "pipe" which results on solidification. Recognizing these facts, it would appear straightforward to improve coke-making and blast furnace operation, install 100% continuous casting, and never let the solid steel cool after casting until it is in finished form.

For various technical and marketing reasons, these are not always easy to accomplish, and this is especially true in a plant which is already built. Short of tearing down the whole plant, which may be economic suicide, and with capital limitations, generating substantial savings of energy comes slowly and at least partly from operational changes. Continuous casting has been a clear contributor to energy saving once product lines were rationalized within a given plant to enable most, if not all, of the liquid steel to pass through the caster. Eventually considerable savings were made in the US (although perhaps more slowly than in Japan for example where the plants were by and large newer) from nearly 40 GJ/ton to 25 or less.

III.3 Environment

The early seventies saw the passage of the first Clean Air Act; the response from the industry was slow. This was not entirely due to recalcitrance although there was a good deal of this. The early standards were somewhat vague and based on little evidence. As a result, credibility was poor. Furthermore, coordination between local, state, and federal standards was lacking.

The technical aspects of reducing particle and specified gaseous emissions were not simple at any reasonable cost. Comparison of the two basic ways of meeting specifications – putting a "box at the end of the pipe" or totally revising the process- rapidly caused the first to be selected as the only solution possible in a reasonable time scale.

This led to some interesting anomalies. When a by-product coke oven is pushed, a large amount of dust and gases escape into the air unless some capture mechanism is installed. Yet the motors driving the large fans necessary to capture all the emissions have to be so large that the emissions from the power plant (not regulated at that time) are greater than those captured at the oven.

The problems became even more complex when balances had to be reached between media, *e.g.*, the water used for quenching coke (to stop burning away after pushing) can be either clean or recycled from previous quenches. Recycled water puts ammonia and some chlorides into the air; clean water can be passed through a bacterial treatment to remove ammonia, phenols, and cyanides but subsequent disposal can put chlorides into rivers or lakes, which can be a problem if the flow is poor. Seashore plants have an advantage here since an increase in chloride is quite tolerable for ocean disposal. The problem of coke plant emissions is still not very tractable and is currently under review.

The alternative was to transfer the hot (1300°C) coke to a closed container, use an inert gas to cool the coke, and extract the energy to produce electricity. This is

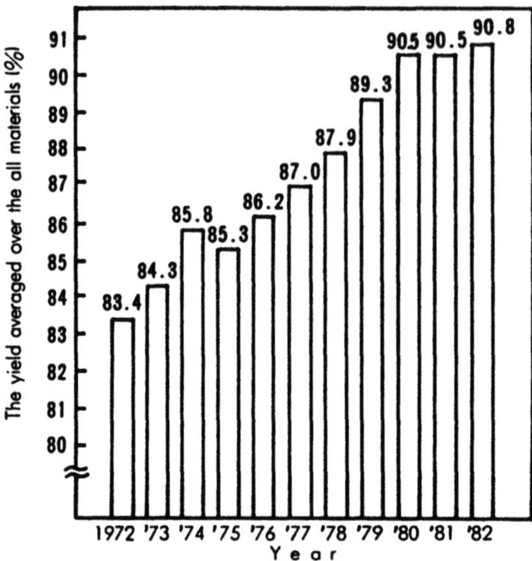

Figure 2. Changes in yield in Japan (average of all products).

a capital intensive process which has only been economically justified in the USSR and occasionally in Japan. Over the last twenty years, environmental constraints have been a net cost. Few, if any, process changes have simultaneously cut costs and improved the environment.

III.4 Yield Improvements

The introduction of continuous casting has been the biggest contributor to the yield, defined for example as the ratio of shipped products to the volume of liquid steel made. However, attention to the details of each stage of the processing also contributes. Technical issues would take too much time to cover here, but we may note that the best yields today are well above 90% whereas with ingots, 75% was often good (Fig. 2).

III.5 Currency Fluctuations

In the period we are discussing, the value of the dollar relative to other currencies has varied significantly. Since most US mills have virtually all their costs in dollars and others have only part, the relative costs of steel as manufactured can fluctuate between countries. When coupled with subsidies of various kinds, which has been common, the resulting pricing in the US causes much friction. More recently, the problem has eased somewhat but has certainly not been eliminated.

III.6 Poor Market Estimates

The predictions of steel shortages in the eighties have been mentioned above. The almost unshakable belief that steel demand would increase at between 2 and 3% per year worldwide (which had historically been true from 1900 to 1980) established a

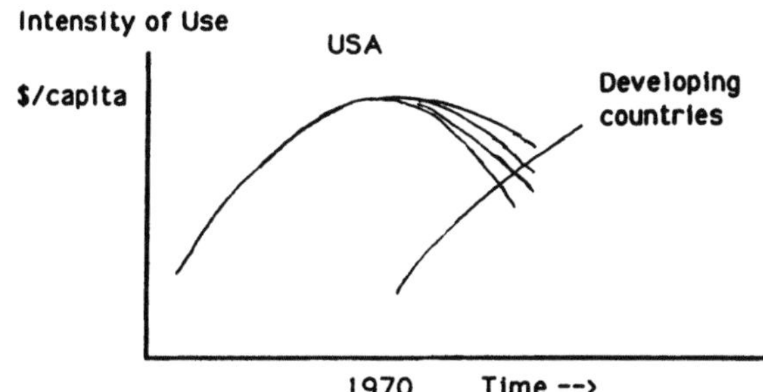

Figure 3. Intensity of use as a function of time and national maturity.

mindset that "things would get better" and prevented steps to both reduce capacity and change the product mix at each plant. An appreciation of the concept of "intensity of use" as summarized by Malenbaum[1] would have permitted an earlier recognition of the inevitability of a decline in consumption (Fig. 3).

Even though the developing world was still increasing its intensity of use, the percentage of the market in the developed world was so high that this increase was almost inconsequential.

III.7 Social Issues

The question of employment in the industry around the world is one of the thorniest and is at the root of much of the international friction and slow responses to difficulties. Third world governments needed employment for stability, and in the developed world, union contracts were drawn with the idea that the industry was expanding. Thus, when it became clear that certain plants would have to shut down and those which remained would be operating with many fewer people to increase productivity, the costs of exit (augmented pensions and lump sum payments) were extremely high - so high that formal announcements were often postponed which led eventually to even more extra costs.

IV. Internal Changes Since 1970

The issues discussed above, while they are covered at some length, are in my view necessary as a backdrop to what has happened on various technical developments in this period, and in what may be possible in the future which is the prime purpose of this paper.

In brief summary, the industry basically employs the same processes which were available in 1970 but the degree of efficiency, control and emphasis on quality has improved almost beyond recognition. The degree of continuous casting employed has contributed to yield and energy reduction in a major way; the sensors and controls introduced have made the quality of cast slab so good that cooling for inspection is usually not necessary allowing hot charging to the further working processes.

Blast furnaces, led largely by Japanese work have become generally bigger, burdened more efficiently with carefully sized self-fluxing sinter and/or pellets, use high quality coke and have simultaneously maximized tons of iron per day and minimized

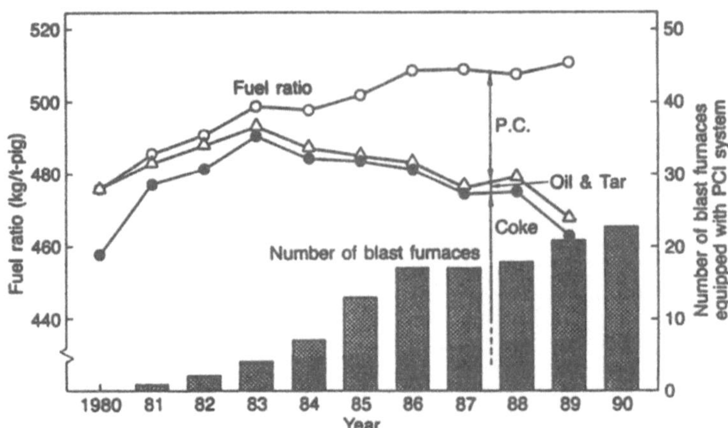

Figure 4. Change in fuel ratio and number of blast furnaces with PCI system (Japan)(2).

fuel consumption. Coke usage reached a minimum around 400 kg/ton until oil injection as a supplement to reduction (10%) had to be reduced for cost reasons and replaced by coke. As coke itself became scarcer because aging ovens which did not meet environmental requirements were not replaced, pressure arose to find other inexpensive reductants while maintaining the mechanical function of coke in acting as a burden support (thereby retaining permeability and keeping the pressure drop within bounds). In the middle 80's substantial pulverized coal injection (PCI) became practical and the amount used has been increasing steadily aided by oxygen injection (Fig. 4).

Because the ironmaking stage is so important to the total process, further refinements to control the composition of the iron (leading to better control and economy when fed to the steelmaking process) are now common. Among these are reduction of Si from several tenths of a per cent to close to 0.1% which keeps slag volumes and hence dissolved and entrapped iron losses low. Sulphur and phosphorus are reduced to lower levels by treatment of the hot metal in transit from BF to steelmaking.

Steelmaking added a new version of the Bessemer converter when Savard and Lee provided a way to blow oxygen from the bottom without ruining the refractory lining - the Achilles heel of Bessemer's original experiments with oxygen! - by injecting endothermic hydrocarbons in an annular pipe around the oxygen inlet. This "Q-BOP" concept was developed on a small scale by Maxhuette in Germany and introduced on a 200 tons scale at the Gary works of USS in 1971. The Q-BOP has certain advantages on its own but for many purposes today, a combination of top and bottom blowing is preferred; the stirring which is beneficial to faster reactions is often accomplished by blowing argon through a set of mini-tuyeres in the refractory bottom. Control adequate enough that the end-point (when the steel has the correct composition and temperature) is automatic began with the invention of the sensor lance at Bethlehem Steel and its continuing refinement by many groups.

A major contributor to control of downstream processing since 1970 is the widespread use of "ladle metallurgy". There are several designs in use but their basic purpose is to be able to remove dissolved gases, change the composition of the steel by appropriate additions, to adjust the temperature prior to casting and to stir the bath to ensure homogeneity.

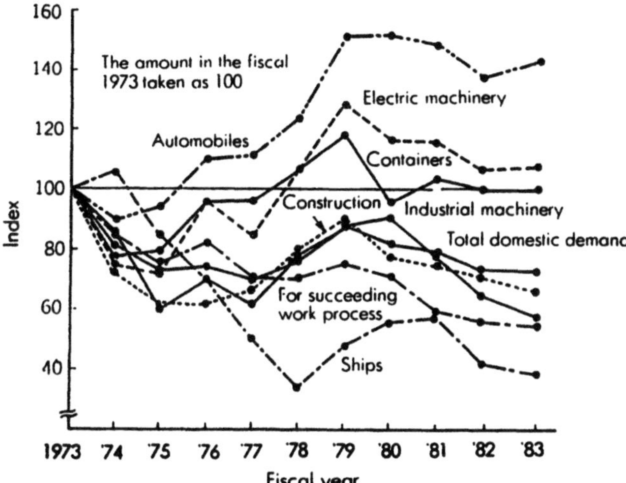

Figure 5. Changes in orders from various industries. The trends to 1991 would not be substantially different.

Continuous casting existed in 1970 but it was almost in an experimental stage. Intensive development has permitted current usage to rise to above 80% in the US and 95% in Japan. Casters coupled with ladle metallurgy stations now run for extended periods with the ability to change mold size and composition of melt with minimum loss of material. The propensity to defects is now low enough that much of the product can be rolled without inspection, a substantial energy saver.

However, the great bulk of casters produce slabs 200–250 mm thick and these must still be hot rolled with advantages to properties but disadvantages in the very substantial capital costs associated with a sophisticated hot rolling mill. There is of course great interest in realizing Bessemer's dream of casting thin sheets and slabs directly; the progress and potential will be discussed later.

Because efficiency, productivity, and quality have become so important in both hot and cold rolling, there have been many developments involving energy efficiency, yield, geometry of product (rolling an accurate rectangle is not an easy task), and both the absolute values of properties and their variance by "heat treatment on the mill".

Finishing operations such as off-line annealing and providing carefully tailored surface layers for cosmetic purposes and corrosion resistance have also been active areas of development. More and more, the trend has been towards developing a continuous process, now available in part at many plants with a suitable product mix.

Product mix leads naturally to discussions of current markets,which have changed noticeably since 1970. Plants changed to reflect these shifts, perhaps more slowly than they should but certainly they shed the image that they could and would supply an over-broad range of products, only a few of which were profitable. The fraction of steel used as coated sheets has substantially increased (as noted above, some important new technology has been developed in this area), whereas the rail and plate business has dropped significantly. Energy products such as seamless and line pipe have been erratic in volume and several bad guesses have occurred in this area on the need for new plants (Fig. 5).

Minimills using local scrap as a feedstock for melting in electric furnaces and producing long products (bars and light beams) for nearby markets became a major competitor to the integrated mills, competing on price, delivery times, and increasingly on quality. Initially using cheap casters, the lack of a huge infrastructure in the minis hit the big mills in two ways. For many purposes the products were "good enough" and, further, the volume lost by the majors with their high capital costs helped to push their operating levels close to or below break-even production levels.

With time and experience, the management and non-union work force (almost all with specifically no previous steel experience), learned to improve quality and began to target higher price markets. For many years, the sheet business, which at its high price end is very technologically intensive, was inaccessible to the minis. However, a courageous gamble by Nucor that they could cast slabs a few centimeters thick by the SMS process and thereby avoid conventional hot-rolling paid off after considerable start-up problems. From this start, they were able to make sheet which is adequate for some purposes. It appears that further development is likely to be able to capture increasing fractions of the sheet market in the same way that much of the bar market was swallowed.

To summarize the needs as seen today, and drawing on the above, "future ferrous technologies", (the title of this paper) will probably focus on the following:

1. Some way of reducing iron ore without using coke, or, in a more extreme form, of reduction without carbon.

2. A possible way to take the fluid (if it is fluid) from 1 and adjust the composition in a continuous process to provide a material which can develop usable properties.

3. A scheme to take the liquid from 2 and solidify it.

4. A system of deformation and/or heating which can develop product properties within acceptable limits.

5. Satisfactory joining and forming technologies without dependence on skilled craftsmen.

6. The ability to recycle products which have served their original purpose.

7. In each of the above steps, to perform the functions without unacceptable environmental effects.

In each of the above cases, we note that while the total "package" may be desirable, in practice the technologies may be developed at different rates and thus initially could be sub-optimal because any single piece will be taking current input to that stage and passing output to the next stage. Only when totally new systems are built will we have the potential of all of the possible gains from new technology. We may perhaps also note that the traditions of the integrated mills, which at one time typically involved vertical integration, have already changed considerably. With more specialization on fewer product lines, it is becoming more common to separate the stages of the current process. It is at least conceivable that as new alternatives become available, and the desire grows to encourage development in poorer countries, that it may make sense to carry out some operations in areas which, for whatever reason, are particularly economical. There are many practical difficulties to this approach, but steel as a world commodity will continue to face pressure from other classes of materials and so must keep total costs low.

V. Future Ferrous Technologies

V.1 "New " Ironmaking Processes

The carbon reduction of iron ore has been carried out for thousands of years with increasing efficiency. Today's blast furnaces with careful burden preparation, good coke and sophisticated control systems have both high thermal efficiency and high productivity especially when compared with existing alternatives such as direct reduction in the solid state. Because our understanding of the complexities of flow of solids, liquids and gases within the furnace, and the accompanying reactions is still not complete, there is some room for improvement. Because reduction costs are a large part of total costs, even modest improvements in efficiency are valuable.

Future threats to the blast furnace come about because the by-product coke oven is under siege as a viable entity under increasingly stiff environmental constraints. The supply of coking coals which are necessary with our current understanding to make good coke is also becoming somewhat scarcer and thus more expensive, but at this time would probably not alone drive out the B.F.

Thus we have to consider whether the B.F. can operate with considerably less coke than formerly (something over 400 kg/ton) to minimize these other concerns. Coke serves two principal purposes; it provides CO as a reductant and also offers mechanical support for the otherwise crushing load of the burden, thereby keeping the pressure drop within bounds and gas distribution across the furnace much better. For some time now, it has been common to provide fluid hydrocarbons (oil or natural gas) at the tuyere level to generate reductants with the amount added limited in part by economics, but also by the effect on solid and gas flow within the furnace.

Because non-coking coal itself can have attractive economics as an injectant, various efforts have been made to find ways to do this for many years. It has been mixed with tar, oil, and water but the current preferred manner is as pulverized coal. This technology has become very active in the last few years, with amounts reaching close to 200 kg/ton on a regular basis. Coal does not displace coke on a 1:1 basis - *e.g.* with a two-week average of 177 kg/ton of injected coal, Thyssen's Hamborn #4 had a coke rate of 295 kg/ton.

The obvious question on which there is a divergence of opinion is whether refinements to B.F. operation can continue and end up with a major variant from current practice. Some such efforts include continuing increases in coal injection, cold blast, nitrogen free operation and the input of plasma energy.[2] Obviously the effectiveness of these will depend on local conditions. For example, some variants produce large quantities of high calorific value top gas for which there may not be a ready use as other downstream uses in the plant disappear with technological changes. However, there is no doubt that extension of the B.F. as a unit will go well into the next century.

The principal competitor, ignoring for the moment the melting of scrap in electric furnaces, is not direct reduction but some updated version of "smelting reduction" whereby partially reduced iron ore, cheap coal and oxygen are reacted in a bath of molten iron. The gas generated in the bath is cleaned and used for further prereduction. The heat and mass balances in this process and the optimum composition of the product are the subject of much research worldwide at the moment. Pooling of industry efforts is common in this research, along with government support, and represents the continuation of a dramatic shift in the philosophy of research especially in the US, which has generally been carried out privately on a company-confidential basis (Fig. 6).

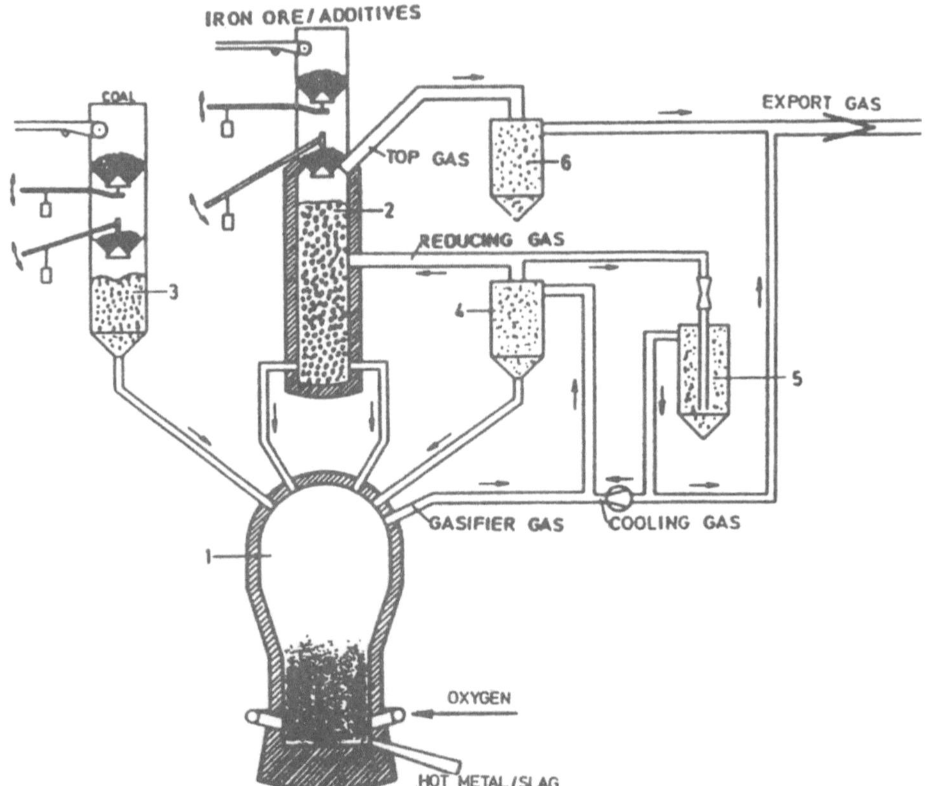

Figure 6. COREX ^{TM} schematic: 1. melter gasifier; 2. reduction shaft furnace; 3. coal feed system; 4. hot dust cyclone; 5. cooling gas scrubber; 6. top gas scrubber.

The potential advantages of this process are the opportunity to use smaller units than the typical B.F., a better chance at a closed system with a reduction in pollution, lower capital costs and the ability to use a range of comparatively inexpensive raw materials. In the present stage, not all these are clearly deliverable but it does seem highly likely that well before 2000, a number of plants will be operational. (In fact, a 300,000 ton/year plant using the COREX^{TM} process developed by the late Willy Korf is in operation in South Africa).

Further ahead in time, we may have to consider the possibility of developing a carbon-free method, since much CO_2 is released to the atmosphere by conventional processes. If the greenhouse effect is real, (or legislated to be real), these may eventually not be legal. There appear to be two options. One would use hydrogen as a reductant but cost may make this unattractive. The alternative may be to consider electrolysis of a molten or dissolved iron salt along the lines of the Hall process for Al. This will take large electric resources, which also cannot be generated from fossil sources so we may be revisiting nuclear fission power.[4]

V.2 Steelmaking

The developments in steelmaking in the last two decades have been considerable. They include the disappearance of the open-hearth process except in Eastern Europe,

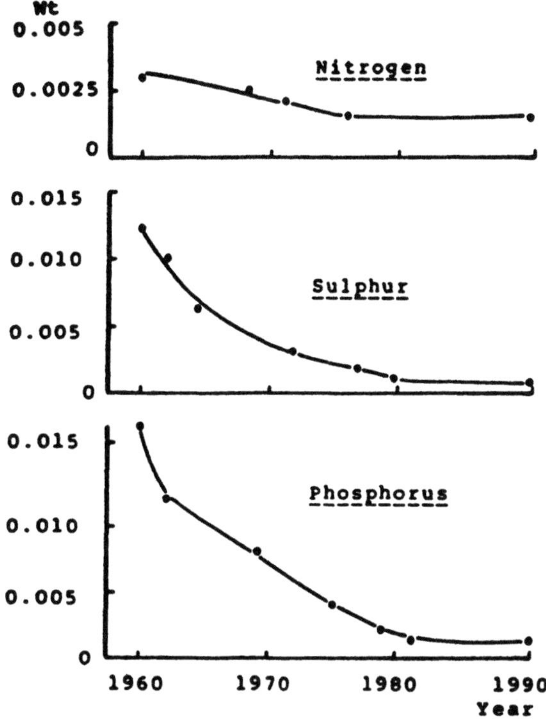

Figure 7. Changes in world best achievable levels of N, S, and P.

an expanded use of combined blowing in oxygen steel-making, and a massive increase (to between 30 and 40%) in the fraction of steel made in electric furnaces. Pre- and post-purification of input and exiting material have allowed greater efficiencies with extremely good control of metal to the caster.

Knowledge of the physical chemistry of the relevant processes at 1600°C has enabled control of levels of important elements such as C, N, S, H, and P to be in the few parts per million range, and thus to enable steels to be produced which are significantly tougher and more ductile at strengths of today's engineering interest. In part, this has resulted from the use of sensors specifically developed for these purposes (Fig. 7).

What then are the limitations of the two important processes and what may we expect over the next decade?

Two important limitations of the BOF are its batch nature and the limited amount of scrap which can be added to take advantage of the heat generated by internal combustion in the bath during blowing. A third desirable economic driver would be the ability to reduce oxides in the vessel, such as MnO to avoid the need to add expensive ferromanganese in the ladle, or iron oxides to provide relatively cheap iron units.

At the present time, there is little we can do about the batch nature (see below) but many schemes have been introduced to be able to handle more scrap, a relatively cheap raw material in most areas. These range from:

1. Pre-heating scrap by exposing it to relatively low-grade waste heat.
2. Pre-heating scrap with gas temperatures up to 1600°C in a shaft furnace.

3. Melting a 100% cold scrap charge in a bottom blown vessel using coal and oxygen.
4. Making partial fuel additions of coal or other combustibles to the molten bath.
5. Avoiding the loss of energy as unburned CO by post-combustion with some of the heat returned to the bath.

Each of these situations obviously must be and has been adapted to local conditions in different countries.

Higher post-combustion ratios (PCR) defined as:

$$\%CO_2/(\%CO_2 + \%CO) \times 100$$

(10 to 20% currently, with perhaps 50% possible) can make a significant difference to the energy available. For example, with a PCR of 40% as opposed to the more usual 10%, about 250 kg/ton extra scrap can be melted. Under the same conditions, about 60 kg/ton of pellets or sinter containing iron as FeO can be smelted. Similar calculations for smelting MnO show that this is generally the cheaper route for realistic slag basicities when ferromanganese is \$500/ton or more.[3]

The dream of continuous steelmaking is of long standing and has not yet been solved. It would not be appropriate to go into detail here, but some of the important issues are:

1. The provision of a continuous supply of iron of appropriate uniform composition and temperature.
2. Adequate containers and nozzles which can be maintained without interference with the operation.
3. Sufficient surge capacity in some form so that short stoppages are not crippling.
4. A system which permits reproducible environmental control of emissions.
5. Last, but not least, a quantitative understanding of the reactions involved so that, with appropriate sensors, automatic control can be obtained.

A brief consideration of these criteria will enable one to understand why this is not an easy task. 2, 3, and 4 are especially difficult.

As noted above, the electric furnace has become a popular route to steelmaking since 1970. Original increases in use came primarily from the desire to serve a local market by melting cheap scrap and using a simple caster to provide material for rods, bars, and small beams. Quality was not a major concern, but price was and the results were most gratifying for these mills. Furnaces were relatively small and used 3-phase A.C.

As the need grew for greater productivity and higher quality, the problems which had to be faced included:

1. Putting more energy into the same volume leading to high refractory wear. Energy is increased by scrap preheating, higher power furnaces, oxygen injection and auxiliary energy sources such as jet burners. The dangers of too high an energy density were reduced by strategically placed water panels (which led to some spectacular explosions when the water and steel met unintentionally) and by operating with a foamy slag from carbon and oxygen injection, which provides better insulation for the arc plasma. Long arcs operating at high voltage and low current are preferred to the shorter arcs operating at low voltage and high current.
2. Refining was at one time conducted in the furnace at low or zero power sometimes with a new slag, and was obviously time consuming. With the emphasis

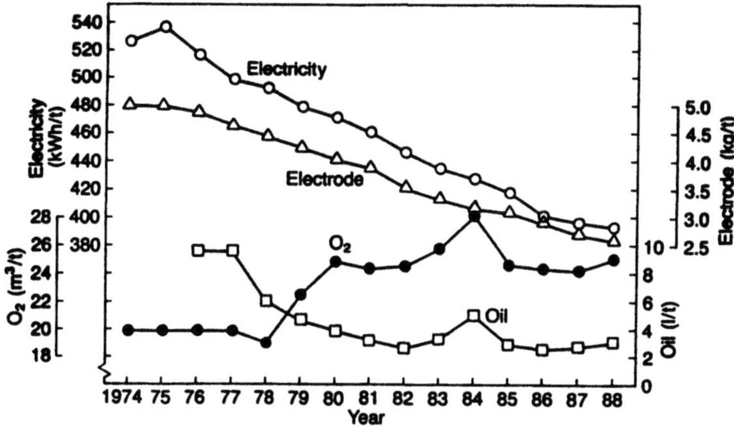

Figure 8. Changes in consumption in Japanese electric furnaces.

on productivity, ladle furnaces were introduced successfully and have enabled a general upgrade in quality as one would expect.

3. Tapping of the heat has been changed from an angle of some 40° to 3 – 5° and has thereby allowed a much greater fraction of the wall to be water-cooled.

These measures have now reduced electrode consumption by about one half to a little over 2 kg/ton. Sprayed water cooling is now used (carefully!) to cool the sides of an electrode and cut down oxidation losses - about 1/3 of the loss (Fig. 8).

An interesting and relatively new development is the introduction of direct current melting. Some of the advantages are more space above the melt, good stirring of the melt, reduced power consumption, and low lining wear. It seems probable that this technology will take over from AC.

One factor which still causes concern in E.F. melting is the disposal of the fume lost to the stacks. The volatile metals will soon no longer be able to be disposed of simply since they have been declared toxic, and they are not sufficiently concentrated to be readily recycled. Regulations and solutions in this area are still being discussed.

The question of whether E.F.s could eventually take over completely from the BF/BOF combination is obviously of interest. Is there enough scrap now in circulation so that re-melting from this pool would provide enough steel to serve the world's needs? Would we experience a progressive deterioration in the amount of undesirable and unremovable elements in solution? At the moment, these answers are not known but the limit of E.F. melting has clearly some way to go.

V.3 Ladle Metallurgy

Once it became feasible to treat freshly refined steel, even though in the early days this might only involve stirring and gas removal by vacuum treatment, advances have continued rapidly. Today's vessels are capable of removing dissolved gases (O, N, and H) along with reaching new lows in S and P, of reacting C and O to produce very low carbon contents, allowing certain inclusions to float out, vastly improving the yield of additives by cutting down side reactions, and of adjusting the temperature for the feed to the tundish and caster. With the necessary improvements in analytical techniques

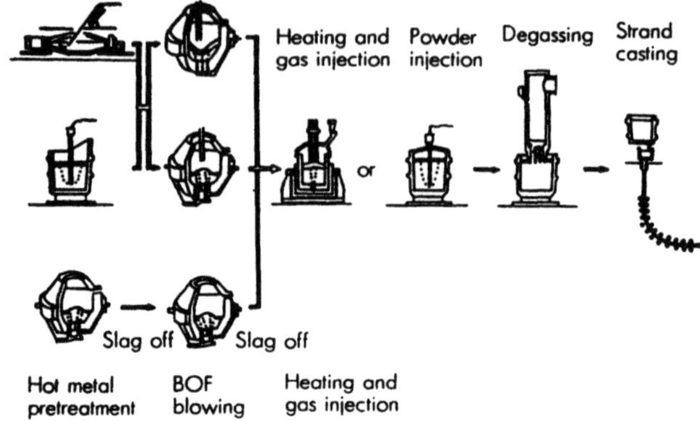

Figure 9. Schematic of system to produce extra-high purity steels.

to track these new low levels, ladle treatment has been and will be a vital component of future technologies. It appears that developments will be incremental but surprises always occur (Fig. 9).

V.4 Continuous Casting

Clearly the biggest change in the technology since 1970 is the development of continuous casting, and its approach to completely taking over the solidification stage for both carbon and special steels. The process itself has many variables which are significant, some independent and some interactive. It is not feasible here to give a complete account; rather some of the interesting areas will be covered along with some guesses as to how new casters will develop.

Some of the important issues common to all casters include:

1. Minimization of the non-metallic inclusions in the steel which can result if the clean steel from the ladle furnace is contaminated by exposure to air while still liquid.
2. Prevention of surface and internal cracks which can result from contamination, hot mechanical deformation, and stresses from the cooling water.
3. Prevention of macro-segregation, usually at the center, because of the transition from dendritic to equiaxed crystals during solidification. For this purpose, and also that of 1 above, electromagnetic fields which can stir or brake fluid flow have eventually been mastered.
4. The ability to change the width being cast, or to change the composition, without stopping the caster and with minimum waste (which must be remelted).
5. The development of sufficient confidence that if the casting parameters are within the proper limits, there is no need to inspect the slab at room temperature prior to further hot working. In part, this has come about because understanding of the mechanical properties of the slab at elevated temperatures, especially ductility, and the effects of trace elements on the grain boundaries has progressed remarkably.

Because of the large economic returns from a good casting operation, much research has been carried out. Some of the ideas and findings can not be easily retrofitted

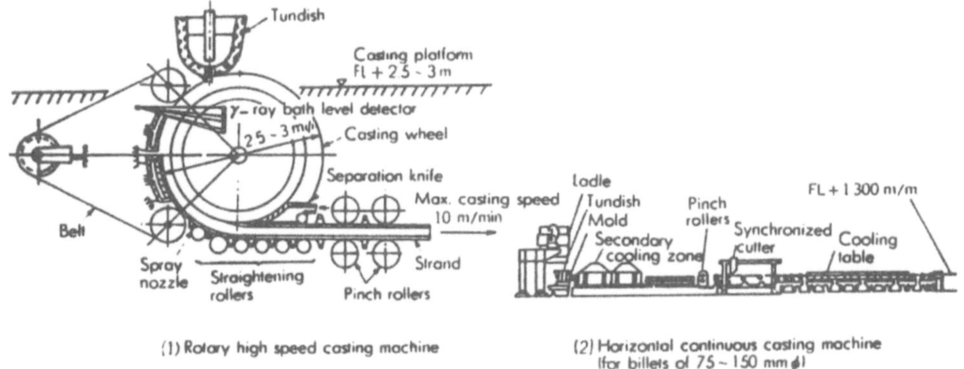

Figure 10. Rotary and horizontal casting machines.

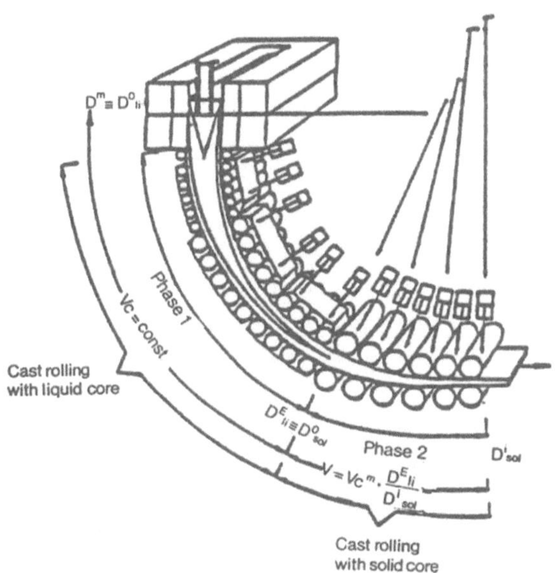

Figure 11. Schematic of a Mannesmann concept for thin slab casting.

and so, often, the effectiveness of a caster depends heavily on installation date. Major rehabilitations or new facilities are often worthwhile and necessary to maintain a competitive productivity. Most casters today cast slabs up to 250 mm x 2000 mm at rates up to 3 MMt per year for a twin slab machine: billet casters 200 mm square can approach 1.2 MMt per year. Other casters such as the horizontal caster producing more special steels in, *e.g.*, twin 150 mm billets, offer a less expensive option for small tonnages, as do a variation on this which permits rotary casting using a copper wheel and a moving belt to contain the liquid steel (Fig. 10).

As we approach these smaller casters, the question of casting steel closer to its finished size in general has always occupied the inventiveness of engineers since Besse-

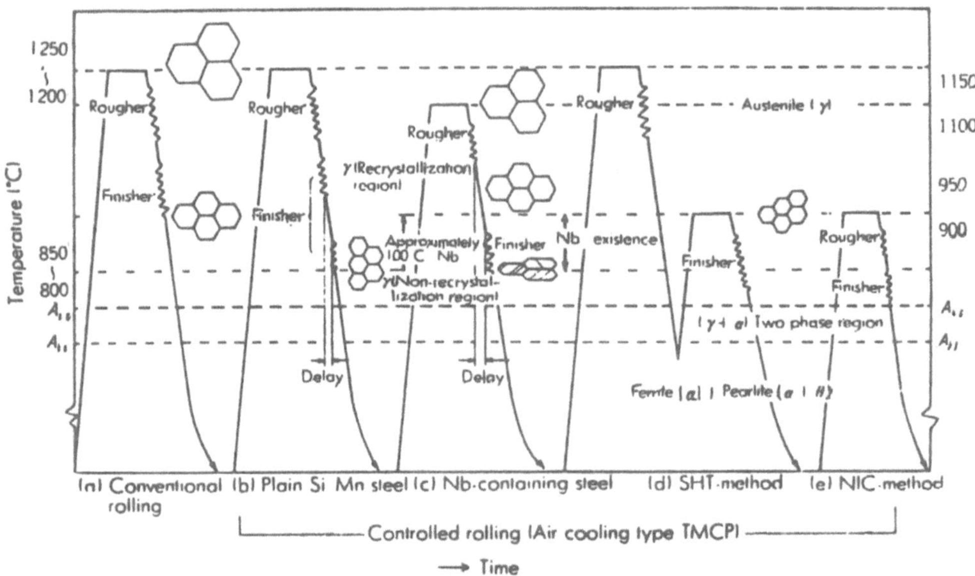

Figure 12. Some examples of controlled rolling.

mer's early attempts. There are two basic choices here - to cast thin slabs perhaps 30 to 50 mm thick and eliminate the need for expensive hot rolling facilities, (Fig. 11) or to proceed to 1 mm sheet directly, a difficult task for carbon steel but rather easier in stainless where Allegheny Ludlum have a working prototype, which will be commercialized soon in association with Voest-Alpine.

Other routes which show promise in the developmental phase involve deposition of droplets of liquid steel on a flat or rotating substrate. Nozzle design, protection from contamination and scale-up to commercial sizes represent some fascinating engineering challenges.

Major difficulties have been encountered in getting sufficient width for all purposes (especially automotive) and adequate quality for demanding applications (such as a cosmetically satisfactory surface). However, the intensity of effort which has been applied to conventional casting has not yet been brought to bear and significant improvements are to be expected. The world interest in this process is high and successful implementation will be an integral part of the future competitiveness of steel.

V.5 Hot Rolling

As discussed above, a major goal is to replace the hot-working step in steel mills, but for the foreseeable future, this will be very gradual. Over the last 15 years or so, hot strip mills and plate mills have become very versatile devices of high productivity which can not only produce good geometries and yields, but can reliably produce desirable properties by "heat treatment on the mill" (Fig. 12). Thus, if the investment in a hot strip mill and conventional casters is still reasonably current, it would take very unusual economics to justify a thin slab caster and atrophy of the front end of the mill for production of substantial tonnages of strip.

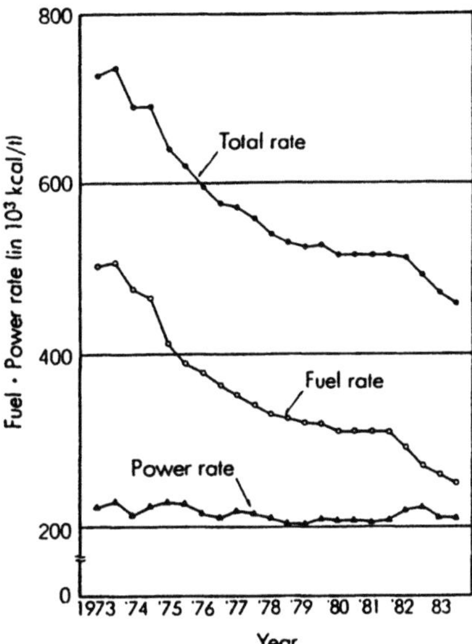

Figure 13. Average changes in hot rolling mill energy consumption in Japan.

Plate mills historically have used ingot product for even moderate thicknesses because to refine the dendritic segregation and give acceptable combinations of properties needed considerable reduction. Even those plate mills which are now sufficiently powerful to reduce the reduction ratio necessary for prime product below 3 (as compared with older figures of 5 to 10), many thickness requirements exist which mandate starting slabs of more than 50 mm or so, and thus rule out "thin slabs".

Some of the developments which have contributed to the development of this competitive position are:

1. Efforts to improve energy efficiency by charging or hot direct rolling. These include improving the basic furnace efficiency, charging slabs at high temperature, lowering the slab extraction temperature, and good scheduling. Fuel consumption can be, and has been reduced by a factor of about 2 (Fig. 13).

2. Improvements of yield by reducing scale loss, avoidance of excessive cropping by new rolling techniques, smoother operations to reduce cobbles and other scrap, and increase of slab weight.

3. The use of high-precision rolling to control accurately thickness, width, and crown to fit the needs of the product as hot rolled or as a feedstock for subsequent cold rolling (which are not the same). Many of the rolling mill developments occurred by joint ventures between steel companies and mill manufacturers in Japan. They include double-chock work roll bending, six-high mills, work roll shifting, large crown back-up rolls, VC (variable crown) rolls, and crossed rolls. All of these work in practice and the choice depends on local conditions (Fig. 14).

4. The development of sophisticated thermomechanical treatments mostly to produce high strength steels without the expense of separate heat treatments and with reduced alloy additions. These treatments are often heavily dependent on the ability of ladle metallurgy to provide accurately controlled composi-

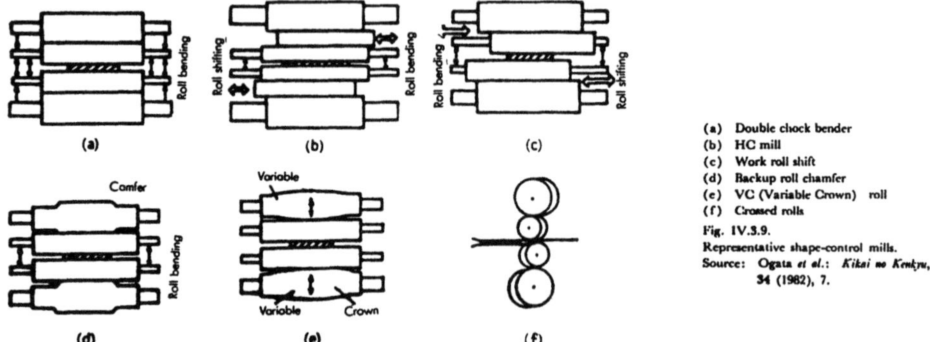

Figure 14. Various representative shape-control mills.

tions and in some cases, very low carbon contents to permit the relatively new bainitic steels. In other cases, the ability to make high strength plates has necessitated the development of robust equipment to control potential distortions of heavy plates during rapid controlled cooling.

5. Substantially increased use of sensors, instrumentation and control so the operator can make on-line corrections, but on most occasions will not have to.

In general, the advances in hot rolling have been so substantial that further refinements seem possible only by evolution, perhaps towards continuing efficiency and customizing individual orders still further. With the tonnages through a typical hot rolling process, small improvements with a full order book are well worthwhile. Plate mills have had excess capacity for some time and the challenges here are to look for new markets and to be able to operate efficiently at very modest capacity utilization.

V.6 Cold Rolling

While much steel is used as hot rolled, the cosmetic demands on sheet which is subsequently cold-rolled and annealed, and possibly coated, have continued to increase. To achieve this, many developments have occurred at all stages. These include:

1. Extension of the now standard HCl pickling lines to introduce partial mechanical descaling as an interim step to perhaps eventual total elimination of today's acid lines. Continual improvements in yield and reduction in energy requirements (steam) are occurring, along with automation especially in surface inspection.

2. Continuing progress towards completely automatic rolling with full control of geometry of the product from front to back, and edge to edge of the coil. This has come about by mill modifications somewhat similar to those listed above for hot rolling (*e.g.* roll bending), and to the development of reliable shape detectors and control systems.

3. Connection of the product exiting the cold mill to a continuous annealing line which dramatically shortens the time formerly occupied by box annealing of individual coils and gives much more reproducible properties. The need for welding coils on a short time cycle to permit continuous operation has been basically solved even for difficult materials by laser welding. The ability to

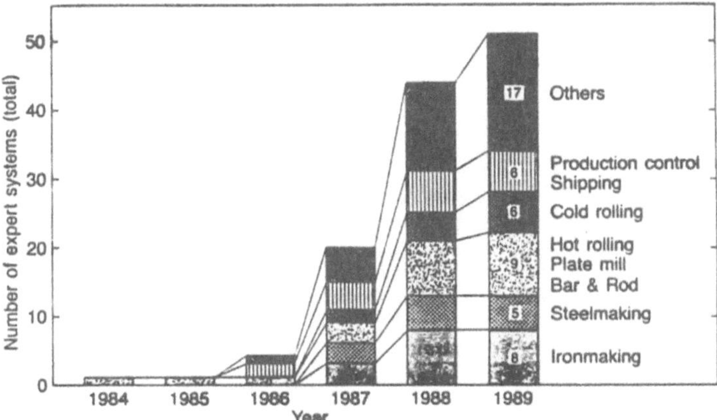

Figure 15. Applications of expert systems in the Japanese steel industry.

meet product formability and weldability requirements is heavily dependent on the physical metallurgy of this annealing process as the last step in the chain which began in the ladle furnace; fortunately this is now well understood as instrumentation developed for basic studies (*e.g.* internal friction) is able to follow such criteria as the amount of interstitials in solid solution at the ppm level.

The above summary covers processes which are common to much, although certainly not all, of the products of today's steel markets. Touched on lightly, or not at all, are such areas as bars and rods, beams, rails, pipes and tubes and specialized materials such as stainless and magnetic materials. Space does not permit these discussions although we may note that many of the principles noted above apply well to general shifts in technology.

V.7 Recapitulation of Process Developments

Perhaps this is a good time to review and summarize the above processes prior to a discussion of potential product developments, which will ultimately depend on the ability to make them to increasingly difficult customer specifications.

1. All the processes which were available in 1970 are recognizable today although the degree to which they have changed is quite remarkable. In particular, these changes have led to much better yields, lower energy consumption, and less environmental problems .
2. The batch nature is still largely present, but after the hot mills, continuous operation is a normal mode in many plants.
3. The ability to control composition, dimensions and properties on-line is changed greatly from 1970 standards. In part, this is due to new and better sensors and automatic controls, but also to the increasing use of such techniques of computer science as artificial intelligence (Fig. 15).
4. More quantitative understanding of the causes of downstream defects has permitted elimination of certain inspection steps and made for much smoother material flow.

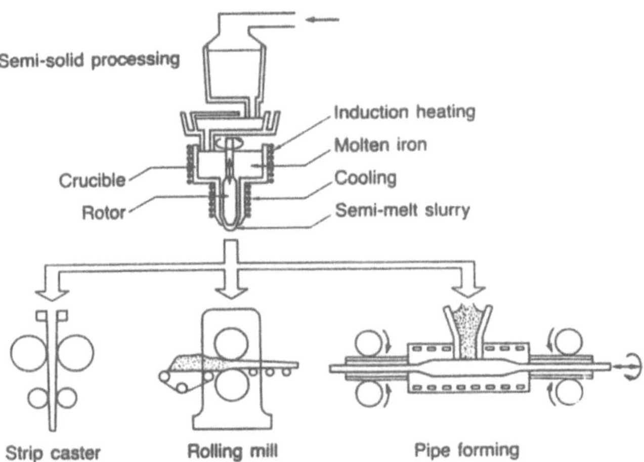

Figure 16. Some ideas for semi-solid processing.

5. Some progress has been made towards flexible manufacturing systems, whereby the customer can order small amounts for rapid delivery. Minimills delivering uncomplicated products have been the leader here. The challenge is much greater for the larger companies, and conscious decisions on product mix and process capability may well be necessary. Much remains to be done in this area, and the two types of companies will probably need to study each other's strengths.

Having now provided this overview of most of the new possibilities in steel processing, we can now properly ask what are pressing and future issues. These appear to be:

1. The stable tonnage of steel produced in the world implies that value added steels will play an increasing role in avoiding further losses to competing materials. However, the need to maintain this competitiveness means that prices will not increase readily and thus cost control will remain a vital issue.

2. The world infrastructure needs massive infusions of structural materials but political and social pressures for public monies make this a "hard sell".

3. The development of a few compositions of steels ("universal grades") which can have their properties adjusted to meet customer demands during the mechanical deformation stage has not made much progress.

4. The ability to recycle coated steels will become increasingly important as the use of these grows and environmental rules tighten.

5. Flexible processing methods which can make small lots of specialized material are likely to be increasingly useful. These may include semi-solid processing (rheo-casting), spray deposition and near net shape casting (Fig. 16).

6. Energy consumption will need to continue to be reduced as its costs rise with time.

7. New iron units for input to steelmaking and subsequent steps will be needed if coke ovens become scarce or non-existent. The transition will involve fuel-augmented blast furnaces, electric furnaces, bath smelting and perhaps currently unexplored ideas such as electrolysis.

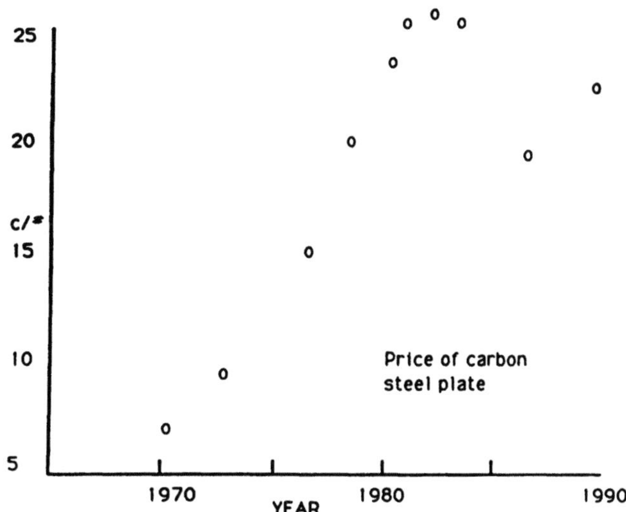

Figure 17. Basic price of carbon steel plate over time.

The author has presented elsewhere[4] a discussion of these topics under the general rubric of :

1. New technical concepts.
2. Materials limitations.
3. Political differences.
4. Markets.
5. Social issues, government actions, regulations, etc.
6. Novel thinking.

The reader may be interested to see how the items discussed above fit into these categories, and the degree to which they help thinking about future developments.

V.8 New Products

No business can hope to survive in the long run if it does not produce products which meet market needs. This can of course be done in a number of different ways.

1. It can produce its conventional product at a lower cost or make it readily available to catch unplanned orders through a better distribution system. This has been a major way in which commercially-driven companies have reacted; the delays in delivery of custom-ordered steel perhaps from offshore are often unacceptable even if the price is attractive.
2. It can produce a clearly better but still basically similar product at the same, or only a mildly increased price. Many of the technical developments of the last 20 years have had the goal of doing this. Fig. 17 shows how carbon steel plate prices stayed constant through the eighties; the quality and delivery improved and if inflation is included, the price actually dropped.
3. Finally, it can produce a qualitatively different product to satisfy an existing market, perhaps at a premium if the customer will accept it, (or has no choice because of regulations) or to open up a totally new application. The most obvious example of this is the introduction of electrolytically deposited zinc

or zinc alloys, either on one side or both sides, of automotive sheet to provide greater corrosion protection while maintaining good cosmetic characteristics. The need for this material came about because of regulations on how long a car body should last. The material was available in the sixties but absent any regulatory pressure, the car makers were not interested.

The obverse of this situation is the loss of the beer and beverage market to aluminum in large part because the aluminum cans could be recycled (thereby saving much energy) and a collection system was developed early. Some of this market may return as steel recycling becomes established since the sheet is now better, the price is a little lower and the manufacturing method only marginally more difficult.

Major new markets and products to satisfy them do not come easily. A study of the US market since 1860 shows that rails, the dominant contributor to the infrastructure as the West was settled with at least 60% of the market, (and are now about 1%) phased out and sheets started to take over as automotive and housing markets became dominant. Structurals and bars have maintained a more or less constant percentage. These generic segments will remain and depending where we are on Malenbaum's curve of intensity of use[1] will in my view not change rapidly even if the long-postponed and sorely needed investment in infrastructure is made.

This is not to imply in any way that product development will stop. The newer processes onstream or in development will allow continuing special steels to be introduced even in small tonnages, and will permit further economies in the large tonnage products. There is still enormous opportunity to apply information technology and computer control at both the local and global level. In fact, just as many problems of physical containment had to await feasible and economical refractories, adequate process control often had to wait for developments in computer hardware and software. The resulting combination of properties at acceptable prices means that ferrous technology will have a reasonably healthy, if not rosy, future. Responses to market challenges can and will be made; longer range social and political issues (especially the need to be able to recycle coated and other inhomogeneous products) can have substantial effects but these will necessarily occur quite slowly.

Acknowledgements

The author is grateful to many colleagues for discussions and information which have helped greatly in the preparation of this paper.

References

1. W. Malenbaum, *World Demand for Raw Materials in 1985 and 2000*, (New York, N.Y., McGraw-Hill, 1978).
2. A. Poos, *Proc. 6th. International Iron and Steel Congress*, Vol. 2, (Tokyo, Japan; Iron and Steel Institute of Japan, 1990), p. 395.
3. R.J. Fruehan, *Proc. 6th. International Iron and Steel Congress*, Vol. 3, (Tokyo, Japan; Iron and Steel Institute of Japan, 1990), p. 73.
4. H.W.Paxton, *Proc. 6th. International Iron and Steel Congress*, Vol. 1, (Tokyo, Japan; Iron and Steel Institute of Japan, 1990), p. 17.

Advanced High Temperature Corrosion Sciences

N. Birks, G.H. Meier and F.S. Pettit

Materials Science and Engineering Department
University of Pittsburgh
Pittsburgh, PA 15261
U.S.A.

Abstract

Advances in high temperature corrosion science are described by considering the current status of developing oxidation resistance in alloys, inhibiting the mixed gas attack of alloys, controlling hot corrosion degradation, and understanding the combined erosion-corrosion of alloys. Recent advances in coatings used to protect alloys from high temperature corrosion are then considered. The high temperature corrosion of intermetallic compounds, ceramics and composites is examined to identify some of the advanced materials upon which this technology is being focussed.

I. Introduction

In a previous paper[1] the important mechanisms for environmentally induced degradation of metallic alloys were described. It was shown that the important forms of degradation consisted of oxidation, mixed gas attack, hot corrosion and erosion-corrosion interactions. The approaches to be used in order to develop improved resistance to these forms of degradation were also described and special emphasis was placed upon the use of coatings for protection. The present paper is concerned with the advances which have been made over the past ten years in high temperature corrosion technology and the materials upon which this technology currently is being focussed.

Advanced Topics in Materials Science and Engineering, Edited by
J.L. Morán-López and J.M. Sanchez, Plenum Press, New York, 1993

II. Advances in the Approaches to Obtain High Temperature Corrosion Resistance

In discussing the advances that have been made to obtain resistance to high temperature corrosion it is useful to consider first the important forms of high temperature degradation, namely, oxidation, mixed gas attack, hot corrosion and erosion-corrosion, and then to examine improved coatings for protection.

II.1 Oxidation

Oxidation is a form of high temperature degradation that is currently of much importance because advanced propulsion systems are requiring higher hardware operating temperatures. The approach to developing resistance to oxidation has not changed. It is necessary to have developed on the surfaces of materials, via selective oxidation,[1] a reaction product barrier through which the reactants diffuse as slowly as possible. The most effective reaction product barriers are still Al_2O_3, SiO_2 and Cr_2O_3 but, as shown in Fig. 1, SiO_2 barriers will result in the lowest oxidation rates at temperatures above about 1300°C. Other oxides have been considered but none appear to be adequate. Beryllia is often mentioned but the available data[2] show that it is not as good a barrier as Al_2O_3 or SiO_2.

The theory for selective oxidation of elements in alloys has not changed substantially over the past ten years. The important factors are,

- The element to be selectively oxidized must form an oxide that is thermodynamically more stable than all other possible oxides.
- Transport of reactants through the oxide must be slow compared to that through other possibly formed oxides.
- The concentration of the element to be selectively oxidized in the alloy must be high enough to permit diffusion processes to be such that a continuous layer of this oxide forms and grows over the surface of the alloy.
- This oxide scale must be as resistant as possible to cracking and spalling caused by stresses, induced thermally or otherwise.

Wagner[3,4] has developed analytical expressions by considering these factors which, for an alloy AB with BO being the more stable oxide, are

$$N_B > \left[\frac{\pi g^*}{2} \frac{N_0^S D_0}{D} \frac{V}{V_{ox}} \right]^{1/2} , \tag{1}$$

$$N_B > \left[\frac{V}{32} \left(\frac{\pi k_p}{D} \right)^{1/2} \right] , \tag{2}$$

where N_B is the mole fraction of B in AB required to form a continuous layer of BO on AB, V is the molar volume of the alloy, k_p is the parabolic rate constant for growth of BO, D is the diffusion coefficient of B in the alloy, D_0 is the diffusion coefficient of oxygen in A, V_{ox} is the molar volume of BO, N_0^S is the mole fraction of oxygen in A at the alloy surface and g^* is an empirical factor usually taken as about 0.3. Eq. (1) describes conditions related to the diffusion of oxygen into the alloy and of B to the alloy surface. Eq. (2) relates diffusion of B in the alloy to transport in BO.

As can be seen from the above equations the diffusion rates of the element to be selectively oxidized and of oxygen, in the alloy,[3] as well as diffusion rates in the

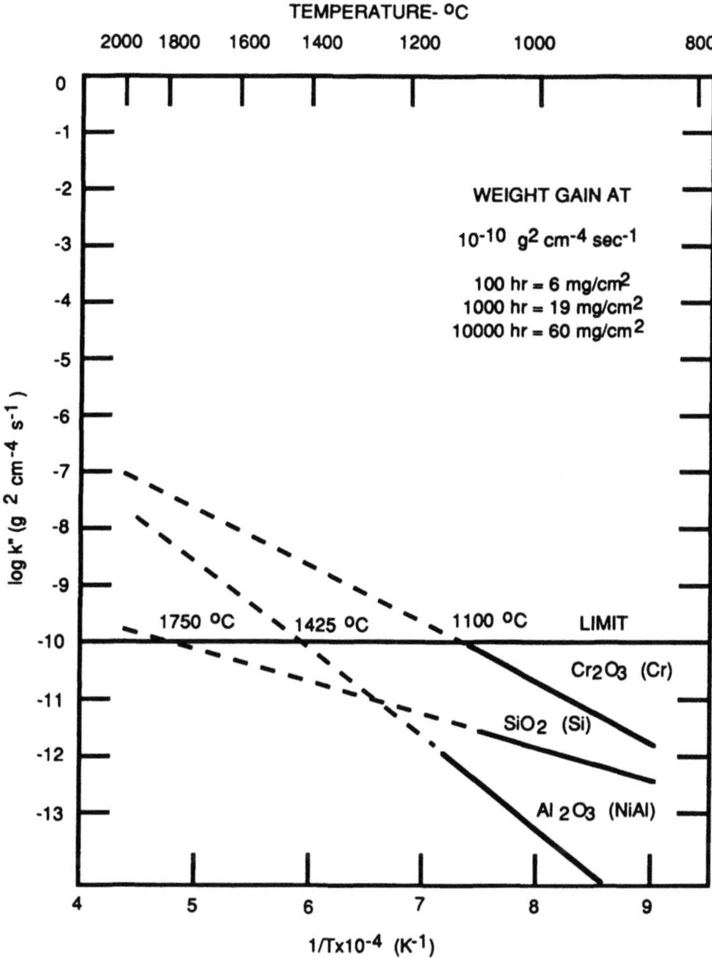

Figure 1. Parabolic rate constants *vs.* temperature for growth of oxides that provide adequate protection at temperatures above 1200°C. A rate constant of $10^{-10}(\text{gm}^2/\text{cm}^4 \text{ s})$ is taken as a limiting rate constant for service.

oxide,[4] are important parameters. The recession rate of the alloy surface during the transient period when the oxide is being selectively developed is also an important factor as discussed by Gesmundo and Viani.[5] As will be discussed subsequently, in the case of some intermetallics the amount of transient oxidation can be high which adversely affects the selective oxidation of aluminum in these types of alloys. In addition to the formation of continuous external scales via selective oxidation, it is very important that this oxide scale is adherent to the alloy substrate and capable of resisting cyclically induced stresses. Oxygen active elements such as Y, Hf, or Ce as well as other similar elements, are widely used to improve the adherence of Al_2O_3 or Cr_2O_3 scales. Over the past five years a number of investigations[6−10] have shown that reduction of sulfur in alloys results in improved adhesion of Al_2O_3 scales on alloys.

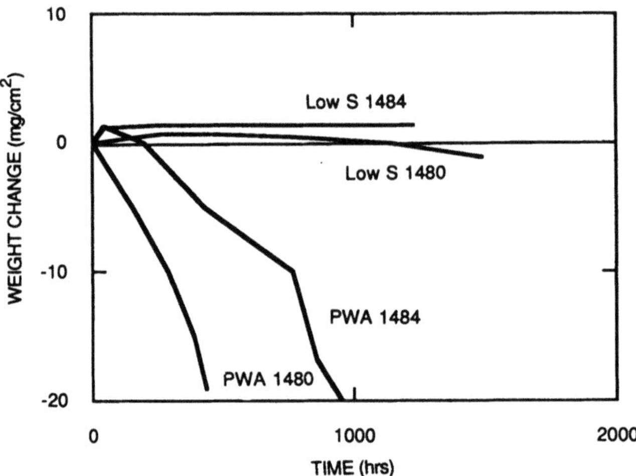

Figure 2. Cyclic oxidation data (each cycle consists of 45 minutes in the hot zone and 15 minutes in the cold zone) for 2 superalloys oxidized in air at 1100°C. The specimens designated low sulfur were exposed to dry hydrogen to remove sulfur.

Smeggil and Funkenbusch[8,9] have proposed that the effect of oxygen active elements such as Y, Ce, Hf is to react with sulfur which inhibits sulfur from diffusing to the Al_2O_3 - alloy interface and adversely affecting scale adherence. This proposal has been supported by some investigators.[6,7] Most investigations show that reduction of the sulfur to very low levels ($\sim$ 1 ppm) in nickel base alloys significantly improves alloy oxidation resistance, Fig. 2. The data available certainly seems to show that the adherence of Al_2O_3 to nickel base alloys is improved as a result of the treatment to remove sulfur. The fact that these treatments improve the oxidation resistance of superalloys even when the superalloys contain sulfur gettering elements such as Hf, suggests that these treatments may produce other effects in addition to reducing sulfur levels.

II.2 Mixed Gas Attack

The approaches to be followed to develop resistance to mixed gas attack have not changed. It is necessary to form protective oxide scales such as Al_2O_3, SiO_2 and Cr_2O_3, and to inhibit the formation of phases composed of other reactants[11] such as sulfur or carbon. Important questions still remain on the mechanisms by which pre-existing oxide scales are broken down. Penetration of oxides by reactants such as sulfur is still not understood. It appears that lattice diffusion is not a likely process. Transport through oxide grain boundaries, microcracks, or channels are considered possible mechanisms.

II.3 Hot Corrosion

Hot corrosion is the increased degradation of alloys which occurs when molten deposits cause oxide scales to lose their protectiveness. As described previously[1] molten deposits can cause oxide scales to become nonprotective via acidic or basic fluxing

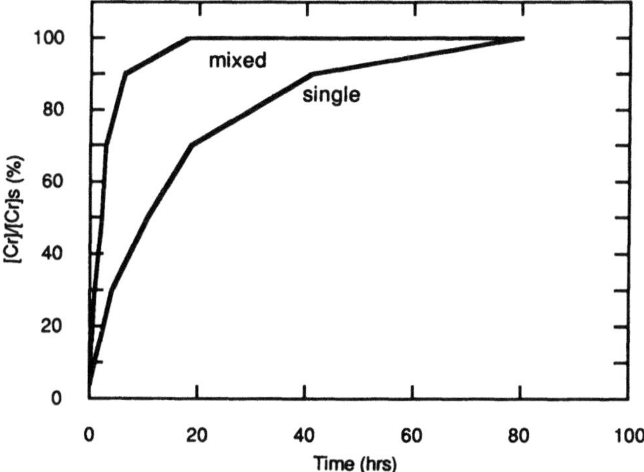

Figure 3. Time dependence of the normalized concentration of dissolved Cr_2O_3 (pure and mixed with Fe_2O_3) in Na_2SO_4 in 1% SO_2 - O_2 gas at 1200 K. Similar results were obtained with Fe_2O_3.

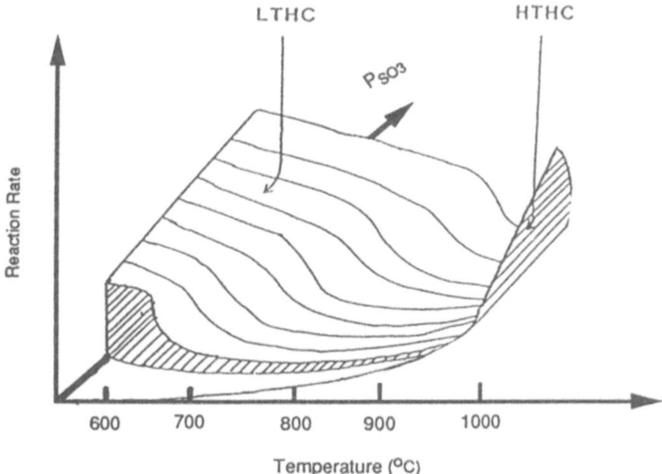

Figure 4. Schematic diagram showing dependence of low temperature hot corrosion rate on temperature and SO_3 pressure.

reactions, and via adverse effects produced by elements such as sulfur or chlorine in some deposits which can accumulate in the surface regions of alloys and coatings. In the fluxing reactions it has been usual to consider rapid attack to occur because the normally protective oxide has been fluxed, but Luthra[12,13] has shown that fluxing of one oxide can prevent another oxide from developing continuity over the surface of an alloy. In the case of some fluxing processes it has been shown that synergistic reactions can be important.[14] For example the solubilities of oxides such as Fe_2O_3, CoO and NiO are controlled by acidic processes, $e.g.$ CoO $\rightarrow$ $Co^{2+}+O^{2-}$, at   melt

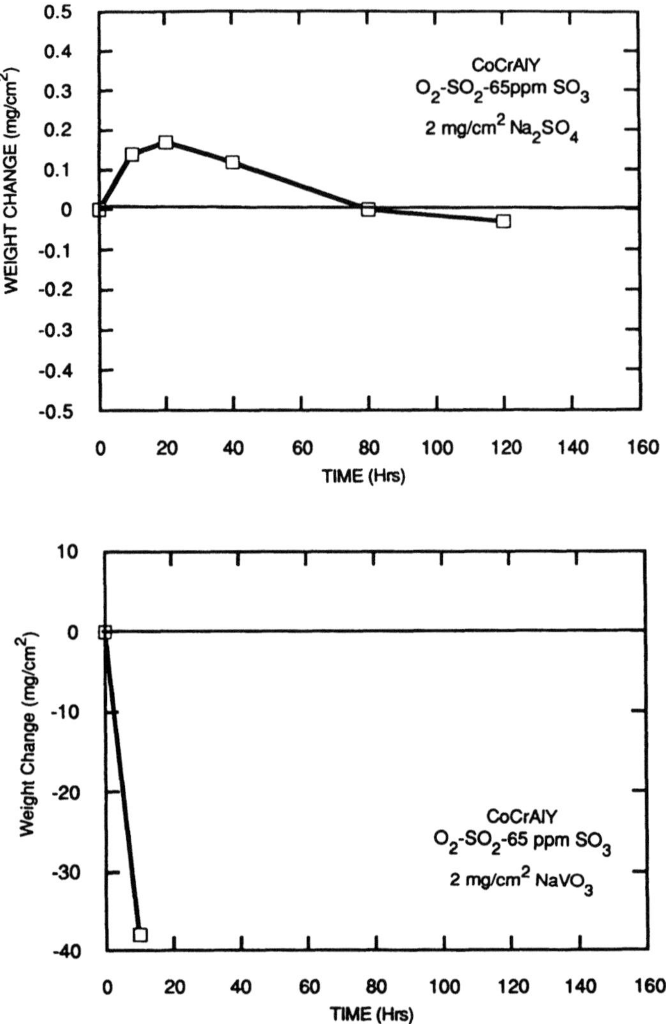

Figure 5. Weight change *versus* time measurements for cyclic (45 minutes at temperature, 15 minutes cold zone of furnace) hot corrosion of CoCrAlY alloy at 900°C using Na$_2$SO$_4$ deposits and NaVO$_3$ deposits in gas mixture containing oxygen and SO$_3$. In the case of Na$_2$SO$_4$ deposits this attack is much less than that observed at 700°C whereas the attack initiated by deposits containing vanadium does not change with temperature.

compositions where the solubilities of oxides such as Cr$_2$O$_3$ and Al$_2$O$_3$ are fixed by basic reactions, *e.g.* Al$_2$O$_3$+ O^{2-} → 2AlO$_2^-$. The acidic dissolution generates oxide ions whereas the basic reaction consumes oxide ions. When either of these two types of reactions are proceeding alone the composition of the melt is modified and the rates of dissolution are affected by diffusion processes in the melt. On the other hand when they occur concomitantly the dissolution rates will be faster because the products of one dissolution process are consumed by the other. Some results obtained by Rapp and Hwang[14] are presented in Fig. 3 which show that the dissolution of Fe$_2$O$_3$ and Cr$_2$O$_3$ take place in Na$_2$SO$_4$ melts at 927°C more rapidly together than separately.

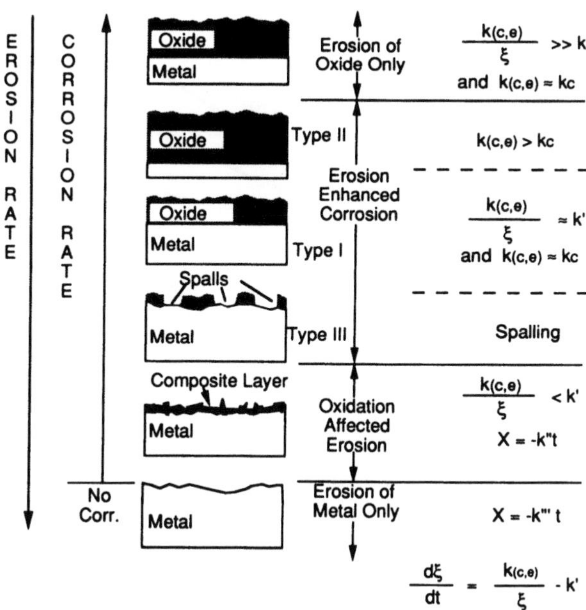

Figure 6. Erosion-Corrosion interaction regimes incorporating three types of erosion-enhanced corrosion behavior.

Synergistic reactions have been proposed to be important in the low temperature hot corrosion of CoCrAlY alloys[15] and also in describing effects produced by impurities during the hot corrosion of silica.[16] In the case of CoCrAlY alloys, the low temperature hot corrosion is much as described by Luthra,[13,14] but synergistic dissolution of CoO and Al_2O_3 may also be important. As temperature is increased, the acidic fluxing of CoO decreases and hot corrosion occurs by the high temperature mechanism which consists of sulfur induced accelerated attack and basic fluxing of Al_2O_3, Fig. 4. When the molten deposits contain $NaVO_3$ - Na_2SO_4 solutions rather than pure Na_2SO_4, the presence of V_2O_5 in the deposits permits acidic fluxing of oxides such as CoO, Co_3O_4 and NiO at temperatures as high as 1000°C. For such conditions the low temperature hot corrosion mechanism extends to much high temperatures, in fact no transition to a high temperature mechanism is evident, Fig. 5.

II.4 Erosion-Corrosion Interactions

In certain systems erosion and corrosion processes act concomitantly and it is necessary to consider how each of these processes affect one another. A number of investigations have examined these interactions.[17-22] It has been established that in the case of erosion-oxidation processes these interactions can be divided into regimes depending upon the relative intensities of the erosive and oxidation components as shown in Fig. 6. When either the erosive or oxidation component is small, degradation of metallic alloys occurs by corrosion or erosion, respectively, Fig. 6. On the other hand when both the erosion and corrosion rates are significant, the interactions can be divided into two different regimes, namely erosion enhanced corrosion, and

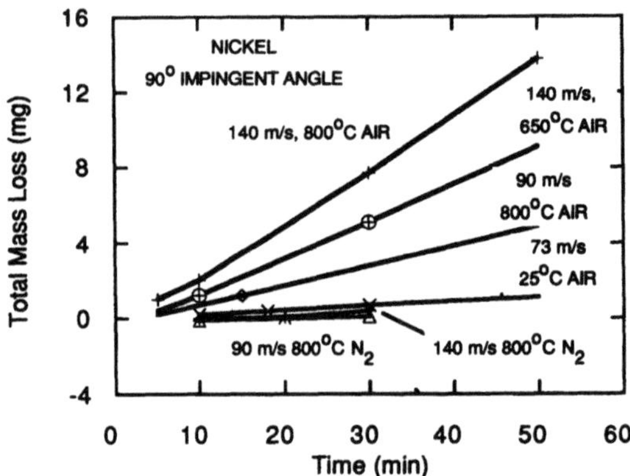

Figure 7. Mass loss *vs.* time for nickel specimens exposed to erosion-corrosion in air and nitrogen at 90° incidence angle. At 800°C much more degradation occurred under oxidizing conditions.

corrosion affected erosion. In the case of erosion enhanced corrosion there are at least three types. One consists of thinning of scales by the erosion process to a point where eventually a steady scale thickness is achieved. Another type also involves thinning of scales but the erosion process also causes transport processes in the scale to be increased. The third type is usually observed when the size of the eroding particles approach the thicknesses of the scales, and then spalling of the scales occur. Spalling of scales induced by particle impacts can occur by a number of different mechanisms. As the intensity of the erosive component becomes large with respect to the still significant corrosive component, the corrosion affected erosion regime becomes important. In this regime corrosion can be difficult to detect due to rapid and efficient removal of the product by erosion. Nevertheless the degradation rate is significantly affected by the corrosive environment, Fig. 7.

II.5 Coatings

Coatings technology[23] has advanced both by the introduction of new, or modified, fabrication processes,[24] and by coatings with new compositions or structures.[25] New processes, such as low pressure plasma spray, permit MCrAlY coatings to be deposited with quality equivalent to electron beam physical vapor deposition, and at lower costs. Greater flexibility is also available to modify the compositions of coatings. For example, elements such as silicon have been added to MCrAlY coatings.[25]

Rapp and coworkers[26,27] have developed pack cementation techniques whereby at least two elements can be diffused into substrates concomitantly. In these processes the vapor pressures of gaseous species arising by use of different activators are important. By using more than one activator, pressures high enough to produce diffusion of at least two elements into alloys can be achieved.

More effective thermal barrier coatings have also been developed.[28] These coatings are usually yttria stabilized zirconia and they are deposited by using plasma spray or electron beam physical vapor deposition (EBPVD). Such coatings have been used on increasingly complex configurations and the use of EBPVD has been especially effective in providing coatings with improved useful lives.

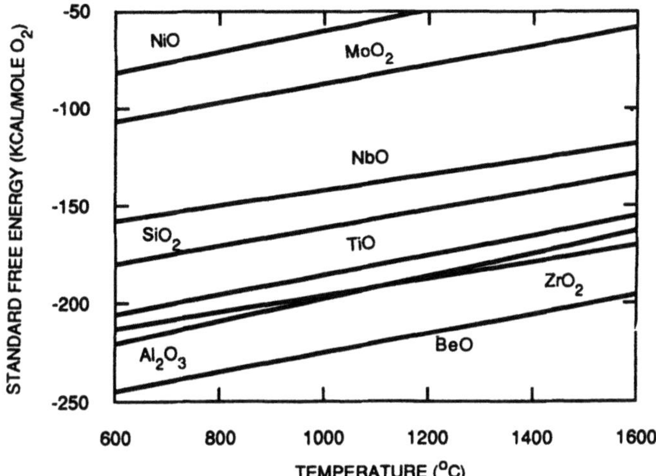

Figure 8. Standard free energies of formation of some oxides relevant to oxidation of intermetallic compounds.

III. Corrosion Resistance of New Advanced Materials

A variety of new materials are being examined for use in advanced propulsion systems, incinerators and coal combustion equipment. The need for new materials arises because of the use of higher operating temperatures for increased power and efficiencies, as well as the constant requirement to reduce costs through longer useful lives of hardware components. The new materials that will be considered in this paper are intermetallic compounds, ceramics, metal matrix and ceramic matrix composites and alloys containing dispersed phases.

III.1 Intermetallics

A number of intermetallic compounds are currently being considered for a variety of high temperature applications because of their possible combination of creep resistance and low density which is superior to that provided by the state-of-the-art, coated, superalloys. The fundamentals of the oxidation of intermetallics have been described by one of the authors of the present paper.[29] In order for intermetallics to develop oxidation resistance, the selective oxidation of elements such as aluminum, silicon or chromium is necessary. The factors that are important for selective oxidation of elements in intermetallics are the same as those for other alloys, but the conditions which exist in some intermetallics are not always conducive for selective oxidation. For example, in the case of intermetallics such as TiAl and Nb$_3$Al or NbAl$_3$, there are not large differences between the standard free energies of formation of niobium oxides and Al$_2$O$_3$, nor titanium oxides and Al$_2$O$_3$, Fig. 8. Consequently the concentration of aluminum must be high if the aluminum is to be selectively oxidized. Weight change *versus* time data for oxidation of NbAl$_3$ is presented in Fig. 9. The rates of oxidation increase with temperature and cracking of the oxide scale is evident at temperatures above about 1250°C. Even though the aluminum concentration is high, the weight change data, as well as metallographic examination of exposed specimens, show that selective oxidation of aluminum is not occurring.

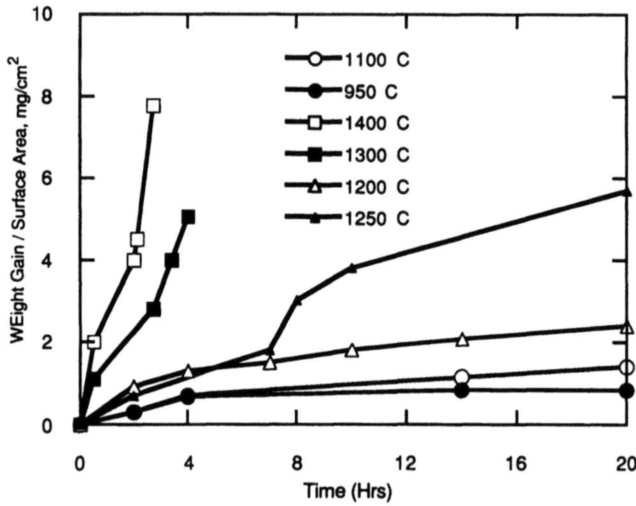

Figure 9. Oxidation of aluminum-rich NbAl$_3$ in O$_2$ at several temperatures.

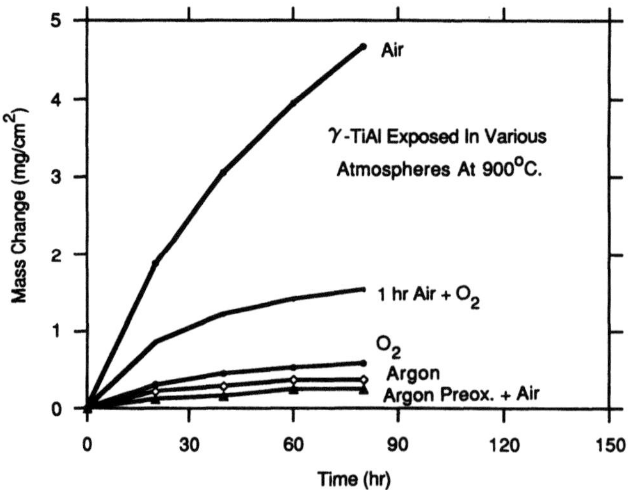

Figure 10. Weight change *versus* time data for oxidation of TiAl at 900°C in different gas mixtures. The argon contained about 10 ppm oxygen.

The oxidation of TiAl has been studied in a number of investigations.[30–32] Even when alloys contain 56 atom % aluminum, selective oxidation of aluminum may not occur at temperatures above about 1100°C. The selective oxidation of aluminum in this intermetallic is also affected by the gas environment, as shown by the data presented in Fig. 10. In pure oxygen at 900°C an Al$_2$O$_3$ layer is formed upon this alloy, but in gas mixtures with nitrogen less protective oxide scales are formed as shown in Fig. 11. Furthermore, the selective oxidation of aluminum becomes less favorable as the nitrogen pressure is increased. In Fig. 12 the oxidation of TiAl as a function of temperature for oxidation in air and oxygen is compared. In oxygen the aluminum

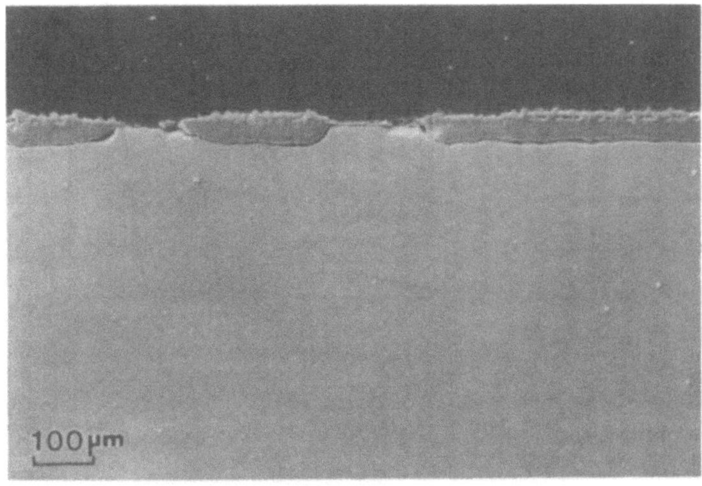

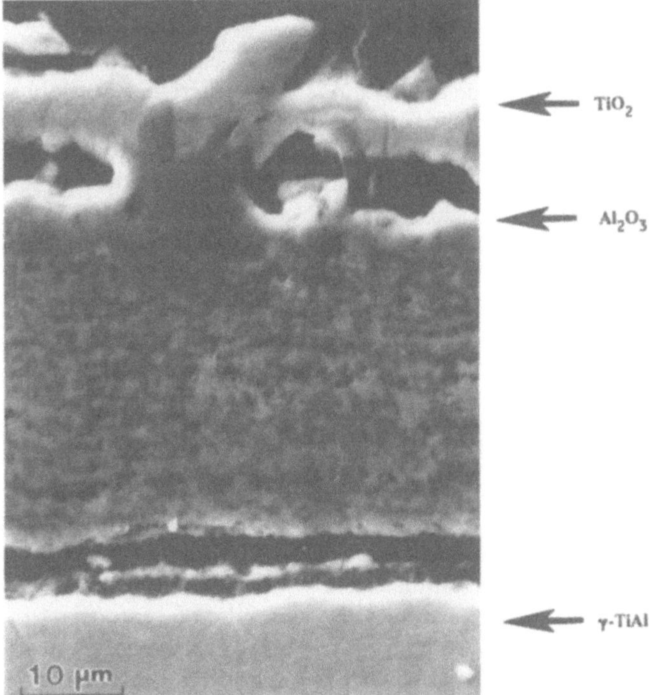

Figure 11. Photomicrograph of TiAl after 80 hours of oxidation at 900°C in air. Protective Al_2O_3 developed in a few areas but most of the surface was covered with discontinuous Al_2O_3 and TiO_2.

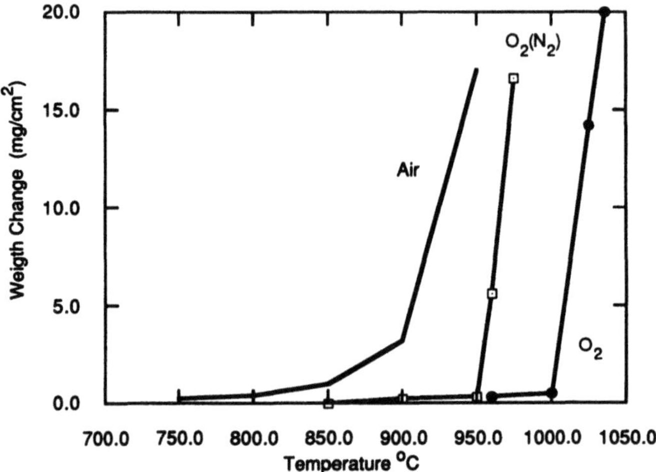

Figure 12. Weight change *versus* temperature for TiAl oxidized for 58 hours in air, oxygen and oxygen containing a small amount of nitrogen.

is not selectively oxidized at temperatures above about 1000°C. The temperature at which aluminum is selectively oxidized in air is reduced to about 800°C. An acceptable explanation for these observations is not currently available. TiAl is not a strong alumina former. As temperatures are increased beyond about 900 to 1100°C, depending upon aluminum concentration, continuous Al_2O_3 cannot be formed in oxygen. At the very beginning of oxidation, titanium nitride has been detected upon the surface of TiAl exposed to air. Moreover such a nitride has been detected in the vicinity of the oxide-alloy interface of specimens upon which continuous scales of Al_2O_3 are not formed.

The intermetallic $MoSi_2$ oxidizes to selectively form a protective scale of SiO_2 at temperatures above about 600°C. As can be seen by the data presented in Fig. 13, at temperatures below about 600°C more rapid oxidation can occur. In Fig. 14 the oxidation of $MoSi_2$, in different fabrication conditions, is compared to the oxidation of pure molybdenum at 500°C. The oxidation rate of the $MoSi_2$ is about one third of that of pure molybdenum and in all cases the kinetics are approximately linear. A tapered section through the scale formed on $MoSi_2$ is presented in Fig. 15. The scale is composed of a mixture of MoO_3 and SiO_2 with some particles $MoSi_2$ which become Mo_5Si_3 near the gas interface. The important point is that, at these low temperatures, silicon cannot diffuse to any great extent and the silicon and molybdenum are converted to oxide *in situ*. The scale therefore consists of a mixture of MoO_3 and SiO_2 in a ratio of about one to three. The kinetics are linear because cracks are formed due to compressive stresses, since the oxidation rate is controlled by the inward movement of oxygen through MoO_3.[33] When cracks are present in $MoSi_2$, as was the case of the cast specimens in Fig. 13, these accelerated oxidation conditions can result in catastrophic degradation called pesting.[34,35]

The point to be emphasized concerning the oxidation of intermetallics, such as $MoSi_2$, is that the oxidation of the base element (*i.e.* Mo) can affect behavior at low temperatures where diffusion of the more active element is slow or negligible. When the base element oxide is not protective, more rapid oxidation can be observed at

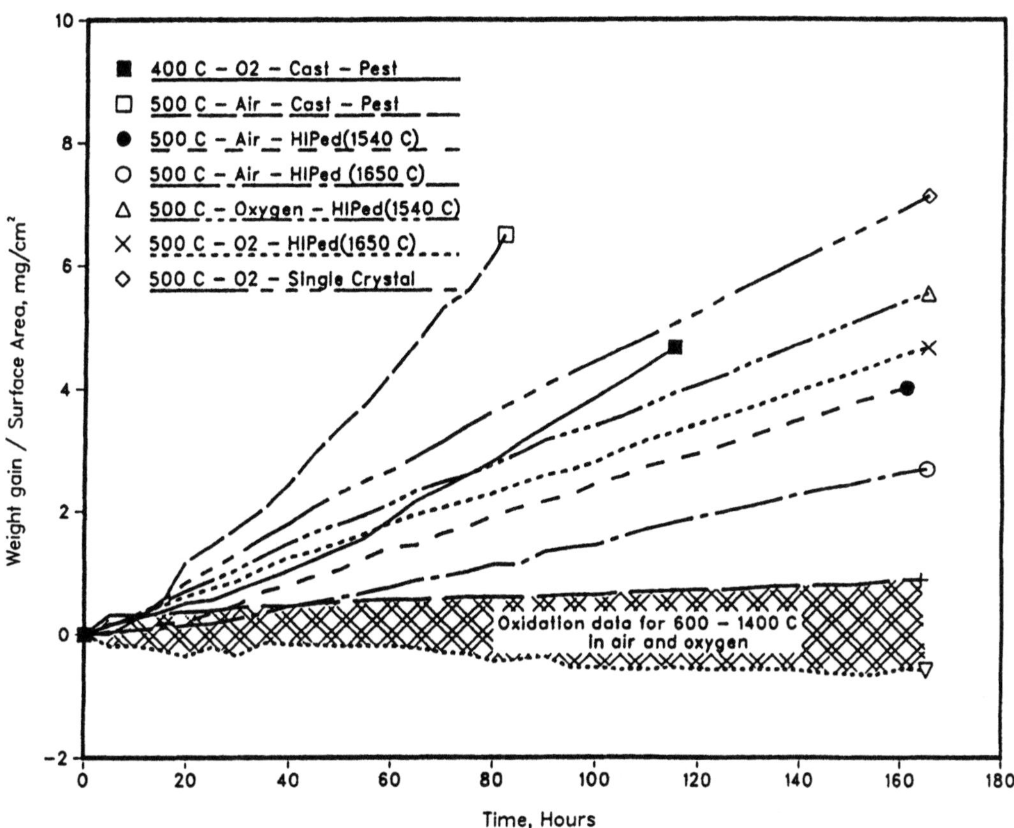

Figure 13. Isothermal weight change *versus* time data for oxidation of MoSi₂ in air and oxygen at temperatures between 400° and 1400°C. The cross-hatched region shows data for temperatures between 600° to 1400°C, regardless of the processing conditions. In this region small weight losses are observed due to volatilization of MoO₃.

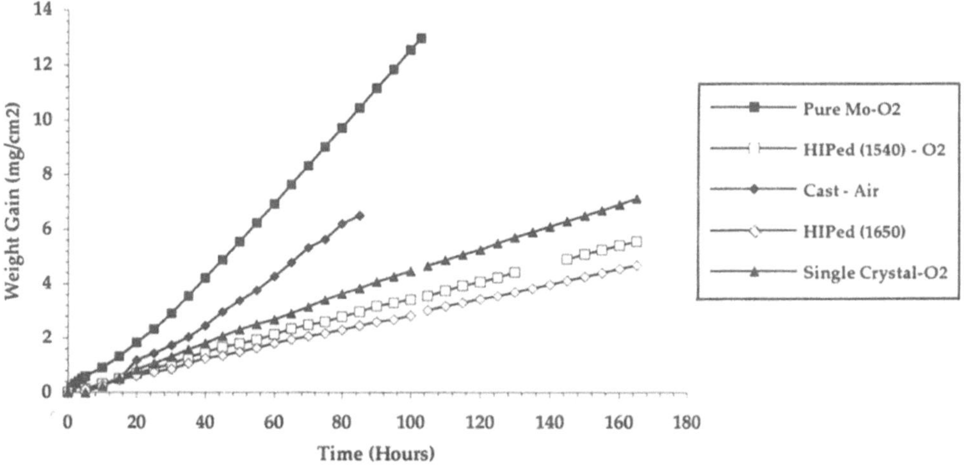

Figure 14. Comparison of weight change *versus* time data for isothermal oxidation at 500°C of pure molybdenum and MoSi₂ in different fabrication conditions.

Figure 15. SEM micrographs of single crystal MoSi$_2$ cross-sections oxidized in O$_2$ for 192 hours and polished at a 17° taper: a. BSE (backscattered electron image), b. SEI (secondary electron image). The oxide scale is composed of a mixture of MoO$_3$ and discontinuous SiO$_2$ with some particles of MoSi$_2$ and Mo$_5$Si$_3$ (near gas interface).

the lower temperatures. It is also important to note that if the base element oxide is volatile, the low temperature accelerated oxidation will not be observed. This occurs in the case of SiC where gaseous carbon oxides are formed and also at temperatures above 600°C in the case of MoSi$_2$.

III.2 Ceramics

In considering the high temperature corrosion of ceramics it is useful to classify ceramics into two types, namely, Type I ceramics which are close to being in equilibrium with the environment, and Type II ceramics which react with the environment and must develop a protective reaction product barrier to become corrosion resistant. Type I ceramics consist of oxides such as Al$_2$O$_3$ or silica which, except for stoichiometry changes, do not react with most industrial gas environments that contain oxygen. The corrosion of Type I ceramics can be influenced by impurities in these ceramics.[16] For

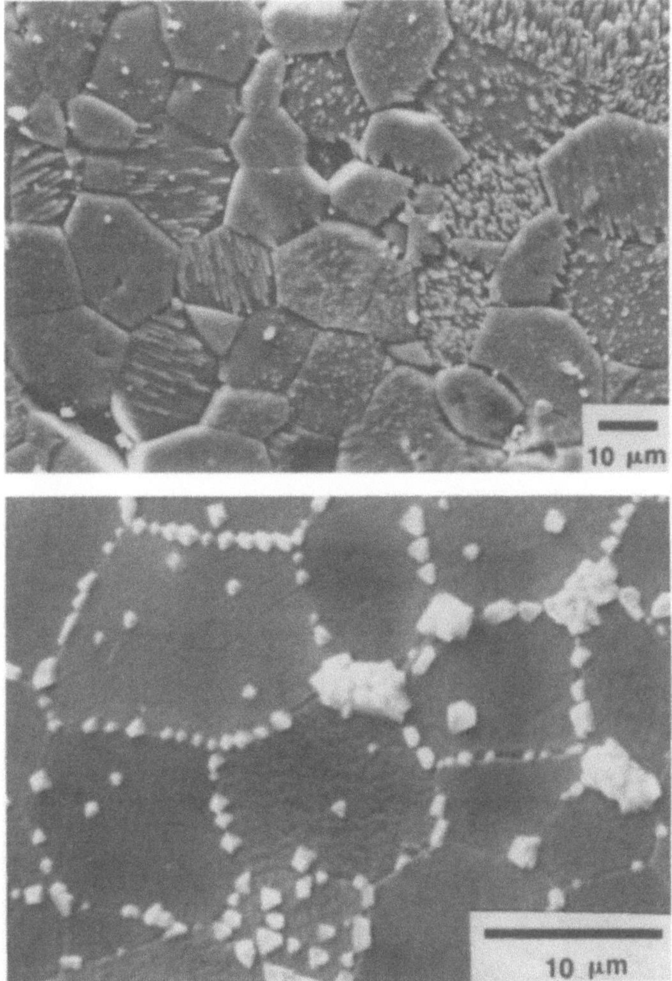

Figure 16. High purity polycrystalline alumina exposed in basic hot corrosion conditions at 700°C for 24 hours (top) and 1000°C for 405 hours (bottom). Oriented silica rich needles (top) and silicate reaction products at grain boundaries (bottom) are shown on washed substrates.

example, impurities such as MgO or CaO can react with SO_3 when present in oxygen to form sulfates. It is also possible that the devitrification of silica glass may occur in different gas mixtures especially at temperatures above about 1000°C

Type II ceramics consist of materials such as SiC and Si_3N_4. Type II ceramics must form reaction product barriers to develop resistance to corrosion. In the case of SiC and Si_3N_4 the protective barrier is SiO_2. These scales are usually very protective but impurities such as MgO[35,36] can adversely affect the corrosion resistance. As mentioned previously, devitrification of the silica can occur in some gas environments, and the formation of SiO (gas) at low oxygen pressures[37] can result in increased degradation rates.

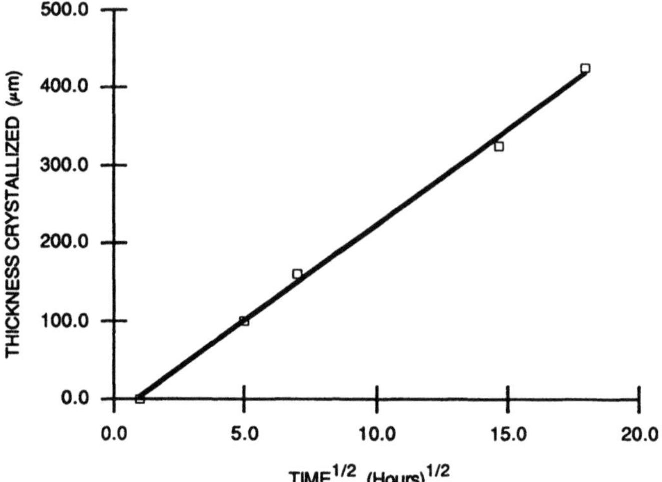

Figure 17. Kinetics of crystallization of silica glass during basic hot corrosion at 1000°C in oxygen.

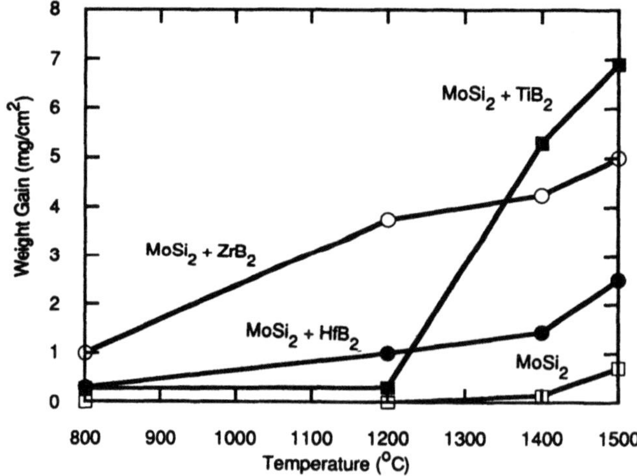

Figure 18. Weight gain of MoSi$_2$ composites after 24 hours in air at temperatures between 800° to 1500°C[41].

Generally speaking, both Type I and II ceramics are more corrosion resistant than metals to corrosive gas mixtures at elevated temperatures. In using ceramics an important form of attack is that induced by molten deposits, in particular, hot corrosion attack.[16] This type of degradation is of concern for both Type I and Type II ceramics but the attack is usually more severe for Type II since the molten deposit can destroy normally protective oxide scales. As discussed previously, molten deposits such as Na_2SO_4 or $NaVO_3$ can dissolve oxides such as Al_2O_3. This dissolution can be affected by impurities, Fig. 16. Silica is resistant to acidic molten deposits, but is rapidly degraded by basic deposits of Na_2SO_4, Fig. 17. In the case of yttria stabilized

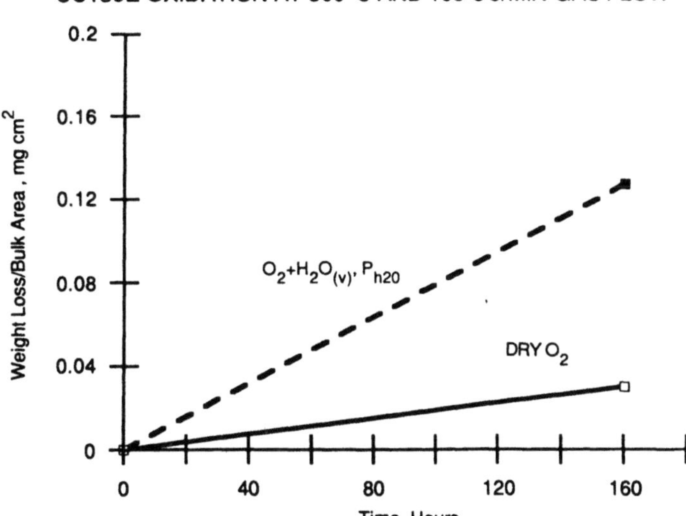

Figure 19. Comparison of the oxidation kinetics of carbon-carbon composits (CC139E) at 300 C in 100cc/min dry oxygen and oxygen/water vapor.

zirconia molten deposits of Na_2SO_4 - $NaVO_3$ can preferentially remove yttria[38] causing cracking of the zirconia under the influence of thermal fluctuations. Jones[39] has attempted to prevent destabilization of zirconia under such conditions by using In_2O_3 as a stabilizer rather than Y_2O_3.

In summary, most ceramics have better corrosion resistance than metals to both gaseous induced corrosion and hot corrosion. When using ceramics in corrosive environments, it is important to be aware of the adverse effects of impurities in the ceramics, and the susceptibility of ceramics to hot corrosion attack.

III.3 Metal Matrix and Ceramic Matrix Composites

The oxidation and hot corrosion of composite materials are just beginning to be investigated. Some typical metal matrix composites are aluminum-base and titanium-base alloys reinforced with carbon or SiC. The oxidation resistances of the metal matrices could be adversely affected by the reinforcing phases. Carbon fibers certainly will be oxidized at catastrophic rates when exposed to oxidizing gases. Reactions at the matrix-fiber interface are also just beginning to be examined,[40] and such reactions may also be important to achieving oxidation resistant systems because some of the products can be susceptible to oxidation.

Ceramic matrix composites in most cases are planned to be used at very high temperatures (> 1200°C) and, at such temperatures, the reinforcing phase may affect the oxidation resistance of the matrix phase. Also some reinforcing phases may be more susceptible to oxidation attack than others, Fig. 18.[41]

Coatings may be required on composites. This certainly is the case for carbon-carbon composites, for which oxidation occurs at temperatures as low as 300°C, Fig. 19. At low temperatures the oxidation rate of carbon composites is controlled by a chemical reaction whereas, at higher temperature, the rates are controlled by the availability of oxygen, Fig. 20. Inhibition of carbon-carbon composites by using boron

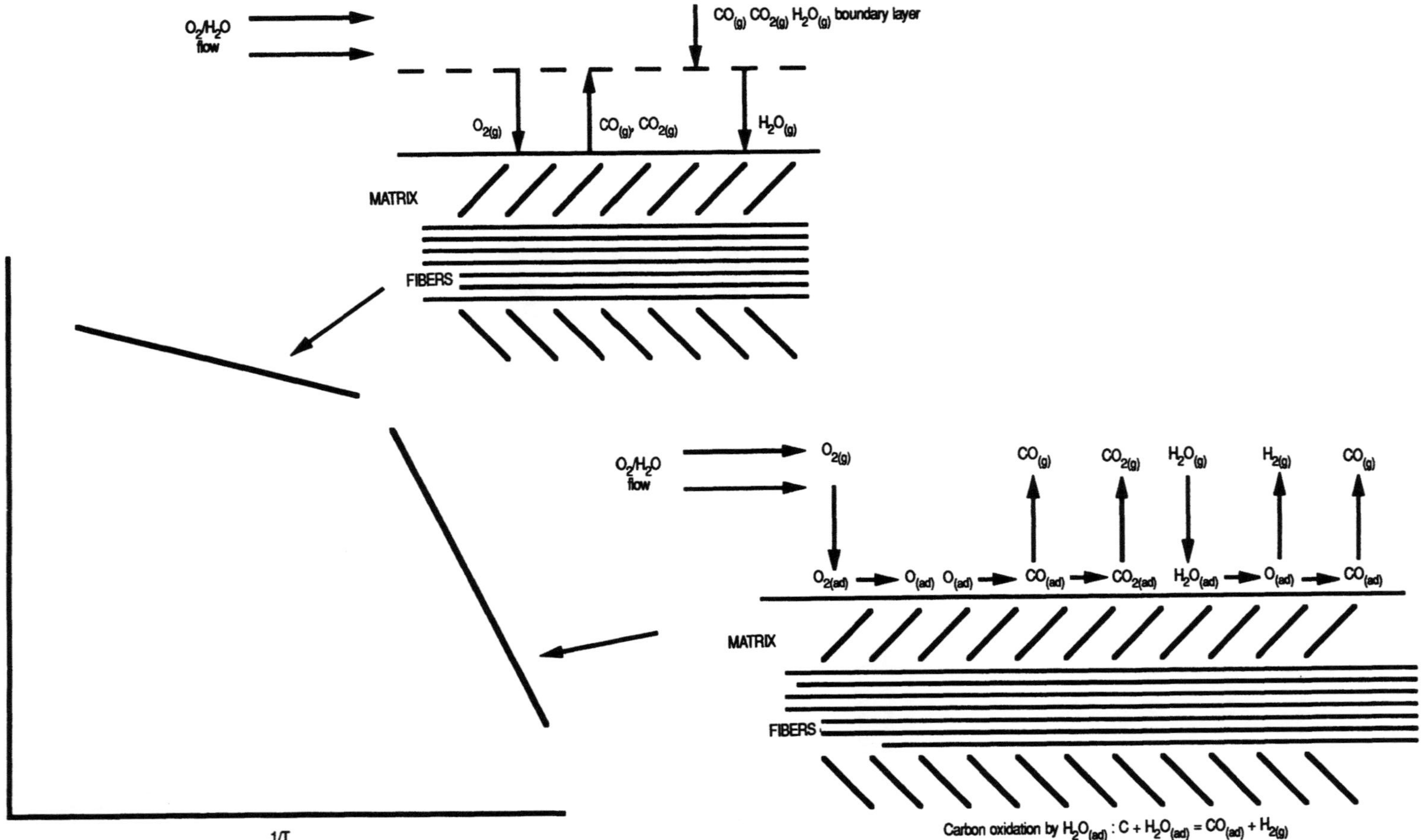

Figure 20. Schematic illustration of adsorption-dissociation-desorption and gas phase diffusion mechanisms of uninhibited carbon-carbon composite oxidation in oxygen/water vapor atmospheres.

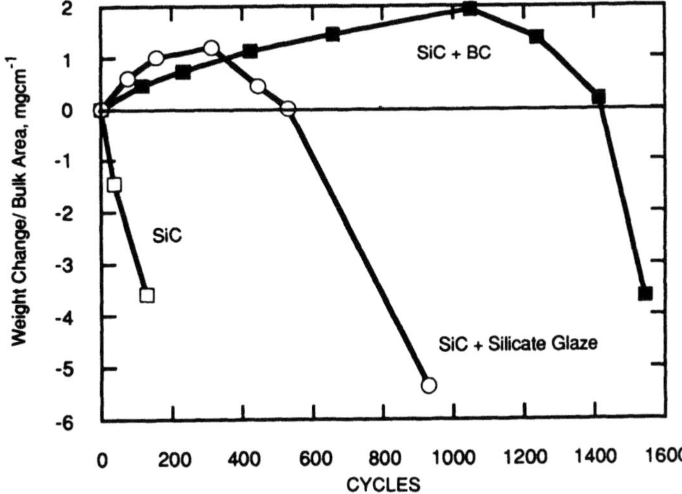

Figure 21. Cyclic oxidation results for the coated carbon-carbon composites at 900°C in still air.

has been found to be a means to decrease the oxidation rate of the carbon reinforcing phase, but not the carbon matrix.[42,43] Coatings such as SiC and Si_3N_4 have been developed to protect carbon-carbon composites but, very often, cracks are developed in these coatings due to differences in the coefficients of thermal expansion, Fig. 21. At temperatures above about 1200°C, these cracks can be sealed via the formation of SiO_2[44] but, at lower temperatures, low melting point glazes must be a part of the coating and such glazes are not highly effective in sealing cracks.[42]

III.4 Alloys Containing Nano-Dispersed Phases

Fabrication techniques are now becoming available by which nanostructured materials can be prepared.[45−47] Certain nanostructured alloys may have some unique oxidation properties. In the case of a nanostructured single phase solid solution, the alloy may lack crystallinity and this could affect oxidation behavior by influencing the diffusion of those elements participating in the selective oxidation process. However, at homologous temperatures greater than 0.5 any effects due to the nanostructure would disappear. In the case of nanostructured multiphase alloys the second or third phases with sizes of about 30 nm would be very uniformly distributed throughout the alloy. Such nanostructured alloys provide a means to alter certain properties of alloys as illustrated in Figure 22.[47] To illustrate this point, it is difficult to develop continuous scales of Cr_2O_3 on Cu-Cr alloys because chromium has a very low solubility in copper. Consequently, chromium cannot be selectively oxidized in Cu-Cr alloys because the copper solid solution in equilibrium with chromium particles does not contain enough chromium in solution. When Cu-Cr alloys are prepared such that the chromium phase is distributed throughout the copper solid solution as 30 nm particles, these particles should provide sources of chromium uniformly distributed over the alloy surface and selective oxidation of the chromium to form a continuous scale of Cr_2O_3 should be possible.

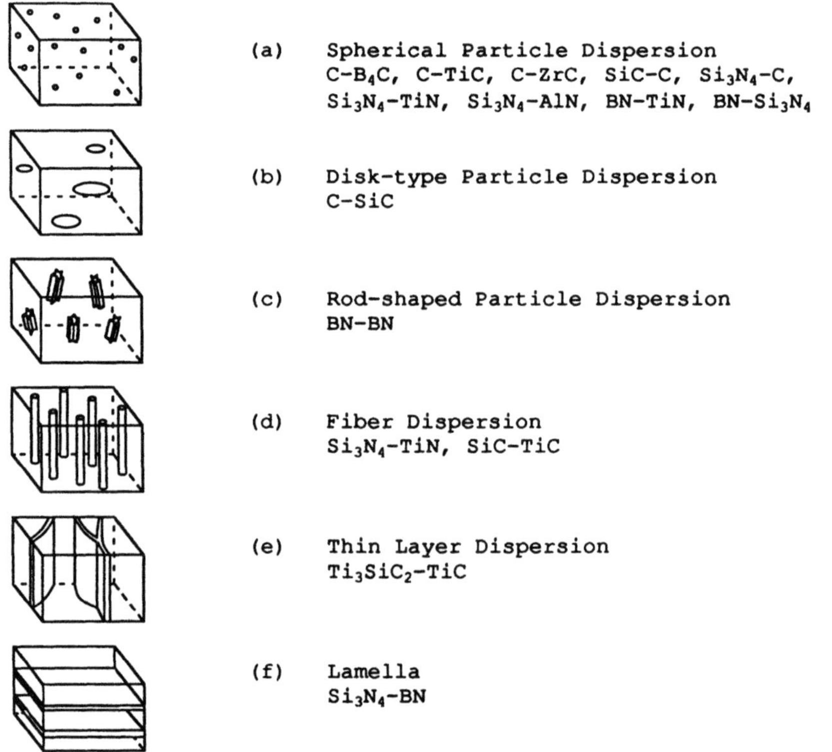

Figure 22. Illustrations of composites containing various nanostructured dispersed phases. The matrix phase is identified first and then the dispersed phase (*e.g.* C-SiC).

Concluding Remarks

The advances in high temperature corrosion sciences over the past decade have not been substantial. The approaches used to obtain resistance to high temperature corrosion still consist of utilizing the concept of selective oxidation and attempting to have these oxide barriers remain protective as long as possible not only in gases containing oxygen but in mixed gases as well as hot corrosion conditions. The current challenge is to apply the selective oxidation approach to advanced materials such as intermetallics, composites and nanostructured materials.

Acknowledgements

The studies on the oxidation of intermetallics are supported by the Air Force Office of Scientific Research. The investigation of the hot corrosion of ceramics was supported by the Office of Naval Research. The program to examine the combined erosion-corrosion of alloys is supported by the National Science Foundation.

References

1. F.S. Pettit and G.W. Goward, in: *Metallurgical Treatises*, edited by J.K. Tien and J.F. Elliott, (The Metallurgical Society of AIME, Warrendale, PA, 1981), p. 603.
2. J.S. Lee and T.G. Nieh, in: *Oxidation of High Temperature Intermetallics*, edited by T. Grobstein and J. Doychak, (TMS Warrendale, PA, 15068 1988), p. 271.
3. C. Wagner, *Z. Elektrochem.*, **63**, 772 (1959).
4. C. Wagner, *J. Electrochem. Soc.*, **99**, 369 (1952).
5. F. Gesmundo and F. Viani, *Oxid. Metals*, **25**, 269 (1986).
6. J.L. Smialek, *Metall. Trans.*, **22A**, 739 (1991).
7. J.L. Smialek, *Metall. Trans.*, **18A**, 164 (1987).
8. J.G. Smeggil, A.W. Funkenbusch, and N.S. Bornestein, *Metall. Trans.*, **17A**, 923 (1986).
9. A.W. Funkenbusch, J.G. Smeggil, and N.S. Bornestein, *Metall. Trans.*, **16A**, 1164 (1985).
10. J.G. Smeggil, *Mat. Sci. Eng.*, **87**, 261 (1987).
11. G.H. Meier and F. Gesmundo, *Oxid. Metals*, **32**, 155 (1989).
12. K.L. Luthra, *Met. Trans.*, **13A**, 1843 (1982).
13. K. Luthra, *Met. Trans.*, **13A**, 1853 (1982).
14. Y. Hwang and R.A. Rapp, in: *Oxidation of High Temperature Intermetallics*, edited by T. Grobstein and J. Doychak, (TMS, Warrendale, PA, 15068, 1988), p. 257.
15. B.M. Warnes, "The Influence of Vanadium on the Sodium Sulfate Induced Hot Corrosion of Thermal Barrier Coating Materials", *Ph.D. Dissertation*, School of Engineering, University of Pittsburgh.
16. J.R. Blachere and F.S. Pettit, "High Temperature Corrosion of Ceramics", (Noyes Data Corp., Park Ridge, New Jersey, 1989).
17. C.T. Kang, F.S. Pettit, and N. Birks, *Met. Trans.*, **18A**, 1785 (1987).
18. S.L. Chang, F.S. Pettit, and N. Birks, *Oxid. Metals*, **34**, 71 (1989).
19. D. Rishel, F.S. Pettit, and N. Birks, *Mat. Sci. and Eng.*, **A143**, 187 (1991).
20. I.G. Wright, G. Nagarajan, and J. Stringer, *Oxid. Met.*, **25**, 175 (1986).
21. A.V. Levy and Y-F. Man, *Wear*, **131**, 39 (1986).
22. G. Zambelli and A.V. Levy, *Wear*, **68**, 305 (1981).
23. J.H. Wood and E.H. Goldman, in *Superalloys II, C.*, edited by Sims, N. Stoloff, and W. Hagel, (John Wiley and Sons, New York, 1986).
24. H. Lammerann and G. Kienel, *Advanced Materials and Processes*, **6**, 18 (1991).
25. D.K. Gupta and D.S. Duvall, in: *Superalloys 1984*, edited by M. Gell, C. Kortovich, R. Bricknell, W. Kent, and J. Radvich, (TMS Warrendale, PA, 1984), p. 713.
26. R.A. Rapp, D. Wang, and T. Weisert, in: *High Temperature Coatings*, edited by M. Khoalaib and R.C. Krutenat, (The Metallurgical Society, Warrendale, PA, 1987), p. 131.
27. V.A. Ravi, P.A. Choquet, and R.A. Rapp, in: *Oxidation of High Temperature Intermetallics*, edited by T. Grobstein and J. Doychak, (The Metallurgical Society, Warrendale, PA, 1988), p. 127.
28. S. Meier, D. Gupta, and K. Sheffler, *JOM*, March (1991).
29. G.H. Meier, in: *Oxidation of High Temperature Intermetallics*, edited by T. Grobstein and J. Doychak, (The Metallurgical Society, Warrendale, PA, 1988), p. 1.
30. N.S. Choudhury, H.C. Graham, and J.W. Hinze, in: *Properties of High Temperature Alloys*, edited by Z.A. Foroules and F.S. Pettit, (The Electrochem. Soc., Pennington, NJ, 1976), p. 368.

31. G.H. Meier, D. Appalonia, R.A. Perkins, and K.T. Chiang, in *Oxidation of High Temperature Intermetallics*, edited by T. Grobstein and J. Doychak, (TMS, Warrendale, PA, 1989), p. 185.
32. G.H. Meier, F.S. Pettit, and S. Hu, *J. de Physique*, in press.
33. M. Simnad and A. Spilners, *J. Metals*, **7**, 1011 (1955).
34. J.B. Berkowitz-Mattuck, P.E. Blackburn, and E.J. Felten, *Trans. AIME*, **233**, 1093 (1965).
35. S.C. Singhal, *J. Mat. Sci.*, **11**, 1246 (1976).
36. S.C. Singhal, in: *Ceramics for High Temperature Applications*, edited by J.J. Burke, Brook Hill Publishing Co., Chestnut Hill, MA, 1974), p. 533.
37. C. Wagner, *J. Appl. Phys.*, **29**, 1295 (1958).
38. B.M. Warnes, S.Y. Hwang, J.R. Caola, F.S. Pettit, and G.H. Meier, in: *Proceedings of the 1990 Coatings for Advanced Heat Engines Workshop*, edited by John Fairbanks, (U.S. Dept. Energy, Washington, DC, 20585), IV-1.
39. R.L. Jones, *Materials for High Temperatures*, **9**, 228 (1991).
40. J.-M. Yang and S.M. Jeng, *JOM*, **41**, 56 (1989).
41. E.W. Lee, J. Cook, A. Khan, R. Mahapatra, and J. Waldman, *JOM*, **43**, 54 (1991).
42. J.C. Schaeffer, "The Oxidation Behavior of Carbon-Carbon Composites and Their Coatings", *Ph.D. Dissertation*, Materials Science and Engineering Dept., University of Pittsburgh, Pittsburgh, PA 15261.
43. J.F. Cullinan, "The Oxidation of Carbon-Carbon Composites Between 300°C and 900°C in Oxygen and Oxygen/Water Vapor Atmospheres", *Master of Science Thesis*, Materials Science and Engineering Department, University of Pittsburgh, Pittsburgh, PA 15261.
44. J.R. Strife and J.E. Sheehan, *Ceram. Bull.*, **67**, 369 (1988).
45. L.E. McCandlish, B.H. Kear, and B.K. Kim, *Mat. Sci. and Tech.*, **6**, 953 (1990).
46. T. Hirai and M. Sasaki, in: *Advanced Structural Inorganic Composites*, edited by P. Vincinzini, (Elsevier Science Publishers B.V., 1991), p. 541.
47. T. Hirai and T. Goto, *Bull. Jpn. Inst. Met.*, **28**, 960 (1989).

Engineering Materials: The Case of Polyelectrolyte Cements

Víctor M. Castaño[1], Irene H. Arita[1], José Saniger[2] and Hailin Hu[2]

[1] *Instituto de Física, U.N.A.M.*
Apartado Postal 20-364
01000 México, D.F.
MEXICO

[2] *Centro de Instrumentos, U.N.A.M.*
Circuito Exterior, CD. Universitaria
04510 México, D.F.
MEXICO

Abstract

A review of the basic properties of the so-called polyelectrolyte cements is presented. Full characterization of these materials was performed by means of microscopy, Fourier transform infrarred spectroscopy, mechanical testing, etc. These studies allow to fully understand the kinetics of formation of the cements and a truly engineered material can then be designed. Some of the practical applications of these investigations are discussed as well.

I. Introduction

Modern Materials Science and Engineering has become a complex and demanding field for both scientists and industry people. Indeed, the degree of sophistication of many practical applications of materials demands a great effort in conveniently applying basic knowledge to the design and fabrication of materials. Thus, the combined use of basic disciplines, specially Physics and Chemistry, represents a must in today's Materials Science. A clear example, among others, of this statement, is the case of the so-called polyelectrolyte cements which are formed by an inorganic constituent (a metal oxide, mineral silicates or ionic glasses) and an aqueous solution of a polyelectrolyte (ionomer polymer) to form a kind of hydrogels. As a result of

Advanced Topics in Materials Science and Engineering, Edited by
J.L. Morán-López and J.M. Sanchez, Plenum Press, New York, 1993

Table I

Qualitative description of the cement samples
(after 48 h of curing).[1]

Sample	ZnO/PAA g/ml	Color	Consistency
I	0.0	Amber	Liquid
II	0.5	Amber	Solid (good)
III	1.0	White	Solid (poor)
IV	1.5	White	Solid (poor)
V	2.0	White	Solid (poor)
VI	2.5	White	Solid (very poor)

the chemical interaction between the organic and the inorganic components of the cement, a crosslinking process takes place to produce a strong and glassy material. An example of such materials are the dental cements.

These materials present several atractive characteristics which are worth studying. First, they are not completely organic nor completely inorganic in nature, but somewhere in between. Second, as we shall see in what follows, by conveniently manipulating the chemistry of the reaction it is possible to design the final properties of the product. Third, since these cements are strong and highly adhesive, the range of possible applications spans from surgical or dental uses to microelectronics, as a packaging material. In this work we present a review of our group's work in the system ZnO/PolyAcrylic Acid in the past few years. The studies performed included mechanical testing as a function of various experimental parameters, microstructure characterization by means of scanning and transmission electron microscopy, the use of Fourier transform infrared spectroscopy for characterizing the kinetics, and so on.

II. Experimental Procedure

All the samples were prepared from ZnO reagent powder and commercial PAA(BASF). The average molecular weight of the PAA was 120,000 in aqueous solution (30% solid contents).

Four different studies have been reported so far for this system, considering synthesis and properties,[1,2] influence of curing environment,[2] porosimetry[3] and characterization of the reaction using Fourier Transform Infrared Spectroscopy (FTIR).[4] The compositions of the samples are expressed in arbitrary units of g/ml, considering the ratio ZnO/PAA. Table I summarizes the compositions considered in the first study,[1] in which a detailed characterization of properties in terms of composition and curing time was done.

As discussed later, the best mechanical properties were found for compositions ranging in a quite narrow interval of compositions, between 0.5 and 0.7 g/ml. Based on that, the authors of the next two studies[2,3] worked within these limits for further characterization of properties. The last and most recent work,[4] deals with the characterization of the reaction from a more chemical point of view, and reports new knowledge about this system. This also opens a trend to follow for other interesting

Table II

Composition of the samples chosen for FTIR studies.[4]

Sample	MO (grams)	PAA Solution (ml)	(MO) mol	(AA) mol
MgO/PAA	0.500	1	12.4	1
CuO/PAA	1.000	1	12.5	5
CaO/PAA	0.700	1	12.5	5
ZnO/PAA	1.000	1	12.3	5

systems of the same kind, *i.e.*, the general case of PAA solution and a metal oxide MO(MgO, CaO and CuO). Table II presents a description of the composition of the samples, based on the number of moles of each component.

In the next section, a complete description of the particular conditions used in each study is given. It will provide a general idea of the work that has been done to the present time and the scope of this research for the near future.

III. Properties and Characterization

III.1 Effect of Composition

Without doubt, this is the most significant parameter to be considered to determine the properties of these materials. As shown in Table I, qualitative observations of the samples after preparation indicate that there is a very limited range of compositions to obtain good samples, in terms of consistency and appearance.

Samples were prepared according to the ASTM[5] standars for mechanical testing. More precise measurements after the preliminary studies revealed that the best samples in terms of mechanical properties and consistency were obtained for compositions between 0.5 and 0.7.

Fig. 1 shows the results of varying the composition of the samples within this range for two different strain rates. The maximum strength was obtained at a ratio of 0.5, independently of curing time.

The former results problably have their explanation in terms of the reaction that occurs between the metal oxide (MO), in this case ZnO, and the polyelectrolitic acid (2HA), *i.e.*, PAA:

$$MO + 2HA \rightarrow MA_2 + H_2O$$

According to this, there must be an optimal value of concentrations, depending upon the specific reactivities for which the reaction can take place to form a strong cement matrix (MA_2). This particular concentration will obviously depend on the metal oxide and polyelectrolytic acid chosen, and on the temperature and pH conditions as well.

Despite the fact that there is one composition for which the mechanical properties, and in particular the compressive strength, are the best, by using multiple correlation numerical techniques it is possible to calculate an empirical relation between the compressive strength and the parameters involved.

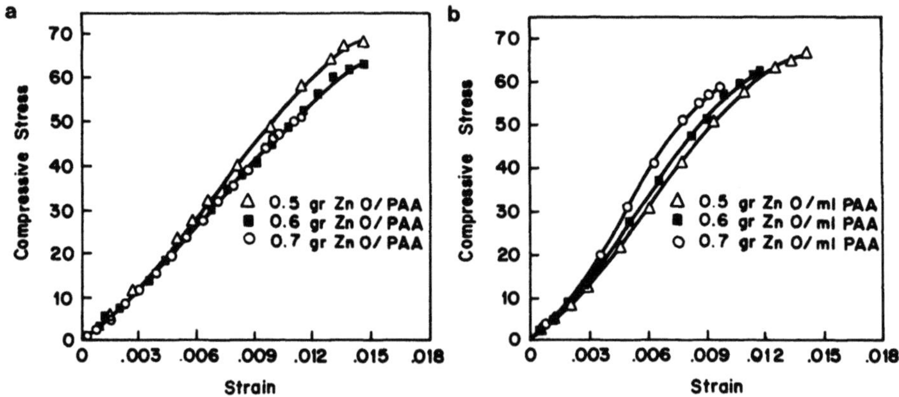

Figure 1. Effect of the ZnO/PAA ratio on the stree-strain curves. (a) 24 h of curing and strain rate equal to 0.2 mm/min. (b) 24 h of curing and strain rate of 0.7 mm/min.[1]

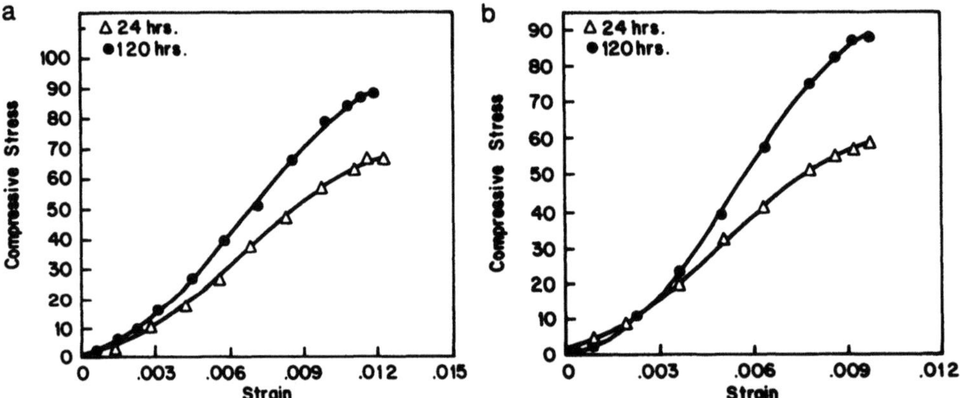

Figure 2. Stress-strain plots to show the effect of curring time. (a) ZnO/PAA= 0.5 and (b) ZnO/PAA= 0.7. In both cases the strain rate was equal to 0.7 mm/min.[1]

In this case, composition and curing time were taken as these parameters, and for values smaller than the optimal ZnO/PAA ratio, the following empirical relation was found:[1]

$$\sigma_f = 68.52 - 25.66X_1 + 0.3387X_2,$$

where σ_f is the compressive strength expressed in kg/cm^2, X_1 is the ZnO/PAA ratio in g/ml and X_2 is the curing time in hours.

Other aspects, like dissolution of part of the PAA in the water present during the reaction, oxide particle size, curing time and environment are to be considered for a more quantitative analysis of the reaction.

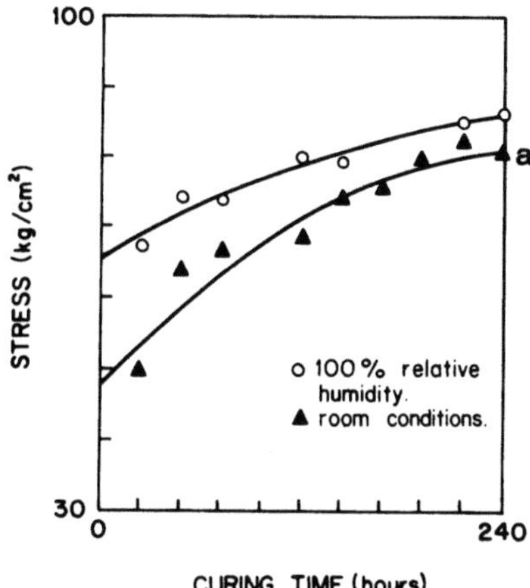

Figure 3. Values of the maximum compressive stress achieved in cements against curing time for (a) samples cured under room conditions and (b) samples cured at 100% humidity conditions.[3]

III.2 Effect of Curing Time

Different curing times were allowed in the samples before mechanical testing. Fig. 2 shows the tests on 0.5 and 0.7 g/ml samples performed 24 and 120 h after preparation in room conditions.

It was found that, for the samples cured for 24 h, there are three different zones in the strain-stress curves: (1) a zone characterized by a small slope and ending at about a deformation of 0.35%, (2) a second (elastic) zone in which the slope increases and that extends up to 0.65% of deformation, and (3) a zone in which the slope decreases back.

For 120 h of curing time, the first zone showed a tendency to dissapear. This behaviour can be explained in terms of the action of uncreated ZnO. It is expected that, the greater the amount of uncreated ZnO, the weaker the material since no chemical bonding has developed, yet. That means that for longer curing times, more uniform materials are expected, with better properties.

In general terms, curing times below 24 h produced very poor quality samples, and after 48 h, the samples did not show any improvement. However, a more detailed study of very short curing times (even in the order of minutes) should provide very useful information about the kinetics of the curing process.

III.3 Effect of Curing Environment

The composition used in all cases in this study[2] was 0.6 g/ml. The compressive strength and elastic modulus were tested as a function of curing time for normal room conditions and for 100% humidity (wet) conditions. The last conditions were achieved by simply curing the samples in pure distilled water.

Fig. 3 shows plots of the maximun compressive stress as a function of curing time for both normal and wet conditions. Regardless of curing conditions, it is observed an increase in strength as the curing time increases, which is consistent with previous results.

Table III

Conditions and number of cycles before fracture in fatigue testing
under different curing and loading conditions.[3]

Curing Conditions	Load (Kg)	Loading Rate 1 (0.4 mm/min)	Loading Rate 2 (0.7 mm/min)
Room + 2 days	40	112	57
	50	68	20
	60	4	–
Room + 3 days	40	174	97
	50	113	53
	60	7	3
Wet + 2 days	40	434	244
	50	167	106
	60	72	29
Wet + 3 days	40	600	392
	50	206	180
	60	96	57

However, it is also found that the wet samples always showed better properties than the samples cured under normal conditions. The same behaviour was observed for the elastic modulus.

It is believed that the reason for this behaviour is that the surrounding water helps control the velocity of the reaction, acting as a medium for the ions to move.

Longer curing times (2 and 3 days) were considered for fatigue experiments and the results from these experiments are summarized in Table III. They show that an increase in the rate of deformation produces a decrease in the number of cycles before failure. This is an indication of an anelastic behaviour of the samples, which means that the energy release has a delay when the load is withdrawn. Again, the wet samples presented more resistance than the samples prepared under normal conditions.

III.4 Strain Rate Studies

The elastic modulus is independent of the composition and curing time, but it was found to increase as the strain rate increases. This can be understood from Nadai's theory.[6]

This model considers the elastic modulus in terms of the chemical bonding of the composite under study. The result is that the elastic modulus increases as the chemical bonds of the composite oppose the deformation as the strain rate increases. This phenomenon corresponds to an adiabatic process in which the strain is so fast that no energy change with the environment is allowed. As a consequence, the temperature would rise, which was actually observed during the experiments. Fig. 4 shows an example of a strain-stress plot for different strain rates, and Table IV summarizes the results from mechanical tests.

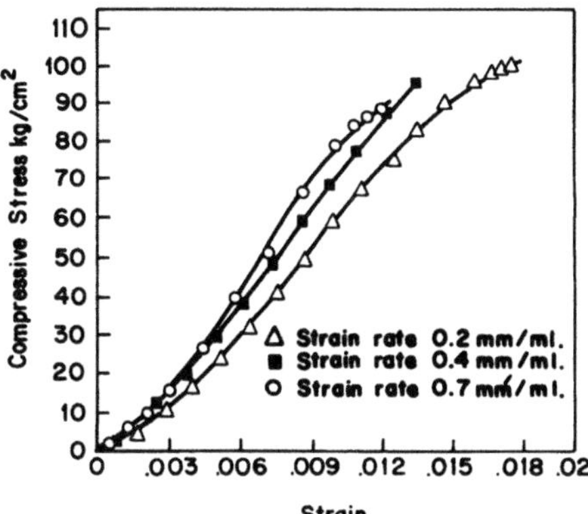

Figure 4. Stress-strain plots for different strain rates for a sample with ZnO/PAA= 0.5 and 120 h of curing time.[1]

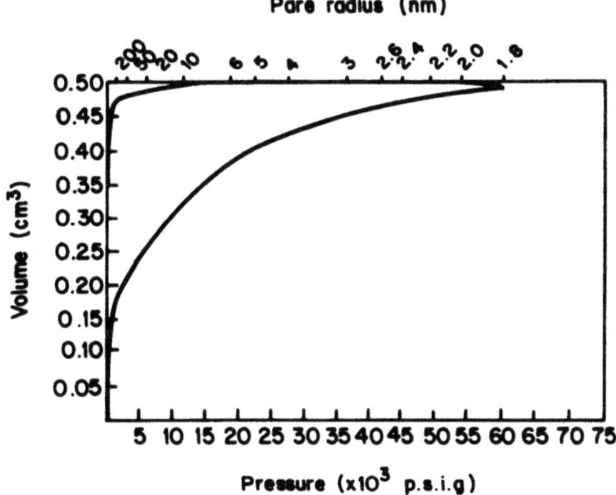

Figure 5. Typical mercury intrusion porosimetry curve of the cements studied. In this case, ZnO/PAA= 0.6 (1psi≃6.9 kPa.).[2]

III.5 Porosimetry Studies

Mercury intrusion porosimetry (MIP) was performed in samples with compositions within the optimum range for best mechanical properties,[3] 0.5, 0.6 and 0.7 g/ml. The samples were prepared according to ASTM standards[5] and the MIP measurements were done in a Quantachrome Autoscan-60 machine.

Fig. 5 shows the MIP plot for 0.6 g/ml composition, which turned out to be the one with the smallest pore volume, as can be observed from Table V. The shape of the porosimetry curves was very similar for all compositions. It is worthwhile mentioning that the composition with the smallest pore volume corresponds to the one with the best mechanical properties according with former discussion.

Table IV

Summary of mechanical testing.[1]

Curing time (h)	ZnO/PAA g/ml	Maximum stress Kg/cm^2	Strain	Elastic modulus Kg/cm^2	Loading rate mm/min
24	0.5	67.93	0.0146	5476.40	0.2
		67.14	0.0146	6107.71	0.4
		67.16	0.0115	7116.12	0.7
	0.6	63.20	0.0139	4758.40	0.2
		62.40	0.0120	6238.27	0.4
		62.40	0.0129	7697.88	0.7
	0.7	51.34	0.0122	5108.12	0.2
		50.53	0.0108	6237.30	0.4
		59.24	0.0100	7333.90	0.7
120	0.5	99.526	0.0174	7292.30	0.2
		94.78	0.0116	8800.25	0.4
		88.46	0.0120	9389.89	0.7
	0.6	84.51	0.0133	8804.25	0.2
		88.866	0.0101	9964.73	0.4
		94.78	0.0104	10674.20	0.7
	0.7	110.6	0.0120	7244.40	0.2
		94.0	0.0120	8393.50	0.4
		88.50	0.0096	11000.00	0.7

III.6 FTIR Studies

Although four different metal oxides (including ZnO, of course) were used in this study,[4] the results in all cases were very similar in terms of the nature of the reaction detected and the structure to be expected from it. This is offering a whole new family of materials. Table II summarizes the conditions of the various samples.

The characterization of the reaction was performed by comparing the FTIR spectrum of unreacted PAA to the corresponding spectra of samples prepared by mixing the metal oxide with PAA at room temperature and conditions until a hard paste was formed.

Unreacted PAA films were prepared by dissolving a drop of the solution in methanol and depositing the mixture onto a KBr window with further evaporation of the methanol. The cured samples of MO-PAA were powdered in an agate mortar and then mixed with powered KBr in a proportion of 1:100 and the resulted mixture was compacted. The spectra were registered between 4000 and 600 cm^{-1} in a Nicolet FTIR.

Table V

Summary of porosimetry measurements.

ZnO/PAA contem (gml^{-1})	pore volume (cm^{-1}g^{-1})
0.5	0.2264
0.6	0.1584
0.7	0.1828

Table VI

$\Delta\nu$ values for the various metal salts of PAA.[4]

Compound	$\Delta\nu$ (cm^{-1})
Na(OH)-PAA	128
MgO-PAA	231
CuO-PAA	228
CaO-PAA	227
ZnO-PAA	244

Fig. 6(a) shows the infrared spectrum of unreacted PAA. Notice the characteristic stretching band of carbonyl groups al 1717 cm^{-1}. Weaker peaks at 1451 and 1403 cm^{-1} can be associated with scissor and bending vibrations of $-CH_2$ and $=CH-CO-$ groups, respectively. The bands at 1235 and 1170 cm^{-1} may be related to the coupling between in-plane $O-H$ bending and $C-O$ stretching of neighboring carboxyl groups.[7,8]

Fig. 6(b) shows the spectra of the reacted PAA with the different metal oxides. In all cases, there is a shift of the carbonyl band towards lower energies, between 1570 and 1550 cm^{-1}. Moreover, the doublet at 1235 and 1770 cm^{-1} present in the unreacted PAA dissapears in these cases, and the same behavior is detected for the band at 797 cm^{-1}. On the other hand, new peaks appear at 1300 and 860 cm^{-1}, indicating a chemical reaction taking place between the PAA and the different metal oxides.

In general terms, the reaction between a polyelectrolytic acid, such as PAA, and a metal oxide would be the formation of the corresponding metal salt. The formation of these salts involves a change of the carbonyl group band arrangement, which will go from a localized $C=O$ bond in its structure to a symmetric ionized structure. This structural change results in the appearence of a doublet near the stretching vibration band of the carbonyl, that is due to the asymmetrical and symmetrical stretching vibrations of the carboxilate anion.

According to the former reasoning, the peaks shown in Fig. 6(b) at 1570 and 1550 cm^{-1} bands would correspond to the asymmetrical vibration whereas the symmetrical vibrations would be located around 1330 cm^{-1}. The absence of the bands at 1235 and 1170 cm^{-1} can be explained as a loss of hydrogen bondings between $O-H$ and $C-O$ groups as a result of the formation of the salts.

The 1451 and 1403 cm^{-1} bands remain present because the $-CH_2$ and $=CH-$ groups are not involved in the reaction. However it is interesting to discuss the changes

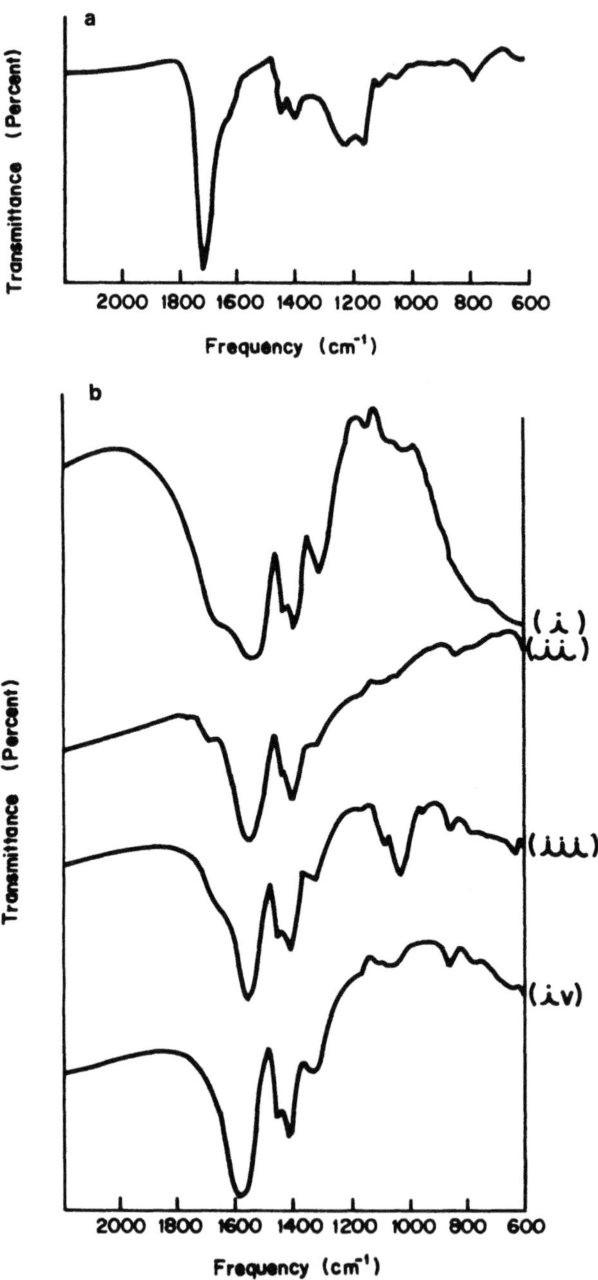

Figure 6. FTIR spectra of (a) pure unreacted PAA and (b) PAA reacted with different metal oxides: i) MgO, ii) CuO, iii) CaO and iv) ZnO.[4]

in their intensities, as well as the change of shape of the $-CH_2$ band, when comparing the MO-PAA spectra to the unreacted PAA. These changes can be explained by the assumption made before, *i.e.*, that the $-CH_2-$ band coincides with the symmetric stretching vibration.

The coordination compounds formed in the reaction could have three different structures: monodendate, bidendate bridging and bidendate chelating. A qualitative determination of the predominant structure can be done by the following considerations. Knowing the values of the wavenumbers of the asymmetric (ν_{as}) and symmetric (ν_s) stretching vibrations, and defining $\Delta\nu = \nu_{as} - \nu_s$, it can be shown that $\Delta\nu$ has its smallest value when the bidendate chelating structure is predominant, an intermediate value for the bidendate bridging structure, and the highest value corresponding to monodendate structure.[7]

Table VI contains a list of $\Delta\nu$ values for the systems studied, and the corresponding value for sodium salt. It is known that this last salt has a bidendate bridging structure, *i.e.*, the intermediate value of $\Delta\nu$, and since all the values determined for the other salts are much higher than this one, it can be concluded that the expected predominant structure has to be the monodendate.

Finally, it is important to consider the stoichiometry of the reaction. Notice that, in all cases, the samples were prepared with a molar ratio ZnO/PAA of about 2.5 (see Table II), obtaining the consequent dissapearance of the carbonyl group band at 1717 cm^{-1}. Other experiments were performed at different concentrations, finding that the minimum molar ratio for consuming all the carbonyl groups was about 1. However, the expected stoichiometric ratio according to the chemical reaction is 0.5, half that experimentally determined. This suggests that the metal oxide does not react completely when mixed with the PAA.

This result can be explained if the relatively large particle size of the starting metal oxides is taken into account. According to this, the final structure of the materials studied would not be homogeneous, but formed by grains of unreacted metal oxide surrounded by a shell of the reacted material. this is supported by X-ray studies of the samples that showed peaks for the metal oxides plus a halo corresponding to an amorphous phase, which can be related to the product of the reaction. No evidence of other crystalline peaks was found in any case.

III.7 Electron Microscopy Studies

Only preliminary electron microscopy studies have been done so far, basically motivated by two aspects of the problem. The first is the influence of the sample geometry in the evolution of the reaction and, the other, the final structure of the materials.

SEM observations of transversal planes in samples prepared for mechanical testing[5] showed unreacted polymer and metal oxide near the centre of the samples. This can be explained by the fact that the reaction is exothermic and, therefore, outer regions of the sample will react more rapidly since they are allowed to have thermal exchange with the environment whereas the inner regions, being thermally isolated, will have slower reaction kinetics. Fig. 7 shows a micrograph of a sample observed by SEM.

TEM observations of thin slices obtained by ultramicrotomy show a film with plenty of holes in the order of microns, as well as some evidence of ZnO remanent particles (See Fig. 8(a) and (b)). However, these observations are not representative since polymer deformation and ZnO pulling-off are expected from the preparation

                                    V.M. Castaño *et al.*

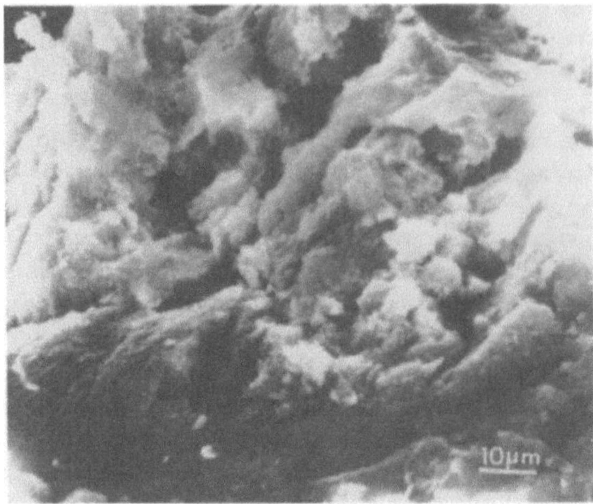

Figure 7. Scaning electron micrograph of a sample with ZnO/PAA= 0.5. Notice the higher concentration of unreacted material near the centre of the sample.[2]

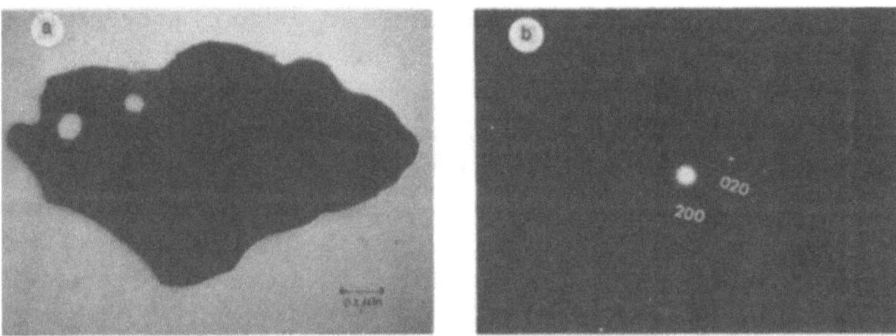

Figure 8. (a) bright field image of a region in a thin slice of ZnO-PAA sample sectioned by ultramicrotomy and (b) diffraction pattern showing the presence of unreacted ZnO.

method. Other methods for thin film preparation are under study at the present time to determine not only the presence and distribution of the ZnO unreacted particles, but also the extent of the reaction around each particle.

IV. Conclusions and Further Work

There are many parameters involved in the final product obtained from the reaction of PAA and ZnO. The most important have been discussed extensively in this paper, namely, the ZnO/PAA ratio and the curing time and environment.

There is a clear evidence of a chemical bond between the polymer and the metal, showing that these materials do not consist of a physical mixture of components. The chemical structure formed during the reaction is expected to be predominantely monodendate, meaning that crosslinking of the carboxyl groups through the metal is produced. However, the fact of whether or not this crosslinking occurs within the same chain is still to be clarified.

From all the testing done it was found that the ratio needed for the best properties, in terms of mechanical behaviour and porosity, is much higher than the stoichiometrical, leading to the conclusion that the metal oxide reacts only partially with the PAA, probably due to a particle size effect. This opens the possibility to the fabrication of novel materials, consisting of oxide particles surrounded by a cement, product of the reaction of the metal oxide at the surface of the particles and the polymer.

According to the former conclusions, there are several trends that can be proposed for further work:

- A better and more reliable characterization of the materials from all points of view, which would include the characterization of the starting materials on their own and of the final products and unreacted polymer and metal oxide. This is important in order to establish standards for preparation and the final products to be expected according to the selection of the initial conditions.
- A more detailed study of chemical reaction, and the subsequent bondings at the interfaces metal oxide-cement and polymer-cement. Other aspects to be considered are the kinetics of the reaction and the influence of sample geometry and external conditions, such as temperature, pH and curing environment.
- In more general terms, it would be relevant the extensive study of other systems of polyelectrolyte-metal oxide, since the chemical reaction involved is the same. This fact, together with the control of the final particle size and the complete characterization of the interfaces involved, is opening the way to develop a new family of composites, which in a near future may be produced by using the so-called Nanophase Technology.[9]
- Finally, applications of this family of materials remain to be studied. In particular, the use of polyelectrolyte cements as biomaterials (specifically as adhesives for dental use, or bone implants) is a current field of interest in our group. Also, the feasibility of using these cements for microelectronics (packaging) represents a rather attractive possibility.

Acknowledgements

The authors are grateful to Dr. A. Padilla, Dr. O. Manero, Dr. Gustavo Vázquez-Polo, and Dr. D.R. Acosta for their valuable discussion and comments. Also, they would like to thank R. Hernández, L. Rendón and R.M. Lima for their skilful technical assistance.

References

1. A. Padilla, A. Vázquez, and V.M. Castaño, *J. Mater Res.* **6**, 2452 (1991).
2. A. Padilla, J.L. Martínez, A. Sánchez, and V.M. Castaño, *Mat. Lett.* **12**, 445 (1992).
3. A. Padilla, A. Vázquez, G. Vázquez-Polo, D.R. Acosta, and V.M. Castaño, *Mater. in Medicine* **1**, 154 (1990).
4. H. Hu, J. Saniger, J. García-Alejandre, and V.M. Castaño, *Mat. Lett.* **12**, 281 (1991).
5. Annual Book of ASTM standards, Volume 08.01 (1985).
6. A. Nadai, *Theory of Flow and Fracture in Solids*, (McGraw and Hill, New York, 1950).

7. K. Nakamoto, *Infrared Spectra of Inorganic and Coordination Compounds*, (John Wiley and sons, New York, 1963).
8. K. Nakanishi, *Infrared Absorption Spectroscopy*, (Holden Day Inc. and Nakodo Co. Ltd., Tokyo, 1969).
9. R.P. Adres, R.S. Averback, W.L. Brown, L.E. Brus., W.A. Goddard III, A. Kaldor, S.G. Louie, M. Moscovits, P.S. Peercy, S.J. Riley, R.W. Siegel, F. Spaepen, and Y. Wang, *J. Mat. Res.* **4**, 704 (1989).

Advanced Textile Structural Composites

Frank K. Ko

Fibrous Materials Research Laboratory
Department of Materials Engineering
Drexel University
Philadelphia, Pennsylvania
USA

Abstract

The need for significant improvements in intra- and inter-laminar strength, damage resistance and the large scale economical manufacturing of structural composites resulted in the reemergence of textile structural composites in the 1980's as a new class of advanced composites. After a review of the worldwide activities in textile structural composites, the role of fiber architecture in the manufacturing of tough, net shape composites will be discussed. The toughening and strengthening of polymer, metal and ceramic matrix composites through the use of 3-D fiber architecture is demonstrated with experimental evidence. Techniques for the modelling of textile structural composite are reviewed and illustrated with examples.

I. Introduction

Textile structural composites represent a class of advanced materials reinforced by textile preforms for primary structural applications. Making use of the unique combination of light weight, flexibility, strength and toughness, textile structures have long been recognized as an attractive reinforcement form for applications ranging from aircraft wings produced by Boeing Aircraft Co. in the 1920's to carbon-carbon nose cones produced by General Electric in the 1960's. Textile preforms are fibrous assemblies with pre-arranged fiber orientation preshaped and often pre-impregnated with matrix for composite formation. The microstructural organization of fibers within a preform, or fiber architecture, determines the pore geometry, pore distribution and tortuosity of the fiber paths within a composite. Textile preforms not only play a key role in translating fiber properties to composite performance but also influence the ease or difficulty in matrix infiltration and consolidation. Textile preforms are the structural backbone for the toughening and net shape manufacturing of composites.

Advanced Topics in Materials Science and Engineering, Edited by
J.L. Morán-López and J.M. Sanchez, Plenum Press, New York, 1993

Table I

Summary of fiber architecture levels.

Level	Reinforcement System	Textile Construction	Fiber Length	Fiber Orientation	Fiber Entanglement
I	Discrete	Chopped Fiber	Discontinuous	Uncontrolled	None
II	Linear	Filament Yarn	Continuous	Linear	None
III	Laminar	Simple Fabric	Continuous	Planar	Planar
IV	Integrated	Advanced Fabric	Continuous	3-D	3-D

When combined with high performance fibers, matrices and properly tailored fiber/matrix interfaces, the creative use of fiber architecture promises to expand the design options for strong and tough structural composites. Recent advances in computer aided design and manufacturing have facilitated the adaptation of many low cost traditional textile processes to create 2-D and 3-D fiber architectures. Making use of these fiber placement or textile preforming methods, a larger family of fiber architecture[1] has been created to address the needs for structural toughening and processing of composites into net or near net shape structural components.

Considering the critical role which the preform plays in composite manufacturing and performance, there is a worldwide revival of interest in the subject of textile structural composites (TSC).[2,3] The subject of textile preform has been detailed in review articles and books;[1,2,4] this paper will focus on the recent advances in preforming concepts. With an emphasis on structural composites, the role of fiber architecture in the design of composites is illustrated through an examination of the structural features and the properties of polymer, metal and ceramic matrix composites reinforced by textile structures. The methods for the modelling of textile structural composites are also reviewed with an emphasis on their role in the integration of design, manufacturing and structural analysis of composites.

II. Textile Preforming Technology

II.1 Classification of Textile Preforms

There are several ways to classify textile preforms.[3] On the basis of structural integrity and fiber linearity and continuity, textile preforms can be classified into four categories: discrete; continuous; planar interlaced (2-D) and fully integrated (3-D) structures. In Table I the nature of the various levels of fiber architecture is summarized according to Scardino.[5]

The first category of textile preforms is the discrete-fiber system. A discrete fiber system such as a whisker or fiber mat has no material continuity; the orientation of the fibers is difficult to control precisely, although some aligned discrete fiber systems have recently been introduced. The structural integrity of the fibrous preform is derived mainly from inter-fiber friction. The strength translation efficiency, or the function of fiber strength translated to the fibrous assembly of the reinforcement system, is quite low.

Table II
A comparison of yarn-to-fabric formation techniques.

	Basic direction of yarn introduction	Basic formation technique
Weaving	Two (0°/90°)	Interlacing (by selective warp and fill) insertion of 90° yarns into 0° yarn system
Braiding	One (Machine Direction)	Interwining (position displacement)
Knitting	One (0° or 90°)	Interlooping (by drawing warp or fill loops of yarns over previous loops)
Nonwoven	Three or more (orthogonal)	Mutual fiber placement

The second category of textile preform is the continuous filament, or unidirectional (0°) system. This architecture has the highest level of fiber continuity and linearity, and consequently has the highest level of property translation efficiency and is very suitable for filament wound and angle ply tape layup structures. The drawback of this textile preform is its intra-and interlaminar weakness due to the lack of in-plane and out-of-plane yarn interlacings.

A third category of fiber reinforcement is the planar interlaced and interlooped system. Although the intralaminar failure problem associated with the continuous filament system is addressed with this textile preform, the interlaminar strength is limited by the matrix strength due to the lack of through thickness fiber reinforcement.

The fully integrated system forms the fourth category of textile preforms wherein the fibers are oriented in various in-plane and out-of-plane directions. With the continuous filament yarn, a three dimensional network of yarn bundles is formed in an integral manner. The most attractive feature of the integrated structure is the additional reinforcement in the through-thickness direction which makes the composite virtually delamination free. Another interesting aspect of many of the fully integrated structures such as 3-D woven, knits and braids is their ability to assume complex structural shapes.

II.2 Preform Fabrication Processes

The conversion of fiber to preform can be accomplished via the "fiber to fabric" (FTF) process, the "yarn to fabric" (YTF) process and combinations of the two. An example of the FTF process is the Noveltex® method developed by P. Olry at SEP*. As shown in Figs. 1a, b and c, the Noveltex concept is based on the entanglement of fiber webs by needle punching. A similar process is being developed in Japan by Fukuta[7] using fluid jets in place of the needles to create through thickness fiber entanglement.

---

* Societe Europeens de Propulsion, Bordeaux, France.

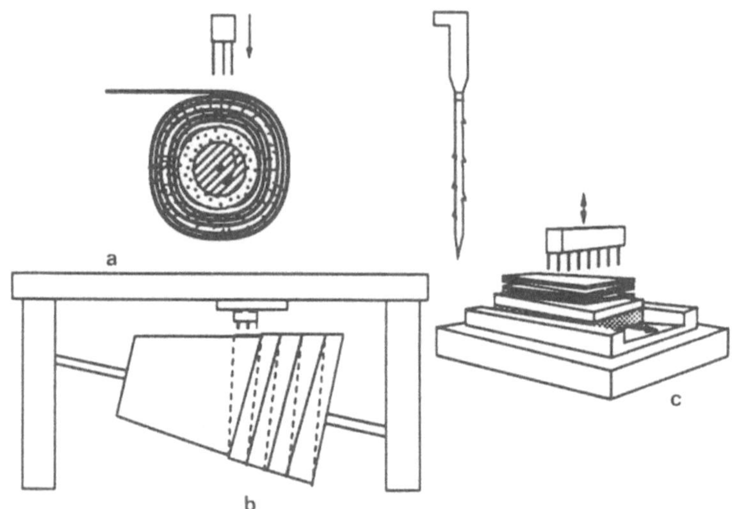

Figure 1. The Noveltex® method.

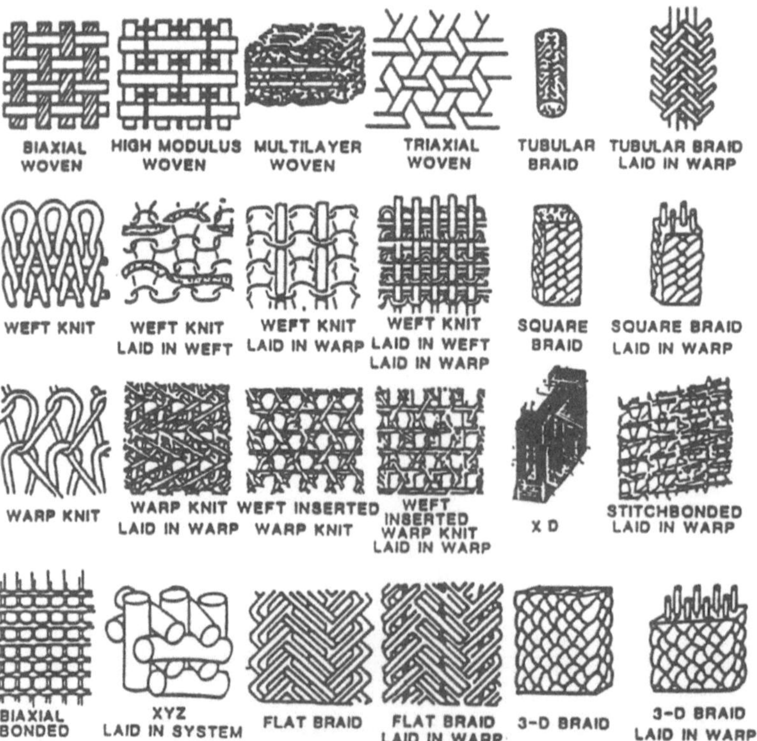

Figure 2. Examples of yarn-to-fabric preforms.

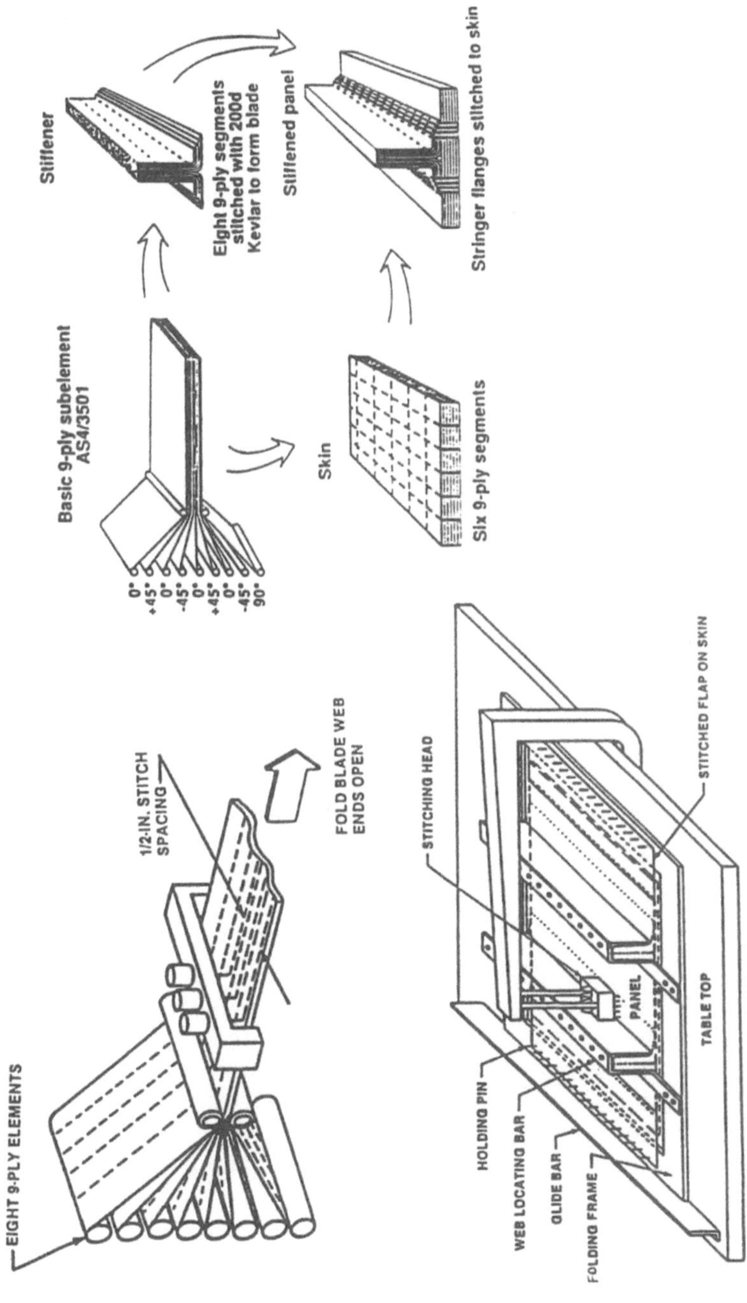

Figure 3. Combination of FTF and YTF processes.

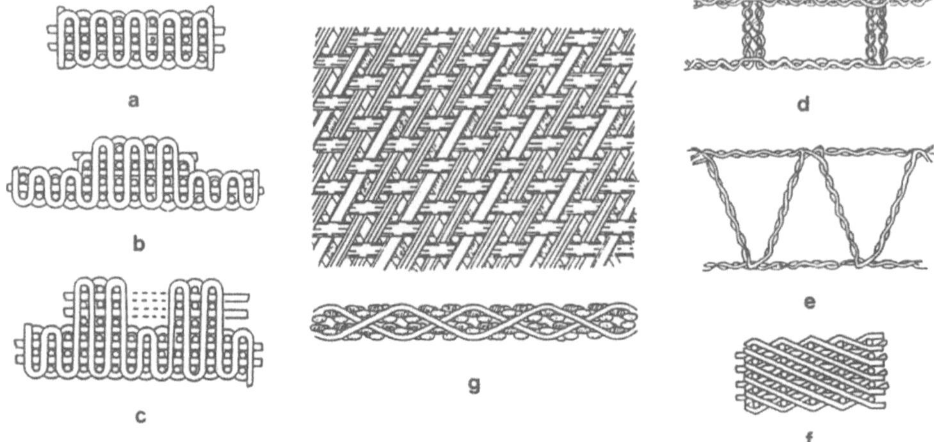

Figure 4. 3-D woven fabrics.

The YTF processes are popular means for preform fabrication wherein the linear fiber assemblies (continuous filament) or twisted short fiber (staple) assemblies are interlaced, interlooped or intertwined to form 2-D or 3-D fabrics. Examples of preforms created by the YTF processes are shown in Fig. 2. A comparison of the basic YTF processes is given in Table II.

In addition to the FTF and YTF processes, textile preforms can be fabricated by combining structure and process. For example, the FTF webs can be incorporated into a YTF preform by needle or fluid jet entanglement to provide through-the-thickness reinforcement. Sewing is another example which can combine or strategically join FTF and/or YTF fabrics together to create a preform having multidirectional fiber reinforcement[8] (Fig. 3).

III. Advanced Preforming Concepts

There is a large family of textile preforming methods suitable for composites.[1] The key criteria for the selection of textile preforms for structural composites are (a) the capability for in-plane multiaxial reinforcement, (b) through thickness reinforcement and (c) the capability for formed shape and/or net shape manufacturing. Depending on the processing and end use requirements some or all of these features are required. In this section the representative structural geometries of 3-D fabrics are introduced according to the four basic methods of textile manufacturing: weaving, orthogonal nonwoven, knitting and braiding.

III.1 3-D Woven Fabrics

3-D woven fabrics are produced principally by the multiple-warp weaving method which has long been used for the manufacturing of double cloth and triple cloths for bags, webbings and carpets. By the weaving method, various fiber architectures can be produced including solid orthogonal panels (Fig. 4a), variable thickness solid

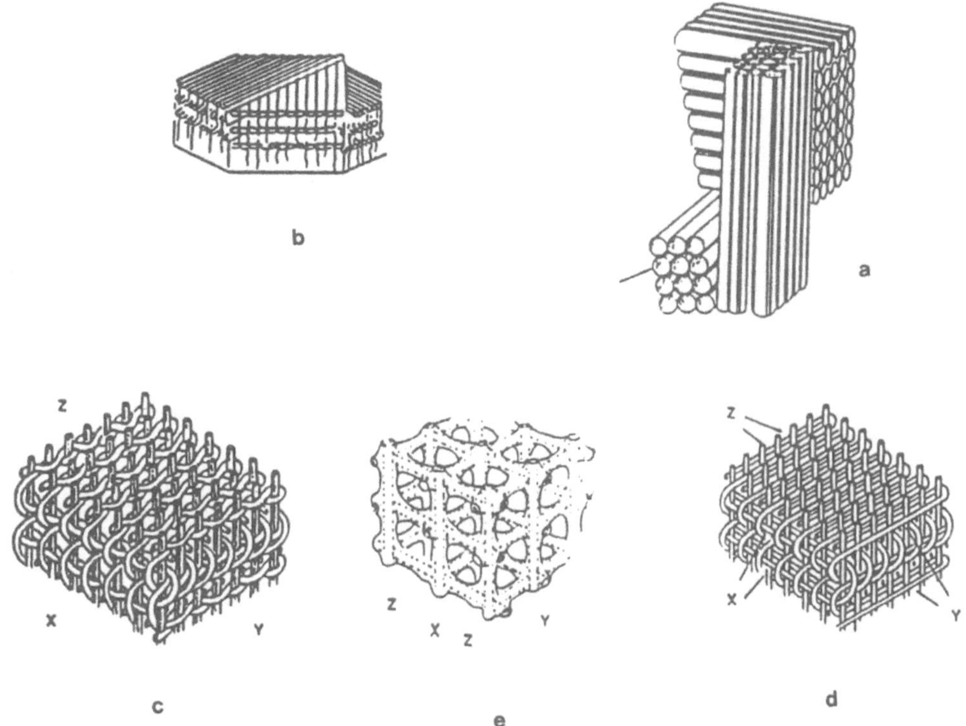

Figure 5. Orthogonal woven fabrics.

panels (Figs. 4b, c), and core structures simulating a box beam (Fig. 4d), or truss-like structure (Fig. 4e). Furthermore, by proper manipulation of the warp yarns, as exemplified by the angle interlock structure (Fig. 4f), the through-thickness yarns can be organized into a diagonal pattern. To address the inherent lack of in-plane reinforcement in the bias direction new progresses are being made in triaxial weaving technology by Dow[9] to produce multilayer triaxial fabrics as shown in Fig. 4g.

III.2 Orthogonal Nonwoven Fabrics

Pioneered by aerospace companies such as General Electric,[10] the nonwoven 3-D fabric technology was developed further by Fiber Materials Incorporated.[11] Recent progress in automation of the nonwoven 3-D fabric manufacturing process was made in France by Aerospatiale,[12] SEP[7] and Brochier[13,14] and in Japan by Fukuta.[15,16]

The structural geometries resulting from the various processing techniques are shown in Fig. 5. Figs. 5a and 5b show the single bundle XYZ fabrics in a rectangular and cylindrical shape. In Fig. 5b, the multidirectional reinforcement in the plane of the 3-D structure is shown. Although most of the orthogonal nonwoven 3D structures consist of linear yarn reinforcements in all of the directions, introduction of the planar yarns in a nonlinear manner, as shown in Figs. 5c, d and e can result in an open lattice or a flexible and conformable structure.

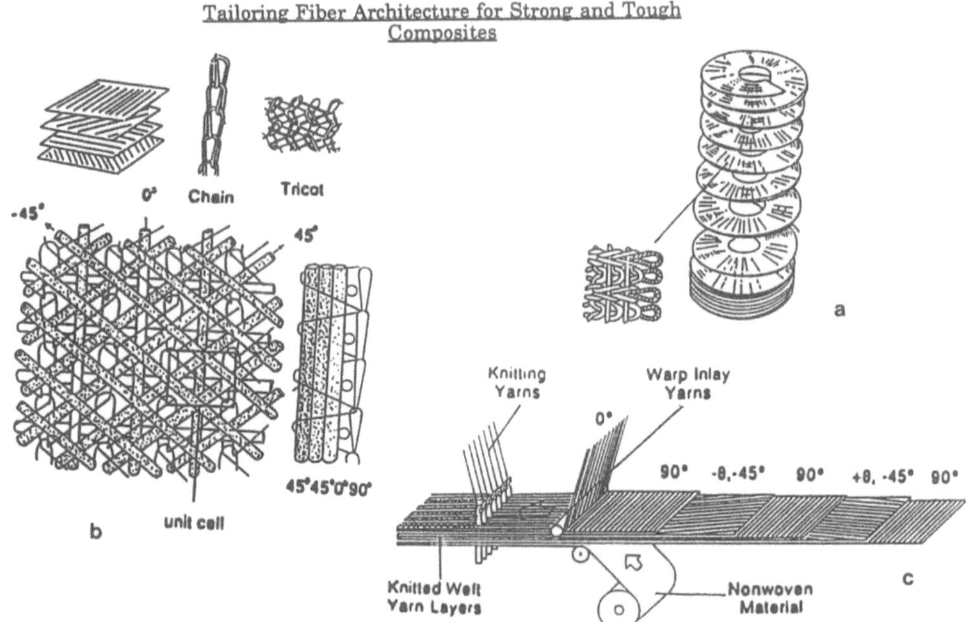

Figure 6. 3-D knitted fabrics.

III.3 Knitted 3-D Fabrics

The knitted 3-D fabrics are produced either by the weft knitting or warp knitting process. An example of a weft knit is the near net shape structure knitted under computer control by the Pressure Foot[R] process[17] (Fig. 6a). In a collapsed form this preform has been used for carbon-carbon aircraft brakes. The unique feature of the weft knit structures is their conformability.[18] While the weft knitted structures have applications in limited areas, the multiaxial warp knit (MWK) 3-D structures are more promising and they have undergone a great deal more development in the recent years.[19,20] From the structural geometry point of view, the MWK fabric systems consist of warp (0°), weft (90°) and bias ($\pm\theta$) yarns held together by a chain or tricot stitch through the thickness of the fabric, as illustrated in Fig. 6b. The major distinctions of these fabrics are the linearity of the bias yarns; the number of axes; and the precision of the stitching process. The latest commercial nonimpaled MWK fabric is produced by the Mayer Textile Corporation utilizing a multiaxial magazine weft insertion mechanism. The attractive feature of this system is the precision of yarn placement with four layers of linear or nonlinear bias yarns plus a short fiber mat arranged in a wide range of orientations. Furthermore, the formation of stitches is done without piercing through the reinforcement yarns at a production rate of one hundred meters per hour.

An example of impaled MWK is the LIBA or Hexcel system, as shown in Fig. 6c. Six layers of linear yarns can be assembled in various stacking sequences and stitched together by knitting needles piercing through the yarn layers. While this piercing action unavoidably damages the reinforcing fiber, it also permits the incorporation of a fiber mat as a surface layer for the composite.

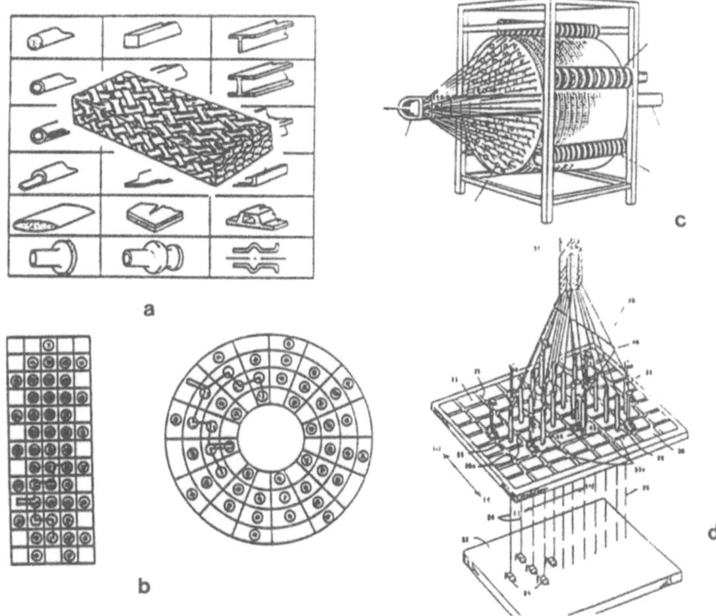

Figure 7. 3-D braided fabrics.

III.4 3-D Braided Fabrics

3-D braiding technology is an extension of the well established 2-D braiding technology wherein the fabric is constructed by the intertwining of two or more yarn systems to form an integral structure. 3-D braiding is one of the textile processes wherein a wide variety of solid complex structural shapes (Fig. 7a) can be produced in an integral manner resulting in a highly damage resistant structural preform. Fig. 7b shows two basic loom setups in circular and rectangular configurations.[21] The 3-D braids are produced by a number of processes including the Track and Column method[22] (Fig. 7c), the 2-step method[23] (Fig. 7d) as well as a variety of displacement braiding techniques. The basic braiding motion includes the alternate X and Y displacement of yarn carriers followed by a compacting motion. The formation of shapes is accomplished by the proper positioning of the carriers and the joining of various rectangular groups through selected carrier movements.

IV. Tailoring Fiber Architecture for Strong and Tough Composites

Strength and toughness are usually considered to be concomitant mechanical properties for traditional engineering methods. By properly manipulation of fiber architecture, the degree of freedom permitted in the engineering of strong and tough composite materials is greatly increased.

The strengthening effect of fiber on polymer matrix composites is well established. The role of fiber in the toughening of ceramic matrix composites is now generally rec-

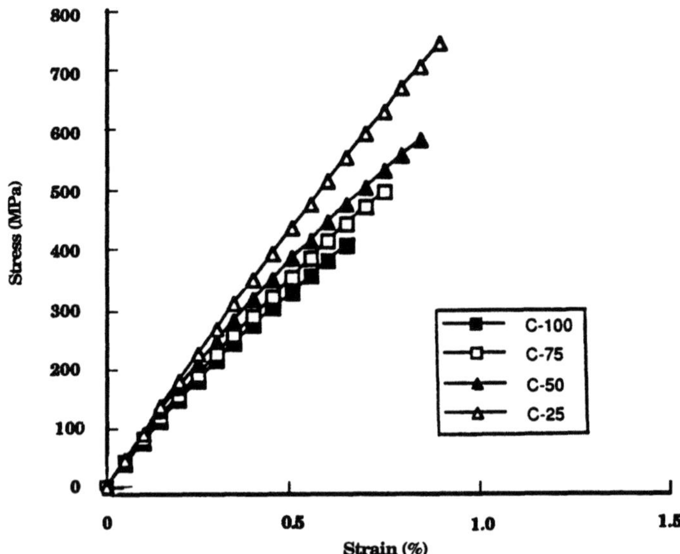

Figure 8. Compressive stress-strain behavior of carbon/epoxy composites.

ognized. A major challenge in advanced composite materials is to achieve a balance of strengthening and toughening effects. While well oriented linear unitape composites provide the maximum strengthening effect when loading is along the fiber direction, the interlaced nonlinear integrated systems tend to maximize damage containment and enhance toughness. Expanding from these strengthening and toughening concepts, one may hypothesize that a combination of the strategically placed linear fibers in a 3-D integrated fiber network would provide the necessary ingredients for a strong and tough composite. The linear fibers and the fibers which make up the 3-D fiber network can be from the same material family but differ in diameter or bundle size. Alternatively, the two material systems can be quite different, each contributes unique properties such as strength, stiffness and thermal stability to the composite structure. To demonstrate this geometric/material hybrid concept, let us examine three examples, including PMC, MMC, and CMC, wherein different proportions of linear fibers were placed in a 3-D braided fiber network.

IV.1 Polymer Matrix Composites

The hybrid concept has been demonstrated in several previous studies for PMC. For example, it was found that a combination of linear fibers in a 3-D braided structure improves the compression after impact strength of PEEK Carbon composites.[24] It was further demonstrated that the linear fibers provide additional flexibility in modifying the failure modes of the composites whereas the 3-D fiber network greatly reduces the damage area caused by impact loading.

In a controlled study,[25] the compressive stress-strain behavior of carbon, glass, and Kevlar fibers in an Epon 828/DETA matrix with linear fiber to 3-D braided fiber ratio ranging from 0/100 to 75/25 were examined. It was found that, in all three cases, both the compressive strength and the elongation to break increase as the proportion of linear fibers increases. This strengthening and toughening effect

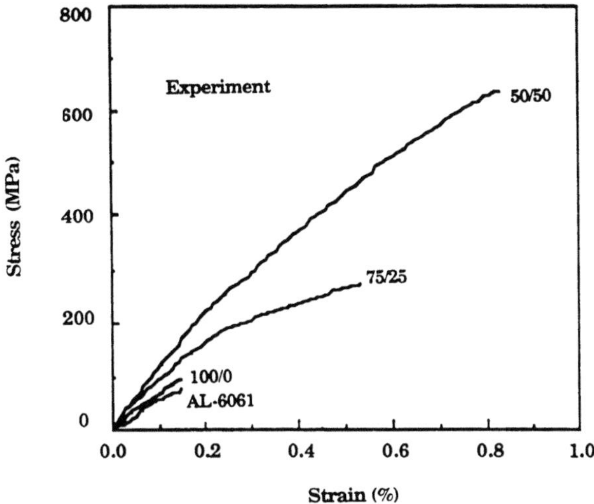

Figure 9. Tensile stress-strain behavior of 3-D braided SiC/Al composites.

can best be illustrated in Fig. 8, which shows that the compressive strength of the carbon/epoxy composites increases from 415 MPa to 760 MPa as the proportion of linear fibers in the 3-D braided structure increases from 0/100 to 75/25. Likewise, elongation to breaking increases from .7% to .9%.

IV.2 Metal Matrix Composites (MMC)

The concepts of geometric and material hybrids for MMC are demonstrated using a combination of SCS-6 SiC filaments in a 3-D braided Nicalon SiC reinforced Al 6061 composites.[26] The hybrid effect was studied by tensile, notched beam three-point bending and compact tension tests of the 3-D braided composites. The hybrid braided SiC/Al-6061 composites studied include a 0/100, 25/75, 50/50 and 75/25 combination of linear SCS with 3-D braided Nicalon. Addition of the strong and stiff SCS filaments to the Nicalon reinforced aluminum strengthens the composites and modifies the composite failure mode from matrix-dominated failure to linear yarns-controlled failure.

Fig. 9 shows the stress-strain curves of the MMC reflecting the nonlinear behavior of the composites. In comparison to the pure cast aluminum sample, the 100% braided composites show only a slight reinforcement effect whereas the hybrid composites show a remarkable improvement in strength (121 to 599 MPa) and toughness (.15% to .81% failure strain) as the percentage of longitudinal SCS-6 layin yarns increases from 0 to 50%.

The effect of geometric and material hybrids on the fracture behavior can be illustrated using the response of the MMC to 3-point bending tests. As shown in Fig. 10, the onset fracture load increases as the percentage of SCS-6 filaments increases. When the propagating crack reaches a bundle of SCS-6 filaments, the crack propagation rate is delayed and gradual failure occurs, as illustrated in the zig-zag pattern of the load-deflection curves.

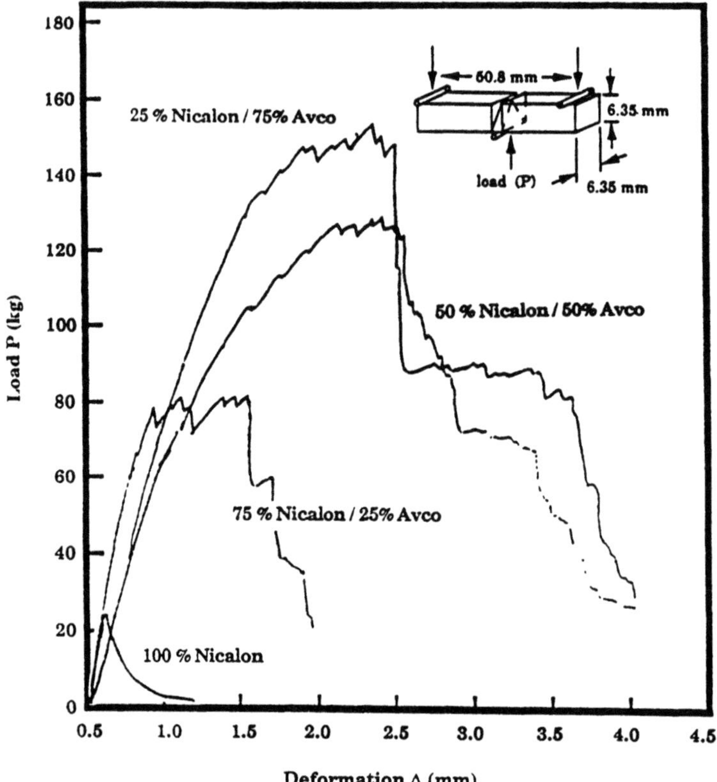

Figure 10. Fracture behavior of 3-D braided SiC/Al composites by 3-point bend test.

IV.3 Ceramic Matrix Composites(CMC)

Employing the same material-geometric hybrid system used in the MMC study, various proportions (ranging from 0/100, 25/75, 50/50, 75/25) of SCS-6/Nicalon 3-D braid were fabricated in a Lithium Aluminum Silicate (LAS III) matrix. Tensile, flexural and fracture tests were carried out on the 3-D braided CMC.[27] Beside enhancing strength, the thermally stable SCS filament played a significant role in preserving the structural integrity of the composite during processing and under high temperature oxidation end use environments. By placing the strong and stiff SCS filaments in the axial (0°) direction in the 3-D braided Nicalon fiber network, a significant improvement in tensile strength as well as the first cracking strength were achieved in the SiC/SiC/LAS III structure (Fig. 11). It is remarkable to observe that the elongation to break of the hybrid composites also increases with the increase of the proportion of SCS filaments resulting in a much strengthened and toughened CMC.

V. Modelling of Textile Structural Composites

Given a large family of fiber architectures which can be generated by an impressive array of textile preforming techniques, it is quite evident that one can tailor composite properties to meet various end use requirements. In order to facilitate the rational selection and stimulate the creative design and development of fiber architecture by

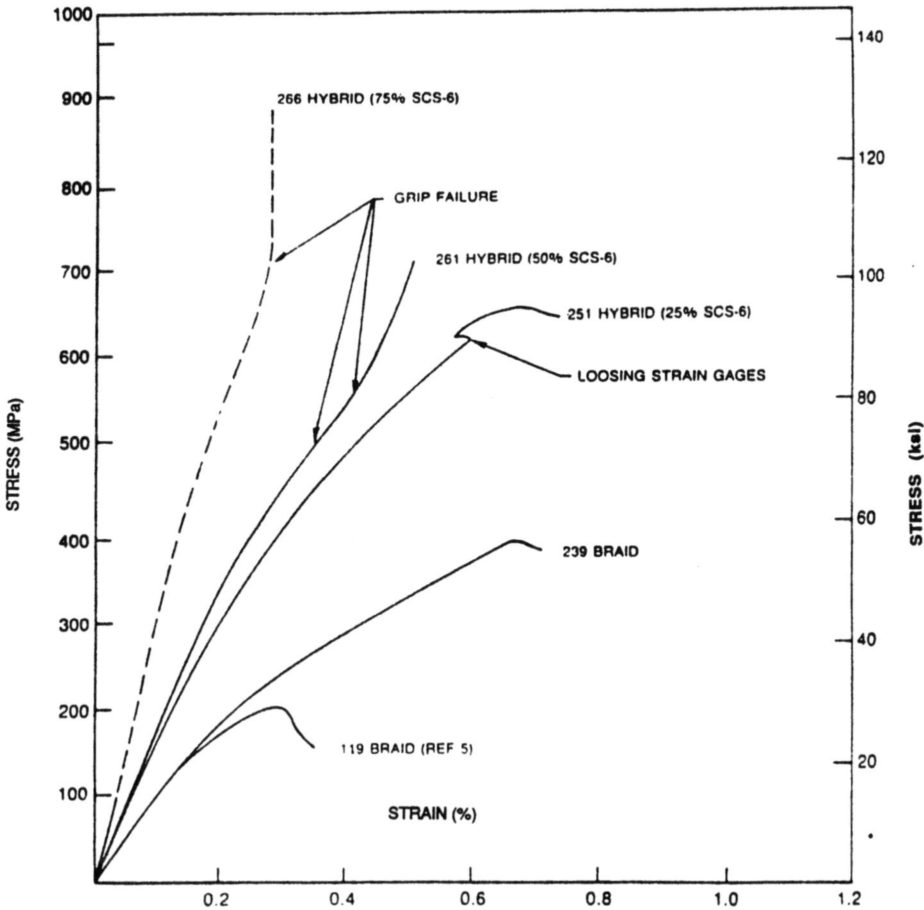

Figure 11. Tensile stress-strain behavior of 3-D braided SiC/LAS III composites.

textile preforming, a science based design framework must be established to bridge the communication gap between textile technologists, composite materials engineers and structural design engineers. This framework must be capable of relating preform manufacturing parameters to fiber architectural geometry as well as material properties. This design framework can be constructed through three levels of modelling including topological, geometrical and mechanical models. The topological model is a quantitative description of the preforming process. The geometrical model is the heart of the design framework which quantifies fiber orientation and fiber volume distribution in terms of fiber bundle (yarn) and fabric structural geometry created by the preforming process. The mechanical model provides the link between the mechanical properties of the material system and the fiber architecture. The product of the mechanical model is the stress-strain response of the textile reinforced composite for a given boundary condition as well as a stiffness matrix reflecting the material and fiber architecture contribution to the properties of the composite system. With the stiffness matrix, the structural designer can perform finite element analysis with meaningful information thus faciliting the integration of material design concepts, manufacturing processes and structural design to product engineering.

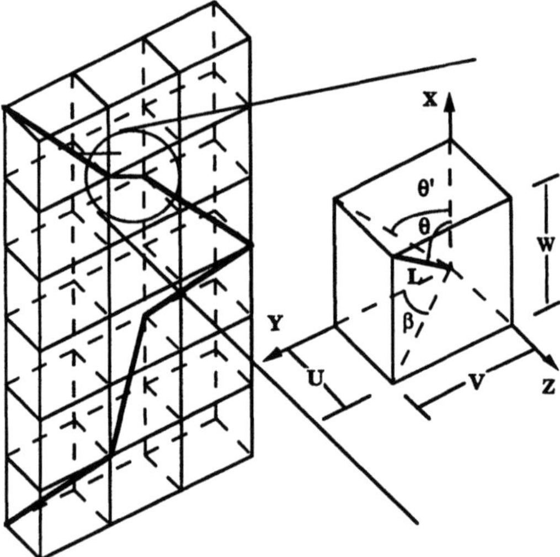

Figure 12. Unit cell geometry for the fabric geometry model.

The mechanical properties of textile reinforced composites can be predicted with
a knowledge of the fiber properties, matrix properties and textile preform fiber architecture through a modified laminate theory approach. Geometric unit cells defining
the fabric structure (or textile preform) can be identified and quantified to form a
basis for the analysis. For 2-D woven fabric reinforced composites, Dow[28] and Chou
et al.[29] have developed models for the thermomechanical properties of plain, twill and
satin reinforced composites. Examples of these include the mosaic, crimp, and bridging models developed by Chou. In the mosaic model, fiber continuity is ignored and
the composite is treated as an assembly of crossply elements. With the crimp model,
the nonlinear crimp geometry as well as yarn continuity is considered. Based on the
geometric repeating unit cell, each yarn segment is treated as a laminar. While the
crimp model was found to be most suitable for plane weave composites, the bridging
model was found to be best for satin weave composites, as it takes the relative stiffness
contribution of the linear and nonlinear yarn segments into consideration.

The modelling of 3-D fiber reinforced composites also begins with the establishment of geometric unit cells. From a "preform processing science" point of view, Ko
et al.[30] developed a "Fabric Geometry Model" (FGM) based on the unit cell geometry
shown in Fig. 12. The stiffness of a 3-D braided composite was considered to be the
sum of stiffness of all its laminae. The unique feature of the FGM is its ability to
handle 3-D braid as well as the other multi-directional reinforcements including 5-D,
6-D and 7-D fabrics with straight or curvilinear yarns. The unit cell for the 3-D braid
can be represented by several yarns running parallel to the body diagonal of the cell.
However, in some instances, yarns are placed in longitudinal (0°) and transverse (90°)
directions of the fabric and referred to as longitudinal and transverse reinforcements
(or lay-ins) respectively.

Using 3-D braided composites as an example, an integrated design and analysis
methodology is presented for 3-D composites. The preform processing parameters are
specifically related to corresponding unit cell geometries. The geometric descriptions
form the basis for a Fabric Geometry Model (FGM) which models a characteristic

volume. Accordingly, the generation of the stiffness matrix through the FGM provides a link between microstructural design and macrostructural analysis.

V.1 Unit Cell Modelling

In order to establish a geometric model and method for analyzing the properties of the 3-D braid, it is necessary to first identify the orientation of the yarns in the 3-D fiber network. Fig. 12 shows a typical 3-D braided structure with an enlarged view of the unit cell. Processing parameters U, V and W, representing the thickness, width and height of the unit cell, are related by the following equation:

$$W = \frac{\sqrt{U^2 + V^2}}{\tan \theta},$$
(1)

where θ, the interior angle, defines the orientation of the yarn with respect to the longitudinal axis of the panel. Given this relationship, it is possible to identify the angle θ associated with a particular fabric system, or to determine the value of W necessary to manufacture a fabric with fiber orientation θ. Once the interior angle θ is identified for a given system, the relation between the desirable fiber volume fraction and the total number of yarns needed in the composite can be established as follows:

$$V_f = \frac{N_y D_y}{C_d A_c \rho \cos \theta},$$
(2)

where

N_y = total number of yarns in the fabric,
$C_d = 9 \times 10^5$, is a constant,
A_c = cross-sectional area of finished composite (cm^2),
ρ = density of fiber (g/cc),
θ = interior angle, defined in Eq. (1),
D_y = linear density of fiber (denier),

and

V_f = fiber volume fraction.

V.2 The Fabric Geometry Model

Once the fabric geometry is quantified, the result can be used together with the fiber and matrix properties to predict the mechanical properties of the composite system through a modified lamination theory.[30] From the geometry of a unit cell associated with a particular fiber architecture, different systems of yarn can be identified whose fiber orientations are defined by their respective interior angle θ and azimuthal angle β, as previously shown in Fig. 12. Assuming each system of yarn can be represented by a comparable unidirectional lamina with an elastic stiffness matrix defined as follows:

$$[C] = \begin{pmatrix} c_{11} & c_{12} & c_{13} & 0 & 0 & 0 \\ & c_{22} & c_{23} & 0 & 0 & 0 \\ & & c_{33} & 0 & 0 & 0 \\ & & & c_{44} & 0 & 0 \\ & & & & c_{55} & 0 \\ & & & & & c_{66} \end{pmatrix},$$
(3)

where

$$c_{11} = (1 - n_{23}^2)E_{11}/K^*$$
$$c_{22} = c_{33} = (1 - n_{12}n_{21})E_{22}/K^*$$
$$c_{12} = c_{13} = (1 + n_{23})n_{21}E_{11}/K^*$$
$$c_{23} = (n_{23} + n_{12}n_{21})E_{22}/K^*$$
$$c_{44} = G_{23}$$
$$c_{55} = G_{13}$$
$$c_{66} = G_{12}$$

and

$$K^* = 1 - 2n_{12}n_{21}(1 + n_{23}) - n_{23}^2.$$

Then the elastic stiffness matrix $[C']$ of this yarn system in the longitudinal direction of the panel can be expressed as

$$[C'] = [T][C][T]^{-1}, \tag{4}$$

in which the transformation matrix[30] is given by

$$[T] = \begin{pmatrix} l_1^2 & m_1^2 & n_1^2 & 2m_1n_1 & 2l_1n_1 & 2l_1m_1 \\ l_2^2 & m_2^2 & n_2^2 & 2m_2n_2 & 2l_2n_2 & 2l_2m_2 \\ l_3^2 & m_3^2 & n_3^2 & 2m_3n_3 & 2l_3n_3 & 2l_3m_3 \\ l_2l_3 & m_2m_3 & n_2n_3 & m_2n_3 + m_3n_2 & l_2n_3 + l_3n_2 & l_2m_3 + l_3m_2 \\ l_1l_3 & m_1m_3 & n_1n_3 & m_1n_3 + m_3n_1 & l_1n_3 + l_3n_1 & l_1m_3 + l_3m_1 \\ l_2l_1 & m_2m_1 & n_2n_1 & m_1n_2 + m_2n_1 & l_1n_2 + l_2n_1 & l_1m_2 + l_2m_1 \end{pmatrix}, \tag{5}$$

where

$$l_1 = \cos\theta \qquad m_1 = 0 \qquad n_1 = -\sin\theta$$
$$l_2 = \sin\theta\cos\beta \qquad m_2 = \sin\beta \qquad n_2 = \cos\theta\cos\beta$$
$$l_3 = \sin\theta\sin\beta \qquad m_3 = -\cos\beta \qquad n_3 = \cos\theta\sin\beta$$

It is noted that the material properties of a unidirectional lamina, $E_{11}, E_{22}, G_{12},\ldots$, can be obtained easily using the well established micromechanical relationships.

V.3 Application of the FGM

In order to determine the stress-strain behavior of the fabric reinforced composites, it is necessary to utilize each of the yarn systems. A model for yarn system interaction has been chosen wherein the stiffness matrices for each system of yarns are superimposed proportionately according to contributing volume to determine the fabric reinforced composite system stiffness.

$$[C] = \sum_i k_i[C']_j, \tag{6}$$

where

$$[C] = \text{total stiffness matrix},$$

and

$$k_i = \text{fractional volume of the } i\text{-th system of yarns}.$$

In order to account for the potentially nonlinear behavior of the materials, the system stiffness matrix should be calculated anew at each strain level. Thus the stress-strain behavior of the composite can be expressed as

$$\Delta\sigma = [C]_s [\Delta\epsilon], \tag{7}$$

where

$\Delta\sigma$ = incremental stress vector (6×1)

and

$\Delta\epsilon$ = incremental strain vector (6×1).

From this, the stress vector can be determined as

$$\sigma = \sigma + \Delta\sigma, \tag{8}$$

where σ = stress vector (6×1).

A failure point for the composite is determined for each system of yarns by a maximum strain energy criterion. If the strain energy on the fiber exceeds the maximum allowable, that system of yarns has failed. Mathematically, if the following expression is true, the system has failed:

$$U_{c,i} \geq U_m, \tag{9}$$

where

$$U_m = \left| V_f(\sigma_{f,u}(\epsilon_{f,u})/2) + (1 - V_f) \int_0^{\epsilon_{f,u}} \sigma_\mu(\epsilon)\epsilon d\epsilon \right|,$$

$\sigma_{f,u}$ = fiber ultimate strength

$\epsilon_{f,u}$ = fiber ultimate strain

$U_{c,i} = \int_0^{\epsilon} \sigma_{c,i}(\epsilon_k)d\epsilon$ = strain energy of the ith yarn system with strain ϵ_k

and

$\sigma_{c,i}$ = stress on the ith composite system.

Using this maximum energy criteria, a failure point for each system of yarns can be found. When a system of yarns fails, its contribution to the total system stiffness is removed. When all systems have failed the composite is said to have failed. In this way, the entire stress-strain curve for the composite can be predicted up to the point of composite failure.

The FGM has been employed with satisfactory results to predict the stress-strain properties of polymer, metal and ceramic matrix components. Shown below in Figs. 13, 14 and 15 are the theoretical and experimental tensile stress strain relatively of Carbon/ Epoxy, SiC/Al and SiC/LAS III composites respectively.

V.4 Finite Element Implementation of FGM

By using the FGM analysis, one can obtain the elasticity matrix $[C]$ of a braided composite. Since in a displacement based finite element formulation the stiffness matrix $[K]$ can be obtained by the equation

$$K = \int B^T CBD_y, \tag{10}$$

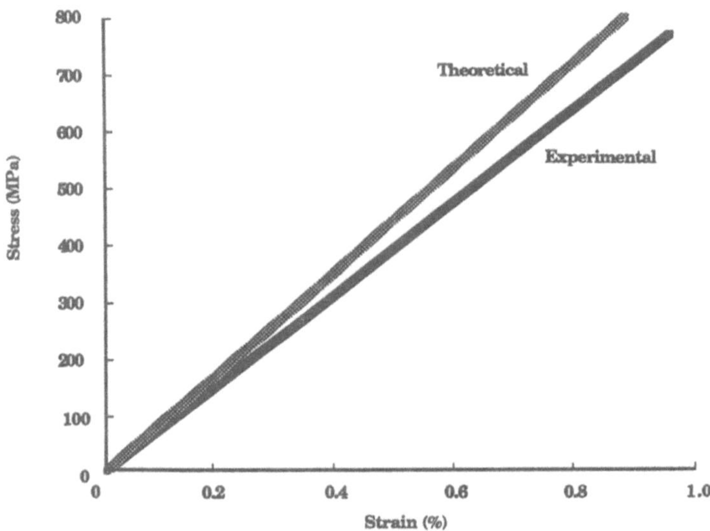

Figure 13. Theoretical and experimental compressive stress-strain relationship of 3-D braided carbon/epoxy composites.

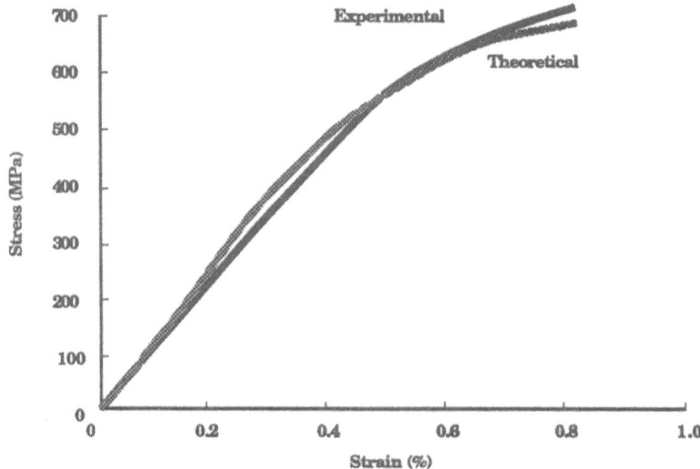

Figure 14. Theoretical and experimental tensile stress-strain relationship of 3-D braided SiC/Al composites.

where $[B]$ is the strain-displacement transformation matrix. Therefore, the FGM can be incorporated into general finite element analysis programs for solving braided composite structures with complex shapes. Materials input to this program includes fiber/matrix properties and braiding parameters, *i.e.*, surface angle, inclined angle, fiber volume fraction and braiding percentage. This methodology has been applied to the design and analysis of various complex shape engine components.[31]

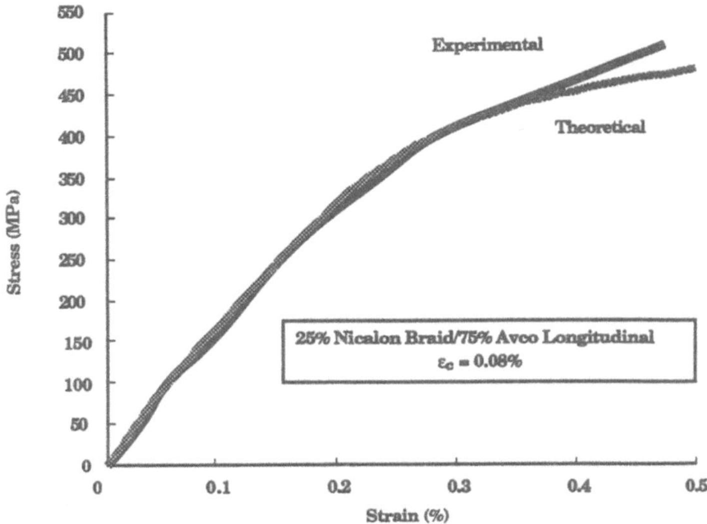

Figure 15. Theoretical and experimental tensile stress-strain relationship of 3-D braided SiC/LAS III composite.

VI. Conclusions

Textile preforms have much to offer in the toughening and manufacture of next generation high performance structural composites. With a large family of high performance fibers, linear fiber assemblies, and 2-D and 3-D textile preforms, a wide range of composite structural performances may be tailored to meet specific requirements.

An examination of the literature indicated that only a limited number of systematic studies have been carried out on fabric reinforced composites. A well established data base is needed in order to stimulate the usage of fabric reinforced composites for structural applications.

The trend towards the use of 3-D textile preforms for structural toughening of composites creates important technical challenges. The first is the question of conversion of brittle fibrous materials to textile structures. As a rule, the higher the temperature capability of the fiber, the stiffer and more brittle it is. This processing difficulty with brittle fibrous structures calls for an innovative combination of materials systems such as the concept of material and geometric hybridization.

The infiltration or placement of matrix material in a dense 3-D fiber network also creates new challenges. It demands an understanding of the dynamics of the process-structure interaction so that questions such as: What is the optimum pore geometry for matrix infiltration? Pore distribution? Bundle size? can be answered.

As the level of fiber integration increases, the opportunity of fiber to fiber contact will intensify at the crossover points. Guidance will be required to select the textile preform and matrix placement method best suited to reduce the incidence of localized fiber-rich areas.

In order to take advantage of the attractive features offered by textile structural composites, there is a need for the development of a sound data base and design methodologies. The fabric geometry models developed thus far establish a necessary but not entirely sufficient first step in the modelling of structural composites. Future work in the modelling of fabric reinforced composites require a better understanding of the dynamic interaction between fiber, matrix and textile preform.

Acknowledgements

Much of the work on textile preforms reported herein and specifically 2-D and 3-D fiber reinforced glass composites have been supported by the Office of Naval Research and the Army Research Office. The assistance provided by Mitchell Marmel of the Fibrous Materials Research Laboratory in the preparation of this manuscript is greatly appreciated.

References

1. F.K. Ko, *Ceramic Bulletin,* **68**, 1989.
2. Y.M. Tarnopol'skiy, I.G. Zhigun, and V.A. Polykov, *Soatiallv Reinforced Composite Materials,* (Moscow Mashinostroyeniye, 1987).
3. T.W. Chou and F.K. Ko, *Textile Structural Composites.,* (Elsevier, 1989).
4. F.K. Ko and C.M. Pastore, *Atkins and Pearce Industrial Handbook on Braiding,* (Atkins and Pearce, 1989).
5. F.L. Scardino, in: *Textile Structural Composites,* edited by T.W. Chou and F.K. Ko, (Elsevier, 1989).
6. K. Fukuta, private conversation, 1989.
7. P.J. Geoghegan, *3rd Textile Structural Composites Symposium,* (Philadelphia, PA., 1988).
8. R. Palmer, *4th Textile Structural Composites Symposium,* (Philadelphia, PA, 1989).
9. R.M. Dow, *Proceedings 21st International SAMPE Technical Conference,* 1989, p. 558.
10. E.R. Stover, W.C. Mark, I. Marfowitz, and W. Mueller, *AFML TR70283,* 1979.
11. J.W. Herrick, "Multidimensional Advanced Composites for Improved Impact Resistance", *10th National SAMPE Technical Conference,* 1977.
12. J. Pastenbaugh, *3rd Textile Structural Composites Symposium,* (Philadelphia, PA., 1988).
13. P.S. Bruno, D.O. Keith, and A.A. Vicario Jr., *SAMPE Quarterly* **17**, 10 (1986).
14. J. O'Shea, *3rd Textile Structural Composites Symposium,* (Philadelphia, PA. 1988).
15. K. Fukuta and E. Aoki, *15th Textile Research Symposium,* 1986.
16. K. Fukuta, E. Aoki, R. Onooka, and Y. Magatsuka, *Bulletin of Research Institute for Polymers and Textiles* **131**, 159 (1982).
17. D.J. Williams, *Advanced Composites Engineering,* 1978, p. 12.
18. G.T. Hickman and D.J. Williams, *Proceedings of the Fourth Annual Conference on Advanced Composites,* (ASM International 1988), p. 367.
19. F.K. Ko and J. Kutz, *Proceedines of the Fourth Annual Conference on Advanced Composites,* (ASM International, 1988), p. 377.
20. F.K. Ko, C.M. Pastore, J.M. Yang, and T.W. Chou, in: *Composites '86: Recent Advances in Japan and the United States,* edited by K. Kawata, S. Umekawa, and A. Kobayashi, (Proceedings, Japan-U.S. CCM-III, Tokyo, 1986), p. 21.

21. F.K. Ko, in: *Textile Structural Composites*, edited by T.W. Chou and F.K. Ko, (Elsevier, 1989).
22. R.T. Brown and C.H. Ashton, *Proceedings 4th Textile Structural Composites Symposium*, 1989.
23. P. Popper and R. McConnell, *32nd International SAMPE Symposium and Exhibition*, (1987), p. 92.
24. PEEK/Carbon, F.K. Ko, J.N. Chu, and C.T Hua, *Journal of Applied Polymer Science: Applied Polymer Symposium 47*, (John Wiley and Sons, 1991).
25. D. Liao, *Ph.D thesis*, (Drexel University, 1990).
26. S. Charles, C. Lei, and F.K. Ko, *ICCM-8*, 1991.
27. Ko, *et al.*, *ICCM-7*, 1989.
28. N.F. Dow and V. Ramnath, (NASA Contract Report 178275, 1987).
29. T.W. Chou and J.M. Yang, *Metallurgical Transaction A* **17A**, 1547 (1986).
30. F.K. Ko, C.M. Pastore, C. Lei, and D.W. Whyte, *Competitive Advancements in Metals/Metals Processing*, (SAMPE Meeting, Cherry, NJ, 1987).
31. T.M. Tan, C.M. Pastore, C. Huang, and F.K. Ko, *Winter Meeting of ASME, Advanced Composites Processing Technology*, Chicago, IL, 1988.

Cracking and Fatigue in Fiber-Reinforced Metal and Ceramic Matrix Composites

A.G. Evans and F.W. Zok

Materials Department
College of Engineering
University of California
Santa Barbara, California 93106-5050
U.S.A.

Abstract

The damage that occurs in unidirectional ceramic and metal matrix composites upon monotonic and cyclic loading involves coupled considerations of mechanics and stochastics. Some of the basic principles are described and models presented that both characterize damage evolution and govern mechanism changes. Comparisons are presented between predictions and experimental data for such phenomena as modulus degradation caused by matrix cracking, fatigue crack growth and tensile strength.

I. Introduction

The fracture and fatigue of fiber-reinforced metal, ceramic and intermetallic matrix composites involve coupled considerations of mechanics and weakest link statistics. Frank McClintock[1] was among the first to address problems that required linkage between these two disciplines. These initial concepts and linkages have grown and become the foundation for understanding some of the mechanical properties of composite systems at the micromechanics level. To provide focus on these issues, a broad view of composite properties is not attempted in this article. Instead, emphasis is given to unidirectional materials, but with the appreciation that the associated mechanical responses provide a fundamental framework for addressing the properties of multi-directional laminated and woven systems.

The three fundamental constituents of fracture and fatigue models for unidirectional composites are schematically represented on Fig. 1. i) Debonding occurs at

Advanced Topics in Materials Science and Engineering, Edited by
J.L. Morán-López and J.M. Sanchez, Plenum Press, New York, 1993

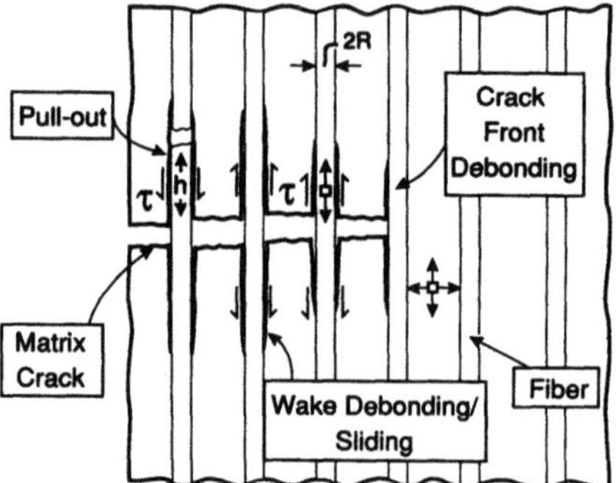

Figure 1. The various mechanisms that accompany mode I matrix crack propagation in unidirectional composites.

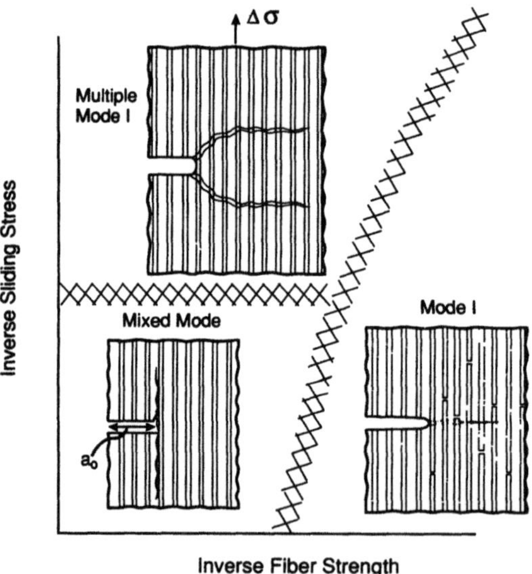

Figure 2. A mapping scheme that displays the various mechanisms of matrix cracking that can occur in notched composites subject to either monotonic or cyclic loading.

fiber/matrix interfaces, requiring an understanding of interface fracture mechanics in mixed-mode.[2,3] ii) Fibers exert tractions on the crack surfaces, requiring a mechanics of large-scale bridging.[3-7] iii) Fiber fracture may occur, usually at locations away from the matrix crack plane, as a result of weakest link statistics,[8-10] resulting in pull-out. The dominant dissipation mechanism that allows fibers to enhance the fracture and fatigue resistance is caused by frictional sliding along previously debonded interfaces.[11] Such dissipation occurs at both intact and failed fibers. However, the extent of the zone that provides dissipation is strongly influenced by the fiber failure site relative to the crack plane, which governs the pull-out length. This phenomenon arises from the stochastics of fiber failure. Large pull-out lengths relative to the crack opening also lead to large-scale bridging (LSB),[13] wherein the nominal crack growth

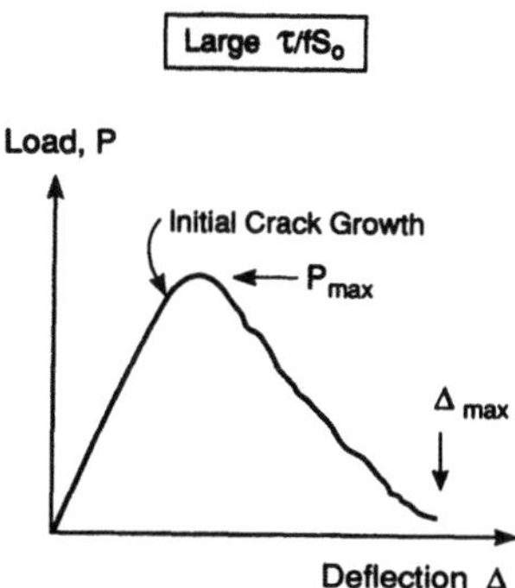

Figure 3. A typical load/deflection behavior for materials with large $\tau/fS_0 (\geq 0.5)$ that exhibit simultaneous matrix crack growth and fiber failure.

resistance depends on crack size and specimen geometry. Consequently, there is an "intimate connection" between the probabilistics of fiber failure and the mechanics of crack growth.

A comprehensive understanding of crack growth and fatigue also recognizes that several crack growth mechanisms are possible, dependent upon the properties of the matrix, fiber and interfaces[14] (Fig. 2). i) Splitting can occur by either mixed-mode or mode II crack growth within the matrix and along the interfaces. ii) A single mode I crack may form, usually accompanied by fiber failure. iii) Multiple mode I cracking may proceed with minimal fiber failure. The same range of behaviors occurs in the monotonic loading of brittle matrix composites and in the cyclic loading of metal matrix composites. This article deals with the two mode I mechanisms and the transition between them for both monotonic and cyclic loading. *The coupling of the stochastics of fiber failure with the mechanics of matrix crack growth is a central theme.* For discussion of basic aspects of debonding and mixed-mode cracking, the reader is referred to other reviews.[3,11,12,14,15]

The macroscopic load/deflection [P(Δ)] characteristics exhibited by unidirectional brittle composites connect with the above mechanisms of matrix and fiber cracking. There are two principal behaviors. i) When cracking occurs in mode I accompanied by fiber failure and pull-out, the loading is essentially linear up to the point of first crack growth. Then, after a small non-linear increase in load, the load diminishes as the crack extends (Fig. 3). The rate of change of load, $-dP/d\Delta$, is governed by the extent of frictional dissipation associated with the fiber pull-out process. ii) When a crack extends across the matrix in mode I *without failing a significant fraction of the fibers*, the crack propagates at constant load.[4,5] The material is also capable of sustaining larger loads and additional mode I matrix cracking occurs as the load increases.[16−18] The cracks reduce the elastic modulus and lead to a permanent strain (Fig. 4). The gradual failure of fibers occurs as the load increases. When a sufficient fraction of fibers has failed, the remaining fibers fracture abruptly, leading to a load drop. In both cases, the important parameters for design purposes are the peak load P_{max}, the failure displacement, Δ_{max}, and the change in compliance with load. The relationships between these composite properties and the *in situ* properties of the matrix, fibers and interfaces are addressed in this article.

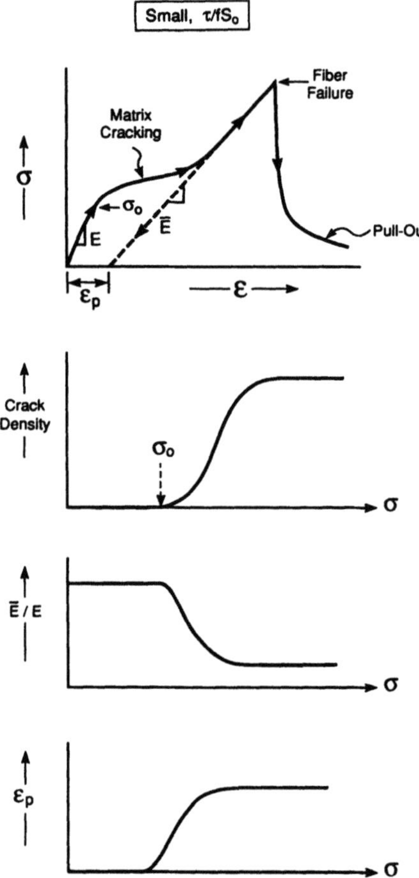

Figure 4. The stress/strain behavior exhibited by materials with small $\tau/fS_0(\leq 0.1)$ subject to multiple matrix cracking. Also shown are schematics of changes in crack density, modulus and permanent strain.

II. Some Basic Mechanics

The approach used to simulate mode I cracking under monotonic loading is to define tractions σ_b acting on the crack faces, induced by the fibers (Fig. 1), and to determine their effect on the crack tip by using the J-integral,[5,19]

$$\mathcal{G}_{tip} = \mathcal{G}_\infty - \int_0^u \sigma_b du, \tag{1}$$

where $\mathcal{G}$ is the energy release rate and u is the crack opening. Cracking is considered to proceed when $\mathcal{G}_{tip}$ attains the pertinent fracture energy. Since the fibers are not failing, the crack growth criterion involves matrix cracking only. A lower bound is given by,[19,20]

$$\mathcal{G}_{tip} = \Gamma_m(1 - f), \tag{2}$$

Table I

Non-Dimensional Parameters

Relative Stiffness	$\xi \longrightarrow fE_f/(1-f)E_m$
Relative Sliding	$T \longrightarrow \left[\frac{f\tau_0\xi}{\sigma}\right]^{1/2}$
Cyclic Sliding	$\Delta T \longrightarrow \left[\frac{f\tau_0\xi}{\Delta\sigma}\right]^{1/2}$
Applied Stress	$\Sigma \longrightarrow \left[\frac{2R\sigma}{f\xi a\tau_0}\right]$
Cyclic Stress	$\Delta\Sigma \longrightarrow \left[\frac{2R(\Delta\sigma)}{f\xi a\tau_0}\right]$
	$\Delta\Sigma_0 \longrightarrow \left[\frac{2R(\Delta\sigma)}{f\xi a_0\tau_0}\right]$
Bridging	$\Sigma_b \longrightarrow \left[\frac{2R\sigma_b}{f\xi a\tau_0}\right]$
Cyclic Bridging	$\Delta\Sigma_b \longrightarrow \left[\frac{2R(\Delta\sigma_b)}{f\xi a\tau_0}\right]$
Characteristic Lenght*	$\delta_c \longrightarrow \left[S_0RL_0^{1/m}/\tau_0\right]^{m/(m+1)}$
Characteristic Strenght*	$S_c \longrightarrow [S_0^m\tau_0L_0/R]^{1/(m+1)}$

*These parameters have dimensions but combine to form
non-dimensional parameters.

with f being the fiber volume fraction and Γ_m the matrix toughness. Upon crack extension, $\mathcal{G}_\infty$ becomes the crack growth resistance Γ_R, whereupon

$$\Gamma_R = \Gamma_m(1-f) + \int_0^u \sigma_b du. \tag{3}$$

A traction law $\sigma_b(u)$ is now needed to predict Γ_R. A law based on frictional sliding along debonded interfaces has been used most extensively and appears to provide a reasonable description of many of the observed mechanical responses. Coulomb friction appears to have a key role in governing the sliding stress, τ, through a relation of the form;[12,21]

$$\tau = \tau_0 - \mu p, \tag{4}$$

where μ is the friction coefficient, p is the residual stress *normal* to the interface and τ_0 represents the sliding resistance afforded by asperities (or roughness). A constant sliding stress assumption, $\tau = \tau_0$, will be used to illustrate the basic mechanics. However, general results based on Eq. (4) are available.[12] In the absence of fiber failure, the sliding distance ℓ is related to the crack surface traction, σ_b, by[4,5]

$$\ell = \sigma_b RE_m(1-f)/2\tau_0 E_f f. \tag{5}$$

Here, R is the fiber radius, E is Young's modulus, and the subscripts m and f refer to matrix and fibers, respectively. The slip length is, in turn, related to the crack opening displacement, such that the traction law becomes,[5,19]

$$\sigma_b = [2\xi\tau_0 E f u/R]^{1/2} - q(E/E_m),\tag{6}$$

where q is the residual axial stress in the matrix and ξ is defined in Table I. When fiber failure occurs, statistical considerations are needed to determine $\sigma_b(u)$, as elaborated below (Eq. (50)).

The matrix fracture behavior can also be described by using the stress intensity factor, K. This approach is more convenient than the J-integral in some cases: particularly for short cracks and for fatigue.[5,22] To apply this approach, it is first necessary to specify the contribution to the crack opening induced by the applied stress, u_∞, as well as that provided by the bridging fibers, u_b. For a plane strain crack of length $2a$ in an infinite plate, these are given by[50]

$$u_\infty = (4/E)\sigma_\infty\sqrt{a^2 - x^2},\tag{7}$$

and

$$u_b = -(4/E)\int_0^a \sigma_b(\hat{x})H(\hat{x}, x, a)d\hat{x},\tag{8}$$

with

$$H(\hat{x}, x, a) = \frac{1}{\pi}\log\left|\frac{\sqrt{a^2 - x^2} + \sqrt{a^2 - \hat{x}^2}}{\sqrt{a^2 - x^2} - \sqrt{a^2 - \hat{x}^2}}\right|,$$

and x being the distance from the crack center. The net crack opening is,

$$u = u_\infty + u_b.\tag{9}$$

The contribution to K from the bridging fibers is obtained using

$$K_b = -2\sqrt{\frac{a}{\pi}}\int_0^a \frac{\sigma_b(x)dx}{\sqrt{a^2 - x^2}},$$

which with σ_b given by Eq. (6) and $q = 0$ becomes

$$K_b = -\sqrt{\frac{a}{\pi}}\left[\frac{(1-\nu^2)fa\xi\tau_0}{R}\int_0^1 \frac{\Sigma_b d\tilde{x}}{\sqrt{1-\tilde{x}^2}}\right],\tag{10}$$

where Σ_b is defined in Table I. The shielding associated with K_b leads to a tip stress intensity factor,

$$K_{tip} = K_\infty + K_b,\tag{11}$$

where K_∞ depends on the loading and specimen geometry.

A *criterion for matrix crack extension* based on K_{tip} is needed. For this purpose, to be consistent with many other *brittle cracking* processes, it has been assumed that fracture is governed by the *stress* in the matrix *ahead of the crack*. The corresponding crack growth criterion for the composite is[5]

$$K_{tip} = (E/E_m)K_m,\tag{12}$$

where K_m is the critical stress intensity factor for the matrix, given by

$$K_m = \sqrt{E_m \Gamma_m},\tag{13}$$

whereupon the crack growth resistance is

$$K_R = (E/E_m)\sqrt{E_m \Gamma_m} - K_b.\tag{14}$$

This criterion would appear to differ from the energy criterion given in Eq. (2). Clearly Eq. (2) is a necessary condition for matrix crack extension, but it may not be sufficient. The criterion given by Eq. (13) is analogous to that used to successfully simulate brittle cracking in steels.[23,24] As demonstrated below, the two criteria lead to essentially the same steady-state matrix cracking stress.

III. Basic Weakest Link Statistics

For fibers failing in a composite, the incidence of failure can be determined using straightforward weakest link statistics, provided that the interface is sufficiently "weak" that fiber fractures do not transmit stress concentrations to neighboring fibers: the so-called "global load sharing" condition.[8-10] This assumption appears to have validity for an appreciable number of brittle matrix composites (ceramic, carbon, glass, intermetallic) that use fiber coatings to impart fiber toughening.[15] Conversely, this assumption is not normally valid for metal matrix composites with "strong" interfaces.[23] For these cases, a "local load sharing" criterion is needed to understand failure.

When fibers do not interact, analysis begins by considering a fiber of length $2L$ divided into $2N$ elements each of length δz. The probability that an element will fail when the stress is less than σ is essentially the area under the probability density curve,[25,26]

$$\delta\phi(\sigma) = \frac{\delta z}{L_0} \int_0^\sigma g(S)dS,\tag{15}$$

where $g(S)dS/L_0$ represents the number of flaws per unit length of fiber having a "strength" between S and $S + dS$. The local stress, σ, is a function of both the distance along the fiber, z, and the reference stress, σ_b. The survival probability P_s *for all elements* in the fiber of length $2L$ is the product of the survival probabilities of each element,

$$P_s(\sigma_b, L) = \prod_{n=-N}^{N} \left[1 - \delta\phi(\sigma_b, z)\right],\tag{16}$$

where $z = n\delta z$, and $L = N\delta z$. Furthermore, the probability Φ_s that the element at z will fail when the peak stress is between σ_b and $\sigma_b + \delta\sigma_b$, *but not when the stress is less than* σ_b, is the change in $\delta\phi$ when the stress is increased by $\delta\sigma_b$ divided by the survival probability up to σ_b, given by,[25-27]

$$\Phi(\sigma_b, z) = [1 - \delta\phi(\sigma_b, z)]^{-1} \left[\frac{\partial \delta\phi(\sigma_b, z)}{\partial \sigma_b}\right] d\sigma_b.\tag{17}$$

Denoting the probability density function of fiber failure by $\Phi(\sigma_b, z)$, the probability that fracture occurs at a location z, when the peak stress is σ_b, is governed by the probability that all elements *survive up to a peak stress* σ_b, but that failure occurs at z when the stress reaches σ_b.[8,27] It is given by the product of Eq. (16) with Eq. (17),

$$\Phi(\sigma_b, z)\delta\sigma_b\delta z = \frac{\prod_{n=-N}^{N}[1 - \delta\phi(\sigma_b, z)]}{[1 - \delta\phi(\sigma_b, z)]}\left[\frac{\partial\delta\phi(\sigma_b, z)}{\delta\sigma_b}\right]d\sigma_b. \tag{18}$$

While the above results are quite general, it is convenient to use a power law to represent $g(S)$,

$$\int_0^{\sigma} g(S)dS = (\sigma/S_0)^m, \tag{19}$$

where S_0 is the scale parameter and m is the shape parameter for the fiber strength distribution. Alternative representations of $g(S)$ are not warranted at the present level of development. Using this assumption, Eq. (18) becomes[8]

$$\Phi(\sigma_b, z) = exp\left\{-2\int_0^L \left[\frac{\sigma(\sigma_b, z)}{S_0}\right]^m \frac{dz}{L_0}\right\}\left(\frac{1}{L_0}\right)\frac{\partial}{\partial\sigma_b}\left[\frac{\sigma(\sigma_b, z)}{S_0}\right]^m. \tag{20}$$

The above results may be applied to various fiber failure problems in composites. Two parameters repeatedly arise in the results (Table I): a characteristic length[9]

$$\delta_c = \left[S_0 R L_0^{1/m}/\tau_0\right]^{m/(m+1)}, \tag{21}$$

and a characteristic strength[9]

$$S_c = [S_0^m \tau_0 L_0/R]^{1/(m+1)}, \tag{22}$$

related by

$$S_c = \tau_0\delta_c/R. \tag{23}$$

A number of important results for fiber bundle failure in brittle matrix composites have been derived using these *statistics*. All of the formulae have been derived by assuming that the interfaces have sufficiently low debond energy and sliding stress that the failed fibers do not concentrate stress.[8-10] In such a case, the matrix influences composite fracture only by transferring load from a failed fiber to *all* remaining intact fibers equally (so-called global load sharing) and by allowing load recovery from a fiber failure site, *along* a fiber, through the sliding stress. Since the load recovery length is related to τ, composite parameters such as the pull-out length and the *in situ* fiber strength have a direct connection with τ.

The nominal *in situ* fiber strength distribution, $G(S)$, may be obtained from measurements of fracture mirrors on fibers after composite tensile testing. The fiber strength S is given by the relationship,[28,29]

$$S \approx 3.5\sqrt{E_f\Gamma_f/a_m}, \tag{24}$$

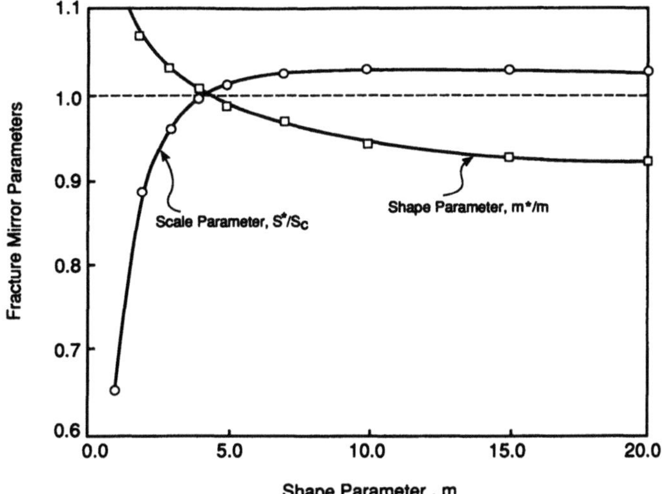

Figure 5. Relationships between fracture mirror parameters and the *in situ* fiber strength parameters.[9]

where Γ_f is the fracture energy of the fiber and a_m is the fracture mirror radius. It is found that $G(S)$ can be represented empirically by,[29]

$$G(S) = 1 - exp\left[-(S/S_\star)\right]^m , \tag{25}$$

where the strengths, S, are subject to an implicit gauge length effect related to τ, through the load transfer length.[9] The details of the preceding statistics depend on the fracture mechanism (multiple cracking or single cracks), as elaborated below.

III.1 Multiple Cracking

Analysis of fiber failure, based on Eqs. (20) and (25), has identified the relationship between the shape parameters $m_\star$ and m. It has also established that the scale parameter $S_\star$ is related to the characteristic *in situ* fiber strength, S_c[9] (Fig. 5). Furthermore, S_c is related to the fiber scale parameters S_0 and L_0 by[9] Eq. (22).

In order to relate S_c determined from fracture mirrors to the strength S_b of the fibers measured in a bundle (without matrix), the difference in gauge lengths must be taken into account. In general, the *in situ* fiber bundle strength at gauge length L is given by,[30,31]

$$S_b \equiv S_0(L_0/Lme)^{1/m}, \tag{26}$$

hence,

$$S_b = S_c^{(m+1)/m}(R/\tau_0 Lme)^{1/m}, \tag{27}$$

where e is the base of natural logarithms. With S_c and m ascertained from fracture mirror data, S_b can be determined from Eq. (27), provided that τ_0 is known.

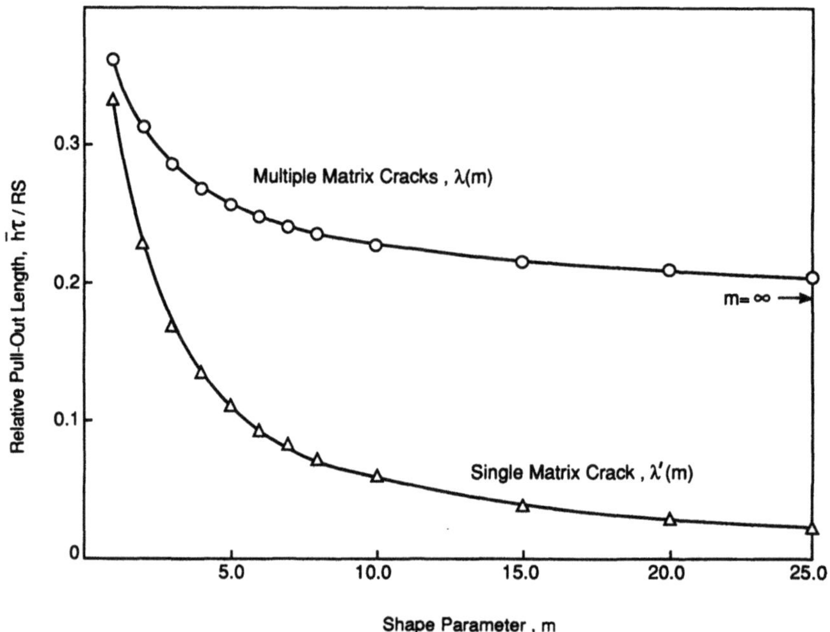

Figure 6. Pull-out parameters for single and multiple matrix cracking.[9]

The sliding stress τ_0 can also be related to the mean pull-out length $\bar{h}$. To determine this relationship, the pull-out lengths are found by placing a randomly-located plane (the fracture plane) over the multiply fractured fibers and finding the fiber failure location relative to that plane by using Eq. (20). The results indicate that $\bar{h}$ and δ_c are related by[9]

$$\bar{h} = \lambda(m)\delta_c/4, \tag{28}$$

or

$$\bar{h}\tau_0/RS_c = \lambda(m)/4, \tag{29}$$

with $\lambda(m)$ plotted on Fig. 6. Consequently, either $\bar{h}$ can be predicted from independent measurement of τ_0 and S_c or $\bar{h}$ can be used to determine δ_c. Another result of importance concerns the ultimate strength of the composite, σ_u, which can be derived from Eq. (20) as,[9]

$$\sigma_u = fS_c \left[\frac{2}{(m+2)}\right]^{1/(m+1)} \left[\frac{m+1}{m+2}\right]. \tag{30}$$

It is of significance to appreciate that σ_u expressed in terms of S_c, through Eq. (30), has an implicit dependence on τ_0, as indicated by Eq. (22), but is independent of the composite gauge length.

The above formulae have been used as follows:[18,32] i) Determine S_* and m_* based on fracture mirror measurements, and then obtain S_c and m from Fig. 5. ii) Evaluate δ_c from $\bar{h}$ using Eq. (28), with m now known. iii) Given S_c and δ_c, determine τ_0 from Eq. (22). iv) With S_c, τ_0 and m known, the fiber bundle strength can be obtained from Eq. (26). v) Calculate the ultimate strength σ_u from Eq. (30).

III.2 Single Cracks

When the fracture mechanism changes to that wherein a single mode I crack extends, accompanied by fiber failure, the fiber is subject to *its maximum stress* between the crack faces and the stress σ_z diminishes as[8,10]

$$\sigma_z = \sigma_b(1 - z/\ell)/f. \tag{31}$$

Inserting this stress into Eq. (20) gives the probability density function,[8,10] $\Phi(\sigma_b, z)$: the basic formula that governs several statistical parameters. Two results are of key interest. i) The effect of z on $\Phi(\sigma_b, z)$ dictates that, on average, the fibers *do not fail* at the location of highest stress on the fiber ($z = 0$, Eq. (31)). Instead, fibers fail at an average distance, $z = \bar{h}$, leading to fiber pull-out and a related frictional contribution to crack growth. The average failure position for all fibers is given by an expression similar to Eq. (28),[8–10]

$$\bar{h} = \delta_c \lambda'(m)/4, \tag{32}$$

where $\lambda'(m)$ is plotted on Fig. 6. ii) The peak stress on the composite when the fibers fail, σ_{max},* is given by,[8,9,33]

$$\begin{aligned}
\sigma_{max} &= fS_c\big[(5m + 1)/5m\big]exp[-1/(m + 1)] \\
&\equiv fS_cG(m).
\end{aligned} \tag{33}$$

IV. Matrix Cracking

The preceding mechanics and stoichastics may be combined to predict matrix cracking subject to either monotonic or cyclic loading, as elaborated in this section.

IV.1 Monotonic Loading

Matrix crack evolution in brittle matrix composites subject to monotonic loading is connected to the incidence of accompanying fiber fracture (Fig. 2). Steady-state matrix cracking with intact fibers results in a reduced tensile modulus and permanent strain[5,18,34] (Fig. 4). This mode of cracking arises when the ratio τ/fS_0 is small,[14] as elaborated in Section IV.1.3. At larger τ/fS_0, fiber fracture may occur simultaneously with matrix crack growth. Then, a single crack occurs and extends subject to diminishing load[13] (Fig. 3). Further analysis of these mode I cracking phenomena is presented below, as well as considerations relevant to the transition criterion.

---

* This stress governs P_{max} in Fig. 3

IV.1.1 Multiple Cracking

The problem of matrix cracking with intact fibers has been comprehensively addressed, commencing with Aveston, Cooper and Kelly[4] (ACK). More recent studies have both verified the ACK solution and introduced considerations additional to those envisaged by ACK.[5,19,20,35] The present understanding involves the following factors. Because the fibers are intact, a steady-state condition exists wherein the tractions on the fibers in the crack wake balance the applied stress. This special case may be addressed integrating Eq. (1) up to a limit $u = u_0$ (obtained from Eq. (6) by equating σ_b to σ_∞), giving[19]

$$\mathcal{G}^0_{tip} = \frac{(\sigma_\infty + qE/E_m)^3 E_m^2 (1-f)^2 R}{6\tau_0 f^2 E_f E^2} .\tag{34}$$

A *lower bound to the matrix cracking stress*, σ_0, is then obtained by invoking Eq. (2) such that,[19]‡

$$\sigma_0 = \sigma_\star - qE/E_m,\tag{35}$$

where

$$\sigma_\star = \left[\frac{6\tau_0 \Gamma_m f^2 E_f E^2}{(1-f)E_m^2 R} \right]^{1/3} .$$

Analogous results can be obtained using stress intensity factors[5,22] (Eqs. (9) to (12)). For the example of a center crack in a tensile specimen ($K_\infty = \sigma_\infty \sqrt{\pi a}$), Eqs. (10) and (11) give the steady-state result at large crack lengths,[22]

$$K^0_{tip} = \sqrt{R}(\sqrt{6}T)^{-1}\sigma\tag{36}$$

where T is defined in Table I and $\sigma \equiv \sigma_\infty - qE/E_m$. When combined with the fracture criterion (Eq. (12)), the matrix cracking stress is predicted to be the same as that given by Eq. (35),[5,22] within 10% in the numerical coefficient.

In addition, this approach may be used to define a critical crack length a_c above which steady-state applies. This critical length occurs at $\Sigma \approx 4$, (Table I) and is given by,

$$a_c/R \approx E_m \left[\Gamma_m(1+\xi)^2(1-f)^4/\tau_0^2 f^4 E_f^2 R \right]^{1/3} .\tag{37}$$

Namely, when the initial flaw size $a_0 > a_c$, cracking occurs at $\sigma = \sigma_0$. Conversely, when the initial flaws are small, $a_0 < a_c$,Eqs. (9) and (10) give,[22]

$$K_{tip} = \sigma\sqrt{\pi a} \left[1 - \frac{3.05}{\Sigma}\sqrt{\Sigma + 3.3} + \frac{5.5}{\Sigma} \right] .\tag{38}$$

---

‡ there is an interplay between the roles of τ and q in matrix cracking. For the case $\tau_0 = 0$, it has been shown that there is an *optimum* thermal expansion misfit at which σ_0 has the maximum possible value.[19]

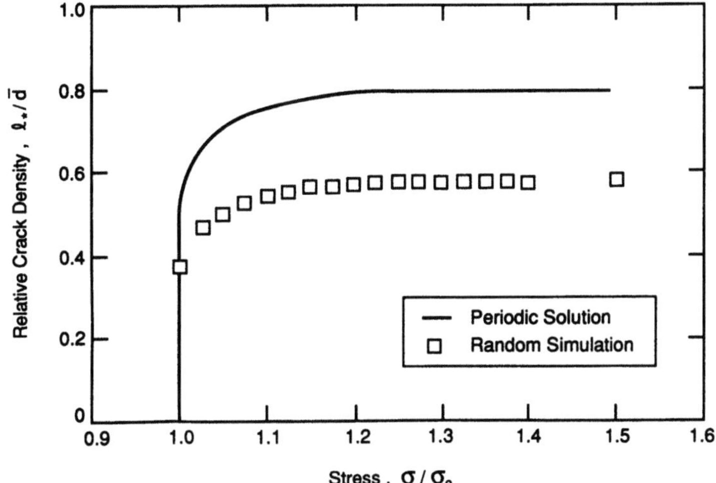

Figure 7. Change in steady-state crack density with applied stress predicted using the mechanics of matrix cracking.[35]

This result, when combined with Eq. (12), gives the matrix cracking stress σ_c for $a_0 < a_c$. In this range, σ_c *exceeds* σ_0.

The evolution of additional cracks at stresses above σ_0 is less well understood, because two factors are involved: screening and statistics.[18,35] When the slip zones between neighboring cracks overlap, screening occurs and $\mathcal{G}_{tip}$ differs from $\mathcal{G}_{tip}^0$.[35] The relationship is dictated by the location of the neighboring cracks. When a crack forms midway between two existing cracks with a separation $2d$, $\mathcal{G}_{tip}$ is related to $\mathcal{G}_{tip}^0$ by[35]

$$\mathcal{G}_{tip}/\mathcal{G}_{tip}^0 = 4\,(d/2\ell)^3, \qquad \text{(for } 0 \le d/\ell \le 1), \qquad (39)$$

and

$$\mathcal{G}_{tip}/\mathcal{G}_{tip}^0 = 1 - 4(1 - d/2\ell)^3, \qquad \text{(for } 1 \le d/\ell \le 2). \qquad (40)$$

Consequently, if matrix cracks develop in a strictly periodic manner, the evolution of the crack density with stress can be predicted by combining Eqs. (12), (34) and (40), leading to the result plotted on Fig. 7. In general, however, non-periodic crack locations exist, resulting in a different distribution of crack spacing. Computer simulations of *random* matrix cracking reveal similar trends[35] (also shown in Fig. 7) and indicate that the average crack density reaches a *saturation* value, $\ell_*/\bar{d}_s = 0.575\ell_*$, when $\sigma/\sigma_0 \ge 1.3$.‡ An *upper bound* to the sliding stress can be inferred from $\bar{d}_s$, given by[35]

$$\tau_0 = 2R(1 - f)(\Gamma_m E_f E_m / f E \bar{d}_s^3)^{1/2}. \qquad (41)$$

---

‡ ℓ_* is the slip length at the onset of matrix cracking. It is given by Eq. (5) with $\sigma_b = \sigma_0$.

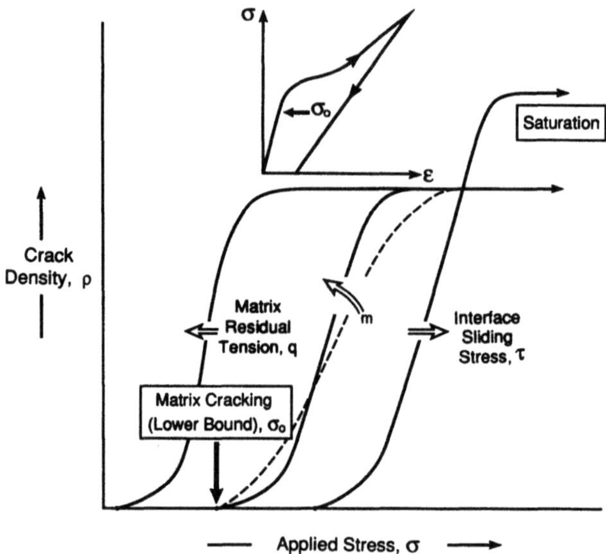

Figure 8. A schematic of general trends in crack density with stress.

The actual evolution of matrix cracks may deviate from the predictions, depending on the spatial and size distribution of matrix flaws. One statistical result of interest is derived by assuming that all matrix cracks initiate from flaws with $a_0 \leq a_c$. The principal result is that significant matrix cracking is not predicted until stresses are appreciably above σ_0, depending on the magnitude of both the shape and scale parameters for matrix flaws.[18,33] A schematic that combines the principal aspects of the mechanics and stochastics is presented in Fig. 8.

The crack spacing relates to the change in Young's modulus and the permanent strain. Analysis of the sliding that occurs in the region between matrix cracks provides explicit predictions:[17] the Young's modulus $\bar{E}$ is given by,

$$E/\bar{E} = E_f R \bar{\sigma} / 2 \bar{d} \xi^2 \tau E, \tag{42}$$

and the permanent strain, ϵ_p, is,

$$\epsilon_p = \frac{R(1-f)^2 q^2}{4 \bar{d} f^2 \tau_0 E_f} \left[\left(\frac{\bar{\sigma} E_m}{qE} \right)^2 - 4 \left(\frac{\bar{\sigma} E_m}{qE} \right) + 2 \right], \tag{43}$$

where $\bar{\sigma}$ is the stress reached upon loading the material.

IV.1.2 Single Cracks

When fracture occurs by a single crack extending with failing fibers, the "weaker" fibers fail in the immediate crack wake, near the crack plane. Upon additional crack extension, more fibers fail in the crack wake, at locations further from the crack plane. The statistics of the process that governs the associated traction laws are as follows.[8-10] For intact fibers, the bridging stress σ_b is given by Eq. (6), while for failed fibers,

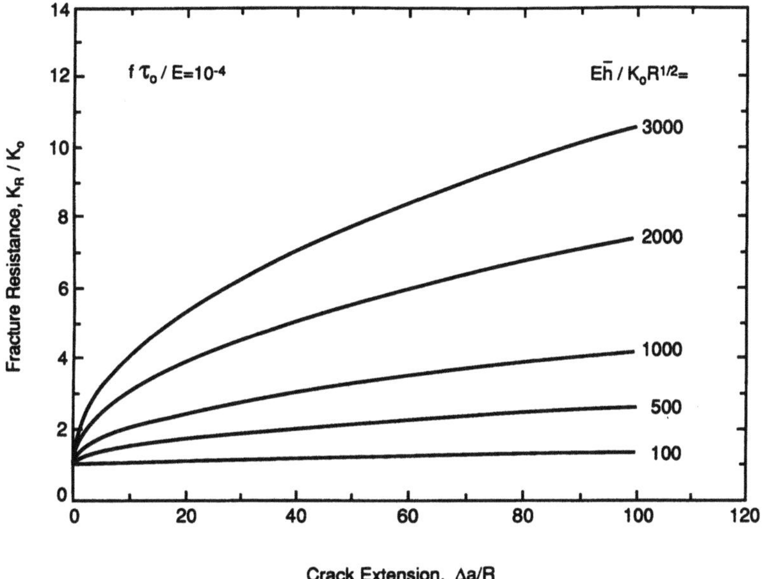

Figure 9. A predicted resistance curve for small-scale bridging.[13]

$$\sigma_b = f(2\tau_0/R)(h - u). \tag{44}$$

The average stress on the fibers thus depends on the fraction of failed fibers f_x which is just the cumulative fiber failure probability at $\sigma_b = S$,

$$f_x = 1 - exp\left[-(S/S_c)^{m+1}/(m + 1)\right]. \tag{45}$$

These formulae may be combined to give the tractions. The result is unwieldy, but the dominant term is[8-10]

$$\sigma_b/fS_c = (u/v)^{1/2}exp\left[-(u/v)^{(m+1)/2}\right](m + 1)^{1/(m+1)}, \tag{46}$$

where

$$v = S_c^2 R/4E_f\tau_0.$$

This fundamental traction law may be used to compute the important fracture properties of the material.

A particularly useful simplification recognizes that the mean pull-out length $\bar{h}$ is typically much larger than the crack opening displacement.[6] Consequently, the traction exerted by the failed fibers is[13]

$$\sigma_b \approx 2\tau_0 f_x \bar{h}/R, \tag{47}$$

with $\bar{h}$ given by Eq. (28) and $f_x = f$, the traction becomes

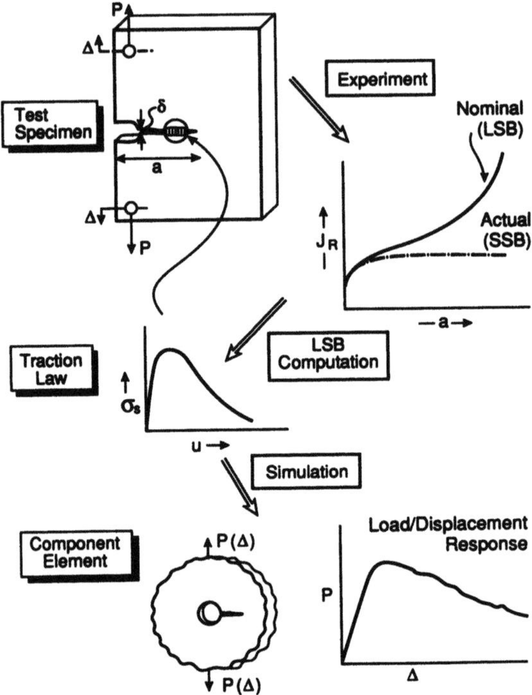

Figure 10. A scheme for calculating the load/deflection response in large-scale bridging.

$$\sigma_b \approx 2fS_c \approx 2f\left[\frac{\tau_0 L_0}{R}S_0^m\right]^{1/(m+1)}. \tag{48}$$

The traction is thus dominated by the *fiber strength*, S_c, and is insensitive to the sliding stress τ_0 and fiber radius, R. The basic traction law (Eq. (48)) can be used with Eqs. (8)–(13) to calculate a resistance K_R given by,[13]

$$\frac{K_R}{K_0} = 1 + 4\sqrt{2/\pi}\left(\frac{f\tau_0}{E}\right)\left(\frac{E\bar{h}}{K_0\sqrt{R}}\right)\left[\sqrt{\frac{\Delta a}{R}} - \sqrt{\frac{2}{\pi}}\left(\frac{K_0\sqrt{R}}{E\bar{h}}\right)\left(\frac{\Delta a}{R}\right)\right], \tag{49}$$

where Δa is the crack extension. A typical result is plotted on Fig. 9. The above formulation applies subject to small-scale bridging (SSB); wherein the bridging zone length is small compared with crack length and specimen dimensions.

Small-scale bridging is often violated and the crack growth resistance can only be deduced using a full, large-scale bridging (LSB) computation. Adequate prediction of the load/deflection characteristic also requires a mechanics of large-scale bridging. The basic approach, schematically illustrated in Fig. 10,[6,36−38] recognizes that the *fundamental property of the composite is the traction law*, $\sigma_b(u)$. ‡ Once this law has been explicitly established, it may be used to predict either $P(\Delta)$ or $K_R(\Delta a)$. The

---

‡ The law appropiate to a particular composite can either be obtained from basic principles of fiber sliding and failure (Eqs. (6) and (46)) or by direct measurement.

procedure is essentially the same as that described in Eqs. (7) to (13), except that the loading function becomes explicit to the geometry of interest. One approach for facilitating this analysis is the use of canonical functions.[37] The choice of function is governed by the characteristic shape of curves of K_{tip}/K_∞ vs. bridging stress. When Eq. (6) applies, the preferred function is found to be

$$K_{tip}/K_\infty = 1/2\{1+$$

$$\tanh\left[\left(1/4 + \lambda_3 exp\left[\lambda_1(\omega - \omega_0) + \lambda_2(\omega - \omega_0)^2\right]\right)(\omega - \omega_0)\right]\}, \qquad (50a)$$

where λ_i are fitting constants for each geometry (tabulated by Cox and Lo),[37] and

$$\omega = \log(\pi\sigma_\infty E_f R/16\xi E\tau w) \qquad (50b)$$

with the parameter ω_0 corresponding to the value of ω when $K_{tip}/K_\infty = 0.5$ and w the width of the component. By using K_{tip} from Eqs. (12) and (13) and by equating K_∞ to the resistance K_R, the nominal composite resistance can be readily derived from Eq. (50). More general numerical procedures have also been devised.[36,39,40]

IV.1.3 The Mechanism Transition

A first order criterion for the mechanism transition from multiple to single matrix cracking is obtained by equating σ_0 (Eq. 35) and σ_{max} (Eq. 33), such that multiple cracking occurs when $\sigma_0 < \sigma_{max}$ and vice versa. This result is unwieldy, but can be presented in an insightful form by neglecting elastic mismatch, leading to:

$$H(f)\left[\frac{E\Gamma_m\tau_c}{RS_0^3}\right]^{1/3} - \left(\frac{\tau_c\ell_0}{RS_0}\right)^{1/m}G(m) = \frac{q}{fS_0}, \qquad (51)$$

where $H(f) = [6/f(1 - f)]^{1/3}$ and τ_c is the critical sliding stress, at the transition. For the case $q = 0$, the critical stress becomes

$$\frac{\tau_c}{fS_0} \approx \left[\left(\frac{S_0^2 R}{E\Gamma_m}\right)\left(\frac{\ell_0}{R}\right)^{3/m}\right]^{1-3/m}\frac{(1-f)}{6}f^{(m-3)/3}G^3(m). \qquad (52)$$

This result has the merit that it identifies τ_c/fS_0 as a key non-dimensional parameter in the transition and establishes the importance of a second non-dimensional parameter, $S_0^2 R/E\Gamma_m$, in the sense that the larger this parameter the greater the critical sliding stress. It is also evident from Eq. (51) that residual tensile stress in the matrix encourages the multiple cracking mode of behavior.

IV.2 Cyclic Loading

Fatigue crack growth in either metal, intermetallic on ceramic matrix composites may be addressed with the same basic formulism used to describe monotonic crack growth.[22,40] Once again, a key factor is the nature of the interface. When the interface is strong, there can be no contribution to the fatigue resistance from frictional dissipation and the composite behaves in approximate accordance with the rule-of-mixtures expected from the matrix and reinforcement fatigue properties.[41] Conversely, when the interfaces are "weak", fibers can remain intact in the crack wake and cyclic frictional dissipation can resist fatigue crack growth.[22] The latter has been most extensively demonstrated on Ti matrix composites reinforced with SiC fibers.[42,43] The essential features of the "weak" interface behavior are as follows: i) Intact, sliding fibers acting in the crack wake shield the crack tip, such that the stress intensity range at the crack tip, ΔK_{tip}, is less than that expected for the applied loads, ΔK. ii) Subject to the reduced ΔK_{tip}, the fatigue crack grows in the matrix, in accordance with the Paris law,*

$$da/dN = \beta(\Delta K_{tip}/E)^n. \tag{53}$$

Using this approach, it has been found that a simple transformation converts the monotonic crack growth parameters into cyclic parameters that can be used to interpret and simulate fatigue crack growth. The key transformation is based on the relationship between interface sliding during loading and unloading, which relates the monotonic result to the cyclic equivalent through[22]

$$\left(\frac{1}{2}\right)\Delta\sigma_b(x/a, \Delta\sigma) = \sigma_b(x/a, \Delta\sigma/2), \tag{54}$$

where $\Delta\sigma$ is the range in the applied stress. Notably, the amplitude of the *change* in fiber traction $\Delta\sigma_b$ caused by a change in applied stress, $\Delta\sigma$, is twice the fiber traction σ_b which would arise in the monotonic loading of a previously unopened crack, caused by an applied stress equal to half the stress change $\Delta\sigma$. *This result is fundamental to all subsequent developments.*[22]

The stress intensity factor for bridging fibers subject to cyclic conditions is

$$\Delta K_b(\Delta\sigma) = -2\sqrt{\frac{a}{\pi}} \int_0^a \frac{\Delta\sigma_b(x, \Delta\sigma)}{\sqrt{a^2 - x^2}} dx, \tag{55}$$

which, with the use of Eq. (54), becomes

$$\Delta K_b(\Delta\sigma) = 2K_b^{max}(\Delta\sigma/2), \tag{56}$$

where the superscript max refers to the maximum values of the parameters achieved in the loading cycle and thus, K_b^{max} is the bridging contribution that would arise when the crack is loaded by an applied stress equal to $\Delta\sigma/2$. Furthermore, since ΔK is linear in $\Delta\sigma$, Eq. (56) is also valid for the tip stress intensity factor:

$$\Delta K_{tip} = 2K_{tip}(\Delta\sigma/2). \tag{57}$$

When the fibers remain intact, a cyclic *steady-state* (ΔK independent of crack length) is obtained when the cracks are long, given by the condition $\Delta\Sigma \leq 4$(Table I)[22] as,

$$\Delta K_{tip} = \Delta\sigma\sqrt{R}(\sqrt{12}\Delta T)^{-1}. \tag{58}$$

---

* The phenomenon has been addressed in terms of stress intensity factors (rather than energy release rates) because of the choise of a Paris law crack growth criterion.

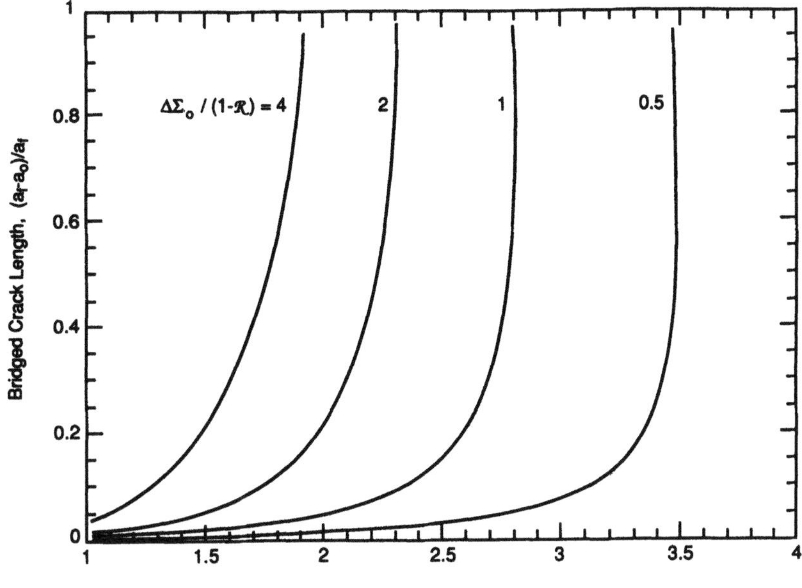

Figure 11. The length of the crack a_f at first fiber failure as a function of fiber strength for a range of stress amplitude, $\Delta\Sigma_0$.

For cyclic loading, the residual stress q does not affect ΔK_{tip}. The corresponding crack growth rate is determined from Eqs. (53) and (58) as[22]

$$\frac{da}{dN} = \beta\left[\frac{\Delta\sigma\sqrt{R}}{\sqrt{6}\Delta T E_m}\right]^n.\tag{59}$$

When short cracks are of relevance ($\Delta\Sigma > 4$),

$$\Delta K_{tip} = \Delta\sigma\sqrt{\pi a}\left[1 - \frac{4.31}{\Delta\Sigma}\sqrt{\Delta\Sigma + 6.6} + \frac{11}{\Delta\Sigma}\right].\tag{60}$$

Consequently, at fixed $\Delta\sigma$, ΔK_{tip} increases as the crack extends, and the crack growth accelerates. However, the bridged matrix fatigue crack always grows at a *slower rate* than an unbridged crack of the same length.

To incorporate the effects of fiber breaking into the fatigue crack growth model, a deterministic criterion has been used:[40] the statistical characteristics of fiber failure have yet to be incorporated. To conduct the calculation, once the fibers begin to fail, the unbridged crack length is continuously adjusted to maintain a stress at the unbridged crack tip equal to the fiber strength, fS. These conditions lead to the determination of the crack length, a_f, when the first fibers fail, as a function of the fiber strength and the maximum applied load‡ (Fig. 11). As an illustration, for an initial condition characterized by $\Delta\Sigma_0/(1 - \mathcal{R}) = 1$ and a strength parameter

---

‡ $\sigma_{max} = \Delta\sigma/(1 - \mathcal{R})$, which defines $\mathcal{R}$ (the "R-ratio" in fatigue parlance).

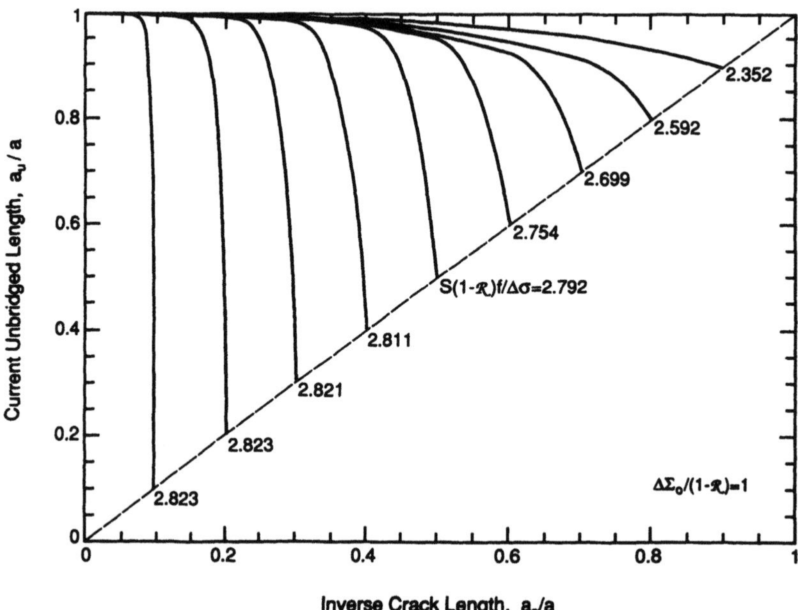

Figure 12. Fiber breaking rate, as manifest in the current unbridged length $2a_u$ as a function of total crack length, $2a$ ($2a_0$ is the initial notch length).

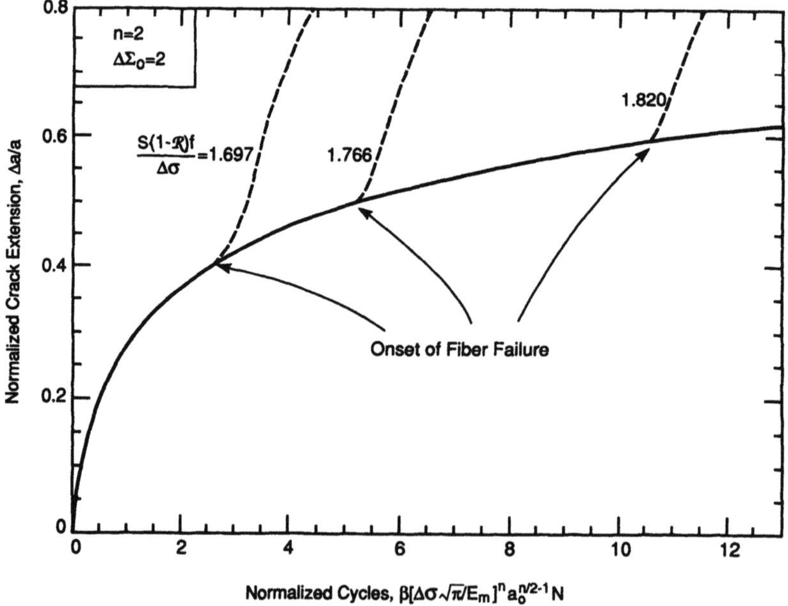

Figure 13. Predicted fatigue crack growth curves: normalized crack extension $\Delta a/a$ as a function of non-dimensional cycles.

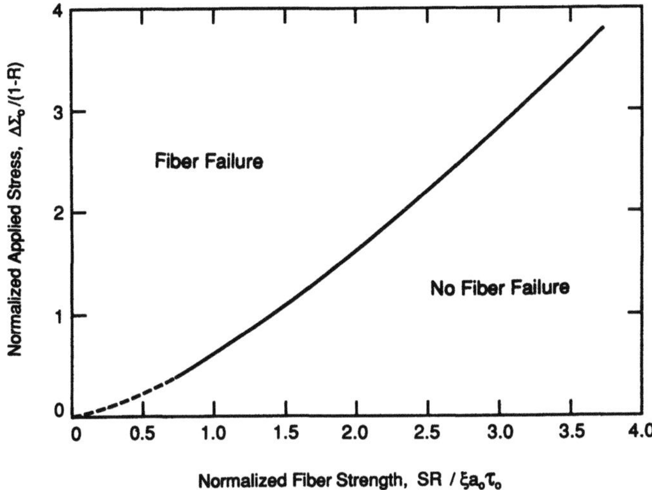

Figure 14. Prediction of the threshold stress range in terms of the fiber strength.

$S(1 - \mathcal{R})f/\Delta\sigma = 2.5$, the onset of fiber failure occurs at $a_f = 1.18a_0$. Note that when either the fiber strength is high or the applied stress is low, no corresponding value of a_f can be identified and the fibers do not fail.

After the first fiber failure, fibers continue to break as the crack grows. Continuing fiber failure creates an unbridged segment larger than the original notch size. However, only the current unbridged length $2a_u$ and the current total crack length $2a$ are relevant, as determined from Fig. 11, with a_0/a_f replaced by a_u/a and $\Delta\Sigma_0$ replaced by $\Delta\Sigma_0/(a/a_0)$. This procedure has been used to compute a_u/a[40] (Fig. 12).

If the fibers are relatively weak and break close to the crack tip $(a_0/a \rightarrow 1)$ the bridging zone is always a small fraction of the crack length. In this case, there is minimal shielding. If the fibers are moderately strong, the fibers remain intact at first. But when the first fibers fail, subsequent failure occurs quite rapidly as the crack grows. The unbridged crack length then increases more rapidly than the total crack length and ΔK_{tip} also increases as the crack grows. When the fibers are even stronger, first fiber failure is delayed. But once such failure occurs, many fibers fail simultaneously and the unbridged length increases rapidly. This causes a sudden increase in the crack growth rate. Finally, when the fiber strength exceeds a critical value, they never break and the fatigue crack growth rate always diminishes as the crack grows. The sensitivity of these behaviors to fiber strength is quite marked (Fig. 12), with the different types of behavior occurring over a narrow range of fiber strength. Some typical crack growth curves predicted using this approach are plotted on Fig. 13. Finite geometry effects associated with LSB also exist.[40]

The results of Fig. 11 can be used to develop a criterion for a "threshold" stress range, $\Delta\sigma_t$, below which fiber failure does not occur for *any* crack length. Within such a regime, the crack growth rate approaches the steady-state value given by Eq. (59), with all fibers in the crack wake remaining intact. The variation in the "threshold" stress range with fiber strength is plotted on Fig. 14. An upper bound to $\Delta\sigma_t$ is obtained when the notch length becomes exceedingly small $(\Delta\Sigma_0 \rightarrow \infty)$, whereupon the threshold condition reduces to

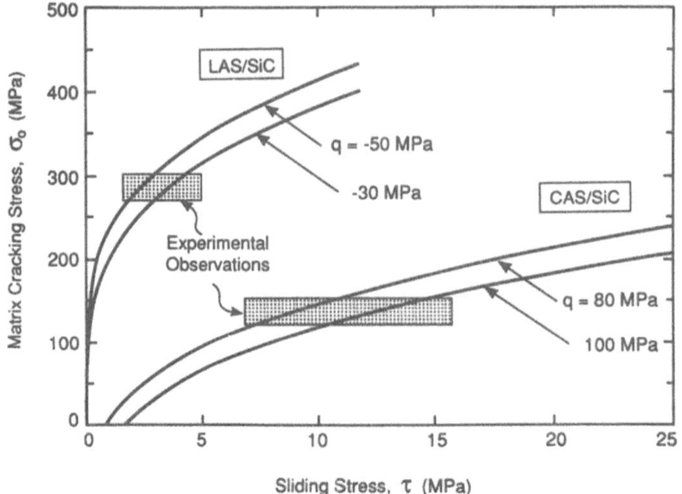

Figure 15. Comparison of predicted and measured lower bound matrix cracking stresses for two CMCs with small τ_0/fS_0: the independently measured values of τ are 1–3 MPa for LAS materials and 15–25 MPa for CAS materials. The residual stress range *measured* in these materials has governed the choices used to plot the predicted curves.

$$\Delta\sigma_t = fS(1 - R). \tag{61}$$

A notable feature of the predictions pertains to the role of the stress ratio R in composite behavior. Prior to fiber failure, the crack growth rate is independent of R (except for its effect on the fatigue properties of the matrix itself). However, R has a strong influence on the transition to fiber failure, as manifest in its effect on the maximum stress, and thus plays a dominant role in the fatigue lifetime. Such trends are evident in both the crack growth curves of Fig. 13 and the threshold stress range of Fig. 14.

V. Structural Performance

V.1 Monotonic Loading

The results in the preceding two sections may be combined to provide predictions of either load/deflection or stress/strain curves that may be compared with experiments and used to assess structural utility. The two features that have been most extensively addressed are the lower bound matrix cracking stress σ_0 (Eq. (35)) and the ultimate strength σ_u (Eq. (30)). To compare with experiments, a substantial range of independent measurements is needed to obtain τ, q, Γ_m, E, m, S_0, etc. The parameter most susceptible to measurement uncertainty is the sliding stress, τ. Consequently, it is judicious to plot predicted properties as a function of τ and compare these with experiments.[17] There are relatively few comprehensive measurements of this type.

Subject to τ/fS_0 being sufficiently small (≤ 0.1) that multiple matrix cracking occurs, the *matrix cracking* results[16−18,34] (Fig. 15) are found to validate σ_0 as a lower

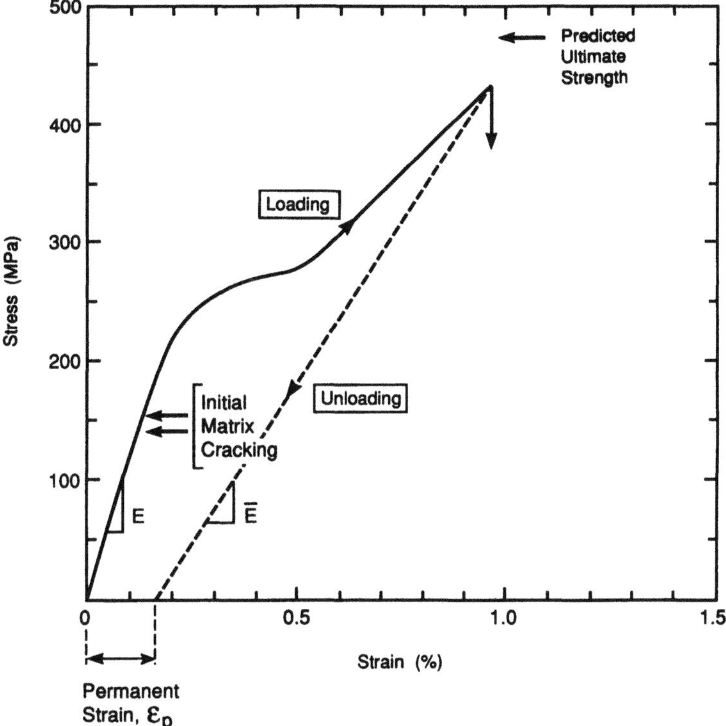

Figure 16. A tensile stress-strain curve for a 1-D CAS matrix composite showing the comparison between the measured ultimate strength and that predicted by the global load sharing analysis[19] ($\tau_0/fS_0 \approx 0.05$). Similar agreement (within 10%) has been shown for other CMCs with small τ_0/fS_0.[9,32]

bound. Results for small diameter Nicalon fibers ($R \sim 10\mu m$) also reveal that the first cracks usually form at stresses close to this bound, in accordance with the following features. The thermal expansion mismatch has a major influence on matrix cracking through its effect on both the residual stress q and the sliding stress, τ. In general, τ is larger in composites subject to positive misfit (residual compression normal to the interface), through the Coulomb term in Eq. (4). But, the effect of larger τ on σ_0 is offset by the influence of the residual tension in the matrix. These effects are evident in the comparison between results for LAS and CAS matrix composites reinforced with Nicalon fibers (Fig. 15).[17] The positive misfit for CAS leads to a larger τ, and a lower predicted matrix cracking stress.‡ Implicit in these findings is that $a_c < a_0$ for small R, because of the scaling of a_c with R (Eq. (37)).

Initial cracking stresses appreciably above the lower bound are generally found for composites containing large diameter fibers ($R \approx 70\mu m$).[44] In such materials, the relatively large matrix flaws needed to satisfy steady-state requirements, $a_0 > a_c$ (Eq. (37)) would generally not occur during careful processing and machining.

---

‡ Explicit measurements of τ by a variety of methods give values in the range 1-3 MPa for LAS and 15-25 MPa for CAS materials: the residual stress q ranges from -20 to -40 MPa and from 80 to 100 Mpa for the two composite systems, respectively.

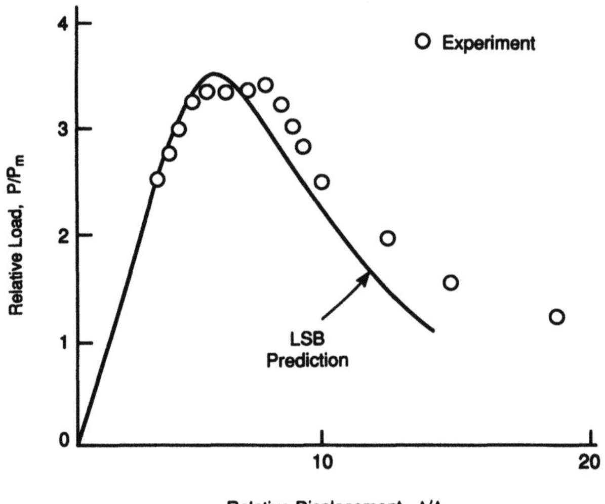

Figure 17. A load/deflection curve for a material that exhibits failure by the growth of a dominant crack. Also shown is a curve predicted from the known bridging traction law.

Statistical information about matrix flaws is needed to correlate the matrix cracking stress in such materials.

The *ultimate strengths* of brittle matrix composites having *small* τ_0/fS_0 also appear to satisfy global load sharing and yield ultimate strengths consistent with Eq. (30).[9,15,29,45] A typical example is shown in Fig. 16.[19] This analysis seemingly applies for unidirectional as well as 2-D (woven and laminated) materials, provided that f refers to the volume fraction of fibers in the loading direction.[32,46]

The occurrence of the transition from global to local load sharing has not been extensively studied, but is found to be qualitatively consistent with Eq. (51): a rigorous comparison has not yet been possible, because of insufficient knowledge concerning residual stresses and matrix fracture energies. Nevertheless, the ultimate strengths at relatively large $\tau_0/fS_0(\geq 0.5)$, are found to be lower than predicted by Eq. (30).[32] Furthermore, fracture involves a dominant matrix crack.[13] In this case, a modelling approach based on crack growth with large-scale bridging, plus some premise concerning the size of the initial flaws, a_0, appears to predict the macroscopic response. A typical LSB, load/deflection curve predicted in this manner and comparison with experiment[39] is shown in Fig. 17. In this case, the initial flaw was introduced to give a well-defined a_0. Otherwise, independent estimates of a_0 are required.

The non-linearity and multiple cracking between σ_0 and σ_u in materials with $\tau_0 < \tau_c$ have yet to be comprehensively compared with models, primarily because of complexities associated with stress corrosion effects. Some typical experimental data[17,18] corresponding with the stress/strain curve of Fig. 16 are presented on Fig. 18.

V.2 Cyclic Loading

The important model predictions that require validation include the fatigue crack growth rates in the matrix both before and after the onset of fiber failure. For this purpose, experiments are conducted on a variety of specimen geometries (charac-

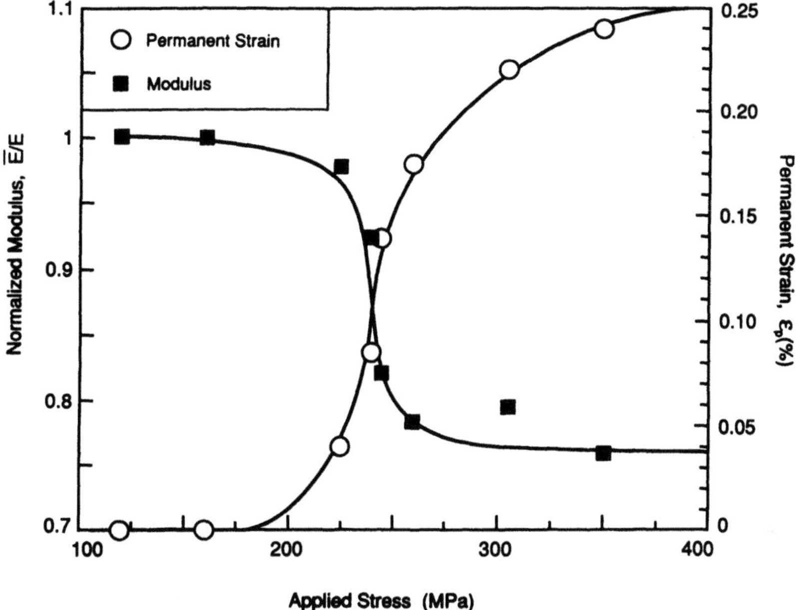

Figure 18. Changes in modulus and permanent strain measured for a 1-D CAS material.

terized by different notch lengths), and the measurements then compared with the model predictions for different values of τ. Following this approach, consistency between experiment and theory can generally be obtained within a narrow range of τ, and the values of τ then compared with those obtained from alternate measurement techniques.

Most experimental studies have focused on Ti/SiC composites, as these exhibit sufficiently weak interfaces to allow matrix cracking to proceed with minimal fiber failure. Examples of crack growth curves in one such material are shown in Fig. 19. Comparisons with model predictions prior to fiber fracture suggest that τ is in the range of 30–60 MPa. Furthermore, the onset of fiber failure is consistent with a fiber strength of 3–4 GPa, which is in the range of values measured on individual fibers.[51] Transitions to fiber failure have been observed with increasing notch length and stress amplitude,[42,52] and found to be in broad agreement with the predictions of Fig. 14.

A notable feature of the crack growth curves pertains to the assumption that τ remains constant throughout the entire loading history. During the early stages, the crack growth behavior is consistent with a relatively high value of τ ($\sim 50 - 60$ MPa). As cracking proceeds, the growth rates in some instances exceed the predicted values (for the same value of τ), suggesting that the interface sliding stress is degraded and thus the shielding contribution ΔK_F is reduced. The data in this regime are consistent with a lower average value of τ, typically ~ 30 MPa.

As a further check, the sliding stress τ has been measured using single fiber push-through tests.[53] Tests have been conducted on both pristine fibers and fibers located adjacent to a matrix fatigue crack. The results indicate that τ does indeed degrade with cyclic loading, from an initial value (obtained on pristine fibers) of ~ 90 MPa to ~ 20 MPa following $\sim 10^5$ loading cycles. These are consistent with the "average"

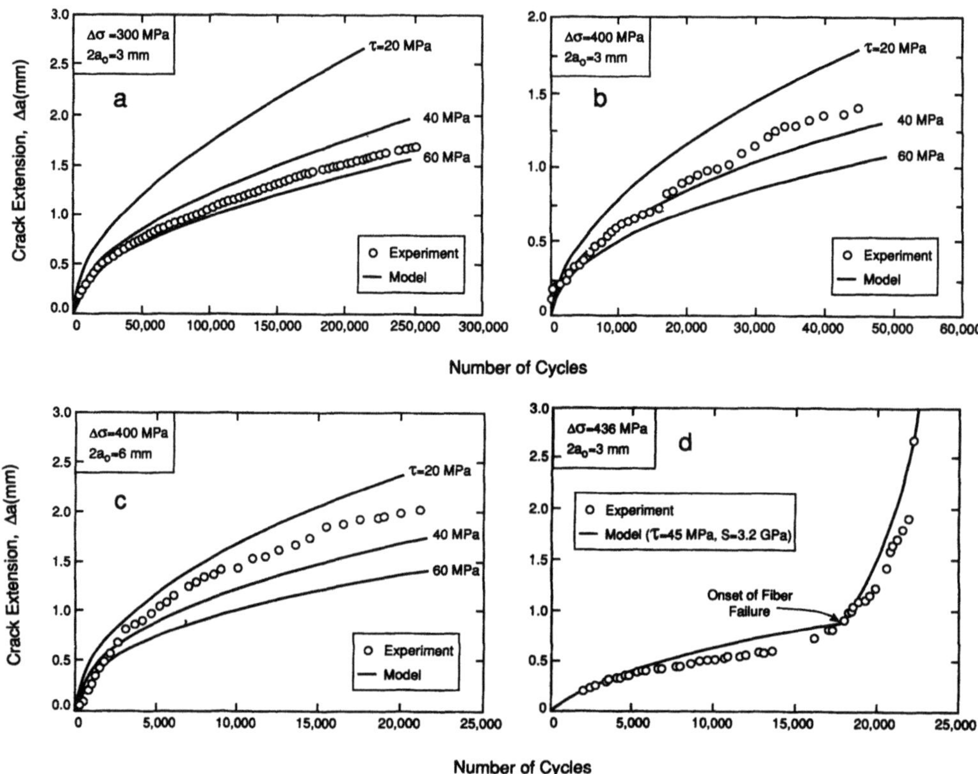

Figure 19. Comparisons of measured and predicted fatigue crack growth curves for a $Ti_{15}V_3Cr_3Al_3Sn$ alloy reinforced with 35% unidirectional SiC (SCS-6) fibers (no fiber failure in the crack wake) for: (a) $\Delta\sigma = 300$ MPa and $2a_0 = 3$ mm, (b) $\Delta\sigma = 400$ MPa and $2a_0 = 3$ mm, (c) $\Delta\sigma = 400$ MPa and $2a_0 = 6$ mm, and $\Delta\sigma = 436$ MPa and $2a_0 = 3$ mm.

values inferred from the fatigue tests (30–60 MPa) and provide confidence in the utility of the micromechanical model described in Section IV.2. It must be emphasized, however, that the process of interface degradation during cyclic loading is *not* yet understood, and thus extrapolation of the model predictions into other regimes (characterized by different values of $\Delta\sigma$, $\mathcal{R}$, a_0, etc.) may provide incorrect trends. It is anticipated that such changes in τ and the associated changes in slip length will also influence both the location of fiber failure relative to the matrix crack plane, as well as the values of σ_b at fiber failure. Additional study is required to establish the important connections between loading history, the statistics of fiber failure and the composite behavior.

VI. Concluding Remarks

The majority of fiber-reinforced composites develop matrix cracks upon either monotonic or cyclic loading. These cracks have a major influence on the macroscopic properties of the composites. The modes of cracking are intimately coupled with the statistical failure properties of the fibers and the nature of the interfaces. This article has summarized present understanding about the coupling between the mechanics and statistics, having origins in the pioneering work of Frank McClintock. A general conclusion is that problems relative to the evolution of individual mode I matrix cracks (in either monotonic or cyclic loading) are rather well understood. Additionally, ultimate failure controlled by the fibers is well-comprehended when the interface are sufficiently "weak" by virtue of their debonding and frictional sliding characteristics that *global load sharing* applies. Phenomena that are not as well understood include: multiple mode I cracking,[35] ultimate failure by local load sharing when the interfaces are relatively "strong" [47,48] and mixed-mode crack growth.[49]

Appendix

Nomenclature:

a	crack length
a_c	critical flaw size for steady-state
a_m	fracture mirror radius
a_0	initial flaw size or notch size
Δa	crack extension
d	crack spacing
$\bar{d}$	average crack spacing
$\bar{d}_s$	saturation crack spacing
E	Young's modulus of composite
E_m	Young's modulus of matrix
E_f	Young's modulus of fibers
$\bar{E}$	Young's modulus of composites with matrix cracks
f	fiber volume fraction
f_x	failed fiber fraction
h	fiber pull-out length
$\bar{h}$	mean pull-out length
K	stress intensity factor
K_f	critical stress intensity factors for fibers
K_m	critical stress intensity factor for matrix
K_0	initiation toughness
K_R	crack growth resistance
ΔK	stress intensity amplitude
n	fatigue growth rate exponent in Paris law
N	number of cycles
ℓ	slip length
L	gauge length
L_0	scale parameter
m	shape parameter for fiber strength
m_*	shape parameter for *in situ* strength
p	residual stresses normal to interface
P	load
P_s	survival probability
q	axial residual stress in matrix
R	fiber radius
$\mathcal{R}$	ratio of minimum to maximum applied stress

S	fiber strength
S_b	fiber bundle strength
S_c	*in situ* strength parameter
S_0	scale parameter for fiber strength
S_*	scale parameter for *in situ* strength
u	crack opening displacement
u_∞	crack opening caused by applied loads
w	width of component
$\mathcal{G}$	energy release rate
$\mathcal{G}_\infty$	applied energy release rate
$\mathcal{G}_{tip}$	crack tip energy release rate
β	Paris law coefficient
σ	stress
σ_0	lower bound matrix cracking stress
σ_b	bridging stress
σ_{max}	peak stress
σ_c	matrix cracking stress
σ_u	ultimate strength of composite
σ_∞	applied stress
$\Delta\sigma$	stress amplitude
$\Delta\sigma_b$	bridging stress amplitude
Δ	displacement
Δ_{max}	maximum displacement
ν	Poisson's ratio
δ_c	characteristic gauge length for fibers
Γ	fracture energy
Γ_f	fracture energy of fibers
Γ_m	matrix fracture energy
Γ_R	crack growth resistance
τ	sliding stress
τ_0	sliding stress caused by asperities
τ_c	transition sliding stress
ϵ_p	permanent strain
μ	friction coefficient
Φ	probability density function

References

1. F.A. McClintock,in *Fracture Mechanical of Ceramics*, edited by R.C. Bradt *et al.*, (Plenum, New York, 1976), p. 33.
2. M.Y. He and J.W. Hutchinson, *Intl. J. Solids Structures,* **25**, 1053 (1989).
3. A.G. Evans and D.B. Marshall, *Acta Metall.,* **37**, 2567 (1989).
4. J. Aveston, G.A. Cooper, and A. Kelly, *The Properties of Fiber Composites JPC*, (1971), p. 15.
5. D.B. Marshall, B.N. Cox, and A.G. Evans, *Acta Metall.,* **33**, 2013 (1985).
6. F.W. Zok and C.L. Hom, *Acta Metall. Mater.,* **38**, 1895 (1990).
7. J. Bowling and G.W. Groves, *J. Mater. Sci.,* in press.
8. M.D. Thouless and A.G. Evans, *Acta Metall.,* **36**, 517 (1980).
9. W. Curtin, *J. Mater. Sci.,* **26**, 5239 (1991).

10. M. Sutcu, *Acta Metall.*, **37**, 651 (1989).

11. A.G. Evans, *J. Am. Ceram. Soc.*, **73**, 187 (1990).

12. J.W. Hutchinson and H. Jensen, *Mech. of Mtls.*, **9**, 139 (1990).

13. F.W. Zok, O. Sbaizero, C.L. Hom, and A.G. Evans, *J. Am. Ceram. Soc.*, **74**, 187 (1991).

14. A.G. Evans, *Mat. Sci. Eng.*, in press.

15. A.G. Evans, F.W. Zok, and J.B. Davis, *Composite Science and Technology*, **42**, 3 (1991).

16. D.B. Marshall and A.G. Evans, *J. Am. Ceram. Soc.*, **68**, 225 (1985).

17. A. Pryce and P. Smith, *J. Mater. Sci.*, in press.

18. D. Beyerle, S.M. Spearing, F.W. Zok, and A.G. Evans, *J. Am. Ceram. Soc.*, in press.

19. B. Budiansky, J.W. Hutchinson, and A.G. Evans, *J. Mech. Phys. Solids*, **34**, 164 (1986).

20. L.N. McCartney, *Proc. Roy. Soc.*, **A409**, 329 (1987).

21. T.J. Mackin, P. Warren, and A.G. Evans, to be published.

22. R.M. McMeeking and A.G. Evans, *Mech. of Mtls.*, **9**, 217 (1990).

23. R.O. Ritchie, J. Knott, and J.R. Rice, *J. Mech. Phys. Solids*, **21**, 395 (1973).

24. A.G. Evans, *Met. Trans.*, **A14**, 1349 (1983).

25. J.R. Matthews, W.J. Shack, and F.A. McClintock, *J. Am. Ceram. Soc.*, **59**, 304 (1976).

26. A. Freudenthal, *Fracture*, edited by H. Liebowitz, (Academic Press 1967).

27. H.L. Oh and I. Finnie, *Intl. J. Frac.*, **6**, 287 (1970).

28. J.F. Jamet, D. Lewis, and E.Y. Luh, *Ceram. Eng. Sci. Proc.*, **5**, 625 (1984).

29. H.C. Cao, E. Bischoff, O. Sbaizero, M. Rühle, A.G. Evans, D.B. Marshall, and J.J. Brennan, *J. Am. Ceram. Soc.*, **73**, 1691 (1990).

30. H.E. Daniels, *Proc. Roy. Soc.*, **A183**, 405 (1945).

31. H.T. Corten, *Modern Composite Materials*, edited by L.J. Brontman and R.H. Krock, (Addison, New York, 1967), p. 27.

32. F.E. Heredia, S.M. Spearing, A.G. Evans, W.A. Curtin, and P. Mosher, *J. Am. Ceram. Soc.*, to be published.

33. W.A. Curtin, *J. Am. Ceram. Soc.*, **74**, (1991), in press.

34. R.Y. Kim and N. Pagano, *J. Am. Ceram. Soc.*, **24**, 1082 (1991).

35. F.W. Zok and S.M. Spearing, submitted to *Acta Metall. Mater.*

36. B.N. Cox, *Acta Metall. Mater.*, in press.

37. B.N. Cox and C. Lo, *Acta Metall. Mater.*, in press.

38. B.N. Cox and D.B. Marshall, *Acta Metall. Mater.*, **39**, 579 (1991).

39. G.R. Odette and B.L. Chao, *Acta Metall. Mater.*, to be published.

40. G. Bao and R.M. McMeeking, *J. Mech. Phys. Solids*, to be published.

41. J.K. Shang, J.L. Tzou, and R.V. Ritchie, *Met. Trans.*, **A18**, 1613 (1987).

42. D. Walls, G. Bao, and F.W. Zok, *Scripta Metall. Mater.*, **25**, 911 (1991).

43. M. Sensmeier and K. Wright, in *Proceedings TMS Fall Meeting*, edited by P.K. Law and M.N. Gungor, (Indianapolis, 1989), p. 441.

44. R. Singh, *J. Am. Ceram. Soc.*, **72**, 1764 (1989).

45. K. Prewo and J.J. Brennan, *J. Mater. Sci.*, **15**, 463 (1980).

46. D. Beyerle, S.M. Spearing, and A.G. Evans, to be published.

47. A.S. Argon, *Treatise On Materials Science*, (Academic Press, 1972).

48. M.S. Hu, H.C. Cao, J.S. Yang, R. Mehrabian, and A.G. Evans, *Met. Trans.*, to be published.

49. Z. Suo and J.W. Hutchinson, *Advances In Applied Mech.*, to be published.

50. H. Tada, P.C. Paris, and G.R. Irwin, *The Stress Analysis of Cracks Handbook*, (Del Research Corp., St. Louis, 1985).
51. K.K. Chawla, *Composite Materials Science and Engineering*, (Springer-Verlag, New York, 1987).
52. D. Walls and F.W. Zok, in: *Advanced Metal Matrix Composites for Elevated Temperatures*, edited by M.N. Gungor, E.J. Lavernia and S.G. Fishman, (ASM, Metals Park 1991n), p. 101.
53. P. Warren, T.J. Mackin, and A.G. Evans, *J. Am. Ceram. Soc.*, in press.

Metallurgy of Permanent Magnet Alloys: Recent Developments

L. Rabenberg

Center for Materials Science and Engineering
The University of Texas
Austin, Texas 78712
U. S. A.

Abstract

The recent rapid expansion of the array of compounds known to possess excellent intrinsic permanent magnet properties has not been matched by an equivalent growth in the ability to process these compounds into magnets having correspondingly high values of the microstructure-sensitive properties such as coercivity and energy product. Since the 1983 discovery of the ternary intermetallic compound, $Nd_2Fe_{14}B$, several other systems, including $Sm(Fe,Ti)_{12}$, $Nd_2Fe_{14}C$, and $Sm_2Fe_{17}N_{3-\delta}$, have been proposed as hard magnet compounds, but none of these are currently being produced as finished magnets. The techniques for the control of the metallurgical development in these systems are simply not yet available. The principles governing the development of coercivity in permanent magnets are briefly reviewed, then several of these newly-proposed systems are examined with emphasis on their potential for the development of high coercive force microstructures.

I. Introduction

The 1982 discovery[1,2] of the ternary intermetallic compound $Nd_2Fe_{14}B$ sparked an extended search for other compounds that might be used as primary constituents in permanent magnets.[3-7] $Nd_2Fe_{14}B$ was the first ternary compound[7,8] to exhibit an outstanding combination of the intrinsic magnetic properties desired for permanent magnet applications. It has a high uniaxial magnetocrystalline anisotropy, high saturation magnetization, and a relatively low cost. $Nd_2Fe_{14}B$ is also the first ternary intermetallic compound to exhibit properties that are substantially different from those available in binary, or pseudo-binary, systems. Its discovery immediately sug-

Table I

Permanent Magnet Phases

	T_c (°C)	$\mu_0 M_S$ (T)	$\mu_0 H_A$ (T)	Theoretical $(BH)_{MAX}$ (kJ/m³)	Reference
SrO 6(Fe_2O_3)	450	0.42		55	9
$SmCo_5$	727	1.14	28.0	259	10
Sm_2Co_{17}	920	1.25	6.5	311	10
$Nd_2Fe_{14}B$	315	1.60	9.1	509	11
$Nd_2Fe_{14}C$	262	1.50	10.1	448	11
$Sm(Fe_{11}Ti)$	311	1.14	10.5	259	4
$Sm(Fe_{10}V_2)$	318	0.72	6.8	162	12
$Sm_2Fe_{17}C$	279	1.24	5.3	306	13
$Sm_2Fe_{17}N_{2.3}$	476	1.54	14.0	471	14,15

Other Candidate Phases
Hydrides of the Fe - rich phases[16]
T_c is generally increased,
H_A is generally decreased.
Compounds based on the Nd_5Fe_{17} phase[17]
e.g. $Sm_5(Fe_{14.5}Ti_{2.5})$[6,18]
$T_c \approx 300°C$
$\mu_o M_s \approx 0.3$ Tesla
$\mu_o H_A \approx 25.1$ Tesla
$(RE)Co_4X$ phases[19], X =B, Al, Ga
$(RE)(Co,Fe)_9Si_2$ Phases[20]

---

gested that there may be other suitable compounds among the relatively unexplored three- or four-component systems involving the ferromagnetic transition metals and the rare earths. Since 1983, there has been a remarkably successful search for new ternary Fe- or Co-rich rare earth compounds. This has resulted in the discovery of several interesting candidate phases such that the list of potential magnet materials (Table I) is longer today than it ever before has been.

Candidate phases that are being considered as permanent magnets include a variety of compounds, all of which are structurally related to known magnet phases such as Sm_2Co_{17} or $Nd_2Fe_{14}B$. All conceivable substitutions in the $Nd_2Fe_{14}B$ structure have been tried; essentially all the rare earths can be substituted for Nd and a wide range of transition metals such as Co, Ni, Mn, Cr, V can be substituted for Fe without seriously changing the stability of the 2:14:1 stoichiometry. C can be substituted for B, but $Nd_2Fe_{14}C$ seems to be significantly less stable[3] than $Nd_2Fe_{14}B$. Pseudo-binary compounds based on the $ThMn_{12}$ structure have been extensively studied; interest has focussed on $Sm(Fe_{11}Ti)$ and $Sm(Fe_{11}V)$, the two compositions having the best intrinsic properties. Sm_2Fe_{17} has been nitrided[14,15] or carburized;[16] interstitials in this compound have been found to dramatically increase both the Curie temperature and the magnetocrystalline anisotropy. Almost all the known permanent magnet phases have been hydrided;[16] hydrogen interstitials generally increase the Curie temperature and decrease the anisotropy field of the Fe-rich phases. A few other Fe- or Co-rich phases, such as $Sm_5(FeTi)_{17}$ (the A2,[17] or ω[18] phase), $(RE)Co_4(Ga,Al,B)$,[19] and $(RE)Co_9Si_2$[20] have recently been discovered and are only now beginning to be characterized. Several of these phases have desirable intrinsic properties (See Table I) and would have been considered to be breakthroughs if they had been discovered before $Nd_2Fe_{14}B$.

None of these new phases have yet been converted into commercially viable magnets. To do so requires that the candidate phase with its favorable intrinsic properties be incorporated into an appropriate microstructure that allows it to display correspondingly favorable extrinsic properties such as coercivity, remanence and energy product. Since each of these new compounds has its own unique metallurgy, economic production and fabrication of magnets having acceptable and reproducible microstructures necessarily requires a significant alloy development effort. Because several different candidate phases are being considered, research efforts have been spread too thinly for rapid progress in any one area. Furthermore, the availability of excellent low cost $Nd_2Fe_{14}B$ magnets may be discouraging investment in research into the development of alternative magnets and processes. Thus, despite the wealth of opportunities represented by these new candidate phases, the metallurgical development aspects have not kept up, and no new materials have been commercialized for several years.

It seems to be an appropriate time to review the metallurgy of permanent magnet alloys. By reviewing the experience gained with past magnets[1-3,21-51] such as Sm_2Co_{17} and $Nd_2Fe_{14}B$, it may be possible to more efficiently begin to exploit the variety of new phases that are currently available or even improve on currently commercialized magnets. Although it will certainly be difficult to improve on $Nd_2Fe_{14}B$, the history of the development of permanent magnets teaches that each successful magnet seems to retain a market niche even when apparently bested by a newer product. Thus, it is reasonable to expect that any new magnet having properties that are unique in one way or another will be able to take a portion of the market.

The purpose of this article is to review the relationships between microstructures, metallurgical development, and coercivity mechanisms in permanent magnets. It is hoped that by examining the metallurgical aspects of successful permanent magnets, insight may be gained into how to proceed with the newer materials. It will be seen that successful magnet microstructures fall into a few classes depending on magnetization reversal mechanism. It will also be noted that the most recent generations of magnet alloys have been blessed with exceptionally favorable metallurgy. In the following sections, the relationships between microstructure, magnetization reversal mechanisms, and coercivity in permanent magnets will be briefly reviewed. Several "ideal" microstructures will be described. The current ideas regarding the development and control of microstructures in commercial practice will be described. Finally, the $Sm_2Fe_{17}N_{3-\delta}$ and the $Sm(FeTi)_{11}$ systems will be examined with regards to possible routes to high coercivity.

II. Magnetization Reversal Mechanisms

It is common to categorize permanent magnets in terms of their magnetization reversal mechanisms. One commonly speaks of either "nucleation controlled magnets" or "domain wall pinning magnets", depending on whether the most difficult step in the process of magnetization reversal is the nucleation of a domain of reversed magnetization or the movement of a domain wall across the magnet material. In reality, the situation is somewhat more complicated; it is necessary to consider a) whether the magnet consists of more than one ferromagnetic phase, b) whether the ferromagnetic grains are larger or smaller than their single domain size, and c) whether exchange interactions cross grain boundaries and/or interphase interfaces. It is especially important to be aware of these complications as technologies that are capable of reducing grains to sizes equal to or smaller than the single domain particle size become more readily available.

Expressions relating magnetic domain wall characteristics to fundamental intrinsic magnetic properties are useful for the following discussion. The simplest possible forms of the essential relations include:

$$\delta \sim (A/K)^{1/2} \tag{1}$$

$$\gamma \sim (AK)^{1/2} \tag{2}$$

$$r_0 \sim \gamma/(Ms)^2 \tag{3}$$

$$\text{Theoretical}(BH)_{MAX} = (1/4)(4\pi M_s)^2 \tag{4}$$

where A is the exchange interaction, K is the first anisotropy constant, δ is the domain wall thickness, γ is the domain wall energy, r_0 is the size of a single domain particle, $4\pi M_s$ is the saturation magnetization, and the theoretical $(BH)_{MAX}$ is the maximum energy product that can theoretically be generated by a specific phase.

The single domain particle size corresponds to the size below which it becomes favorable for an isolated ferromagnetic particle to exist as a single domain; that is, below this critical size the magnetostatic energy supported by the particle when completely magnetized is smaller than the energy of a single domain wall through the particle. For hard magnetic materials such as $Nd_2Fe_{14}B$, r_o is on the order of 20 nm.[51,52] Below the single domain particle size, domain wall processes, such as nucleation of reversed domains or domain wall migration, are not viable; magnetization reversal processes must occur by coherent rotation, and in such a case, the coercivity is expected to be very high - essentially equal to the anisotropy field of the magnet.[53-55] This is not achieved in practice; all known magnets show coercivities that are much lower than their anisotropy fields. Heterogeneous nucleation of reverse domains and/or easy migration of existing domain walls allow the magnetization to change without large spins being oriented in the hard direction. Because even the best commercial magnets show coercivities far below their anisotropy fields and exhibit $(BH)_{MAX}$ values that are between 1/2 and 2/3 of theoretical, there certainly is room for improvement.

The characteristics of magnets having various reversal mechanisms are indicated schematically in Fig. 1. For magnets having a single ferromagnetic phase, grain sizes and the degree of magnetic isolation among grains are important parameters. For alloys having more than one ferromagnetic phase, the sizes and distribution of the particles of the "soft magnetic" phase as well as the extent of magnetic interaction between phases are important parameters. The figure refers to three field intensity parameters, H_A, H_N, and H_P. H_A, the anisotropy field, is the magnetic field intensity required to rotate the magnetization in a single crystal from its easy axis to its hard

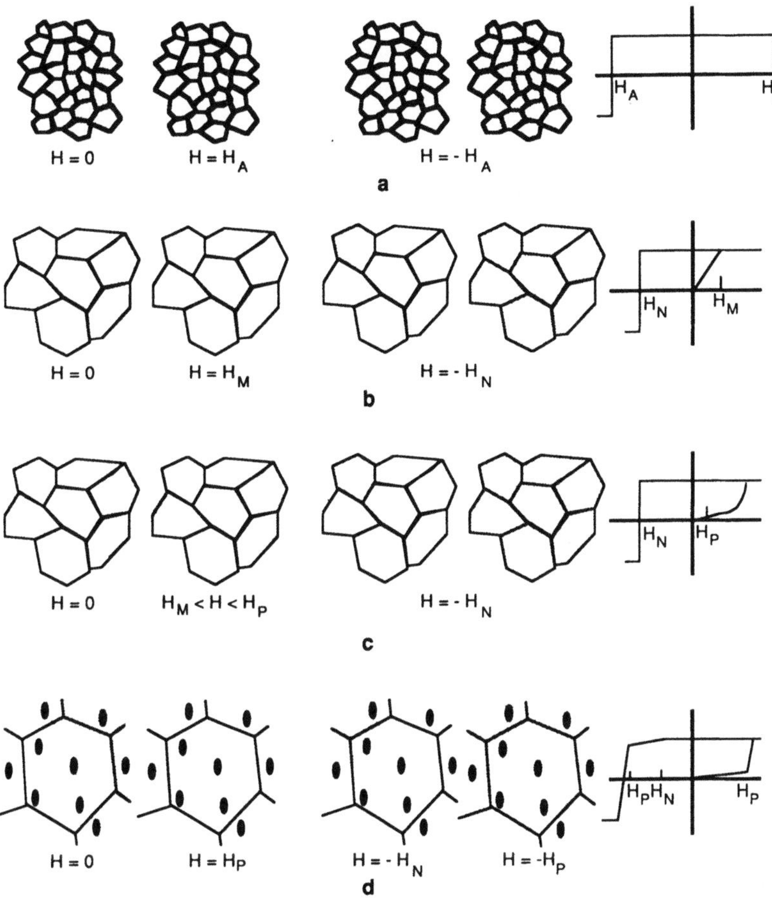

Figure 1. Schematic sketches showing reversal mechanisms in hard magnetic materials having one ferromagnetic phase. In these sketches, shaded and unshaded areas represent regions magnetized in opposite senses. Grain boundaries drawn as heavy lines represent boundaries across which exchange interactions are negligible. Each sequence of five sketches shows (i) the initial, demagnetized state, (ii) the magnet at, or near, positive saturation, (iii) the initial reversed domain at the coercive field, (iv) the magnet at negative saturation, and (v) the corresponding hysteresis loops. (a) Isolated single domain grains. (b) Large grains with exchange interactions across grain boundaries. (c) Large grains with exchange interactions across grain boundaries and some pinning of domain walls by grain boundaries. (d) Very large grains with nonmagnetic inclusions pinning the domain walls.

axis of magnetization. H_N, the reversed domain nucleation field, is the field intensity required to cause the nucleation of a reversed domain in an otherwise completely magnetized grain. Although one can use classical nucleation theory and measured values of bulk intrinsic properties to calculate H_N for homogeneous nucleation, actual H_N's are generally much smaller because of heterogeneous nucleation at magnetic defects. H_P, the depinning field, is the magnetic field intensity required to displace a magnetic domain wall away from a pinned position in the material. Because of structural heterogeneities, the energy of a domain wall varies throughout the solid; the depinning field is essentially the field required to displace a magnetic domain wall over such an energy barrier.

The ideal microstructure for substances having a single ferromagnetic phase of high magnetocrystalline anisotropy is probably that shown schematically in Fig. 1a. In this hypothetical microstructure, all grains are smaller than the single domain particle size; they are perfectly aligned such that their easy magnetization axes are exactly parallel; they are isolated from each other in the sense that exchange interactions do not cross the grain boundaries. Starting from an initially demagnetized state, the magnetization distribution does not change until the fields acting on a grain exceed the field required for coherent rotation, the anisotropy field H_A. When H_A is exceeded for any grain, its magnetization reverses suddenly , and its reversed field acts to reverse the magnetization in adjacent grains. The entire sample reverses by the sequential reversal of grains along an advancing boundary. Once fully saturated, the magnetization is not changed until a reversed field exceeding the anisotropy field is applied.

For magnets having grains larger than the single domain size, or for magnets with strong exchange interactions across grain boundaries, domain wall migration and reversed domain nucleation become effective processes. As seen in Fig. 1b, when a field is applied to such a material in its demagnetized state, domain walls migrate until the magnet is saturated. Upon reversing the field, nothing happens until the reversed field exceeds H_N. Once this happens, the entire magnet reverses by the rapid migration of the domain walls. This is a classic reverse domain nucleation limited magnet.

If crystallographic defects, grain boundaries, or nonmagnetic inclusions impede domain wall motion such that H_P is nonzero, the virgin magnetization curve will tend to become concave-upward as shown schematically in Fig. 1c. Fig. 1d represents the situation for which H_P exceeds H_N, and the pinning centers are uniformly distributed throughout the material. The pinning sites can be any of a variety of defects, interfaces, or nonmagnetic inclusions. This is a typical domain wall pinning magnet for an alloy having only one ferromagnetic phase.

In alloys having more than one ferromagnetic phase, easy reverse domain nucleation and/or effective domain wall pinning may occur depending on the size, distribution, and intrinsic properties of the second ferromagnetic phase. Figs. 2a–2d represent the types of behaviors expected depending on the relative values of H_{N1} (the hard phase), H_{N2} (the soft phase), and H_P.

Fig. 2a shows what is expected if the two ferromagnetic phases having large grains are magnetically isolated from each other. With no exchange interactions across interphase interfaces, $H_P > H_{N1} > H_{N2}$, this magnet is a simple composite of two nucleation-limited magnets. If exchange interactions are present such that $H_{N1} > H_P > H_{N2}$, and if the soft magnetic phase is in the form of grains larger than their single domain size, reversed domain nucleation occurs first in the soft grains, then the domain walls migrate across the interphase interfaces (Fig. 2b). If H_P is very small, such that $H_{N1} > H_{N2} > H_P$, domain walls move easily from one phase to another, and the magnet is only as good as its worst phase (Fig. 2c).

If an alloy has two ferromagnetic phases, but the soft magnetic phase is finely divided such that each particle is smaller than its single domain particle size, then it can be a very effective domain wall pinning magnet. In such an alloy, a pre-existing domain wall will lie in the soft phase and will need to overcome a significant energy barrier in order to be displaced into the hard phase. Since nucleation of reversed domains within the soft magnetic phase simply will not occur, the presence of the

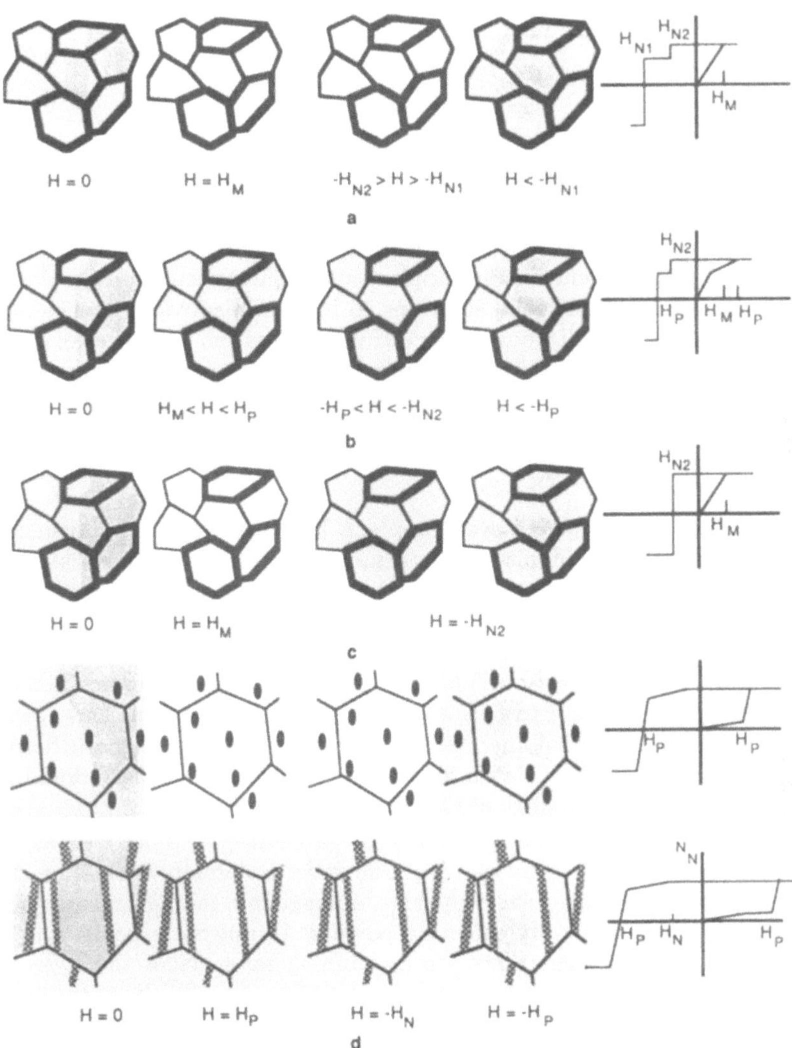

Figure 2. Schematic sketches showing reversal mechanisms in hard magnetic materials having two ferromagnetic phases. In these sketches, the grains outlined in very heavy black lines represent the soft magnetic phase. Grain boundaries drawn as heavy lines in (a) represent boundaries across which exchange interactions are negligible. Each sequence of five sketches shows (i) the initial, demagnetized state, (ii) the magnet at, or near, positive saturation, (iii) the initial reversed domain at the coercive field, (iv) the magnet at negative saturation, and (v) the corresponding hysteresis loops. (a) Two different, magnetically isolated phases. (b) Two interacting phases for which the pinning at the grain boundaries, H_P, is greater than the reverse nucleation field for the soft phase. (c) Two interacting phases with little pinning at the grain boundaries. (d) Pinning magnets containing uniform dispersions, elliptical and platelike, of a finely-divided ferromagnetic phase.

soft phase will not be deleterious to the coercivity of the overall system. As noted in Fig. 2d, the most effective pinning is to be expected for alloys whose soft magnetic phase has a high saturation magnetization and forms as continuous films parallel to the domain walls of the hard magnetic phase.

III. Successful Magnetic Materials

A brief review of the historical development of permanent magnet materials shows that only three of the microstructural classes described above have been successfully exploited for permanent magnet applications. This is true despite the fact that materials ranging from cubic solid solutions to highly anisotropic intermetallic compounds have been used. All known successful magnets fall into the following three microstructural classes: (1) pinning magnets based on a single ferromagnetic phase, (2) pinning magnets based on two or more ferromagnetic phases, and (3) nucleation magnets based on a single ferromagnetic phase. This is true despite the wide range of dissimilar materials systems and microstructural development processes that have been used.

The first permanent magnet was simply a hardened steel;[9] by transforming a steel to a martensitic or a bainitic structure, a high concentration of structural defects is introduced, and the steel becomes a magnetically hard pinning magnet of the type sketched in Fig. 1d. Continuous phase separation in Alnico[9,56,57] and FeCrCo magnets[58-61] results in continuous domain wall pinning by microstructures having two interpenetrating ferromagnetic phases. (Fig. 2d). Ordered, stoichiometric PtCo (or PdCo) magnets have an intricate distribution of orientational and translational variants (twins and antiphase domain boundaries)[62,63] that pin magnetic domain walls (Fig. 1d); nonstoichiometric, Co-rich alloys may also have antiphase domain boundaries coated with continuous films of Co (Fig. 2d).[64,65]

The hard oxide magnets based on Ba- or Sr-ferrite,[9,66,67] as well as the original rare earth magnet,[68-71] $SmCo_5$, are true reverse domain nucleation controlled magnets. These compounds have very high magnetocrystalline anisotropies; useful microstructures are obtained by orienting powders in magnetic fields, and sintering (Fig. 1b). The "2:17" magnets are strong pinning magnets of the type described in Fig. 2d; in these magnets, the soft magnetic phase, $Sm(Co,Cu)_5$, forms a continuous honeycomb-like network, subdividing the hard magnetic phase, $Sm_2(Co,Fe)_{17}$, into 150 nm rhomboid-shaped particles.[22-27] Magnets based on $Nd_2Fe_{14}B$ fall into two distinct classes depending on their processing and the resulting microstructures. Conventional nucleation-controlled magnets (Fig. 1b) are made by orienting, then sintering, powders;[1,34-44] excess Nd is added in order to coat the grain boundaries with a nonmagnetic Nd-rich phase. Microcrystalline magnets can be prepared by hot compaction of Nd-rich melt-spun ribbons.[48-51] These are pinning magnets in the sense that a magnetic domain wall, once created, can not easily propagate through the dense network of grain boundaries coated with Nd-rich material (Fig. 1d).

All known successful magnets fall into the following three microstructural classes: (1) pinning magnets based on a single ferromagnetic phase, (2) pinning magnets based on two or more ferromagnetic phases, and (3) nucleation magnets based on a single ferromagnetic phase. It can also be noted that each successful magnet relies on some metallurgical mechanism that either creates a uniform distribution of pinning sites or isolates grains from each other.

A wide variety of favorable metallurgical phenomena have been exploited for the development of permanent magnets. From the eutectoid decomposition/martensitic transformation in steels, to the spinodal decomposition in AlNiCo and FeCrCo magnets, to the ordering reaction in PtCo, most of the classical solid state phase transformation phenomena have been represented in permanent magnet technology.

With the development of phases having high magnetocrystalline anisotropies, the emphasis has shifted to grain boundary phenomena and the extent to which grains are magnetically isolated from each other. Sintering aids in ferrite magnets and residual oxides in $SmCo_5$ (both of which tend to coat grain boundaries) at least partially isolate grains from each other. Because of the high ratio of anisotropy field to magnetization in these two magnets, the single domain particle sizes are relatively large; even so, commercial polycrystalline magnets in their thermally demagnetized state tend to contain many single domain grains that are, in fact, larger than the critical grain size for a truly isolated particle.

Pinning magnets based on Sm_2Co_{17} (2:17's) and all magnets based on $Nd_2Fe_{14}B$ (2:14:1) have been particularly blessed with highly favorable, and not-to-be-routinely-expected metallurgy. The cellular microstructure in the 2:17 magnets is generated by an allotropic transformation that not only creates a dense distribution of nucleation sites but constrains the resulting precipitates to grow into continuous three-dimensional networks.[72,73] Useful magnets based on $Nd_2Fe_{14}B$ rely entirely on the fact that $Nd_2Fe_{14}B$ "equilibrates" with a metastable phase that aggressively wets 2:14:1 grain boundaries.[39,41,48,49,74,75] Because these materials are most closely related to the next generation of magnet phases, they will be examined at greater length in the next section. It should be emphasized that each successful magnet type has relied on its own, peculiar metallurgy.

IV. $Sm_2(Co,Cu,Fe,Zr)_{17}$ and $(Nd,RE)_2(Fe,Co)_{14}B$

The metallurgical development processes in the 2:17 magnets and in the 2:14:1 magnets are worth further examination for several reasons. These two types of magnets are similar to the new candidate materials in that they are all based on highly anisotropic intermetallic compounds involving rare earth metals. The two essential phases, Sm_2Co_{17} and $Nd_2Fe_{14}B$, have been successfully exploited as permanent magnets largely because of system-specific metallurgical phenomena. Finally, within these two general systems, each of the three useful microstuctural types is represented.

The 2:17 magnets, $Sm(Co,Fe,Cu,Zr)_z$ ($7.5 \leq z \leq 8.0$), develop their coercivity by domain wall pinning in a fully coherent two-phase distribution of $Sm(Co,Cu)_5$ and $Sm_2(Co, Fe, Zr)_{17}$. In optimized magnets, the 1:5 phase takes the form of thin (8 nm) films lying parallel to the pyramid planes in twinned, rhombohedral 2:17.[22–27,32,76] A third phase, best represented as $(Zr,Sm)_1(Co,Cu,Fe)_3$, forms as very thin (0.8 – 4.8 nm) plates parallel to and coherent with the basal planes of the two primary phases.[24,32,76,77] This is nearly an ideal microstructure; the habit of the 1:5 phase films is such that it naturally forms a continuous, cellular network in three dimensions (Fig. 3) even though it occupies as little as 8 volume percent of the alloy. Domain walls are trapped in the energy minima created by the continuous Cu-rich 1:5 phase.[28,29]

The cellular microstructure develops as the result of a series of transformations[72,73] in nonstoichiometric $Sm_2(Co,Fe,Cu,Zr)_{17}$. For compositions used in permanent magnets, a single phase solid solution based on the hexagonal form of 2:17 is stable at very high temperatures. During cooling, this phase simultaneously undergoes an allotropic

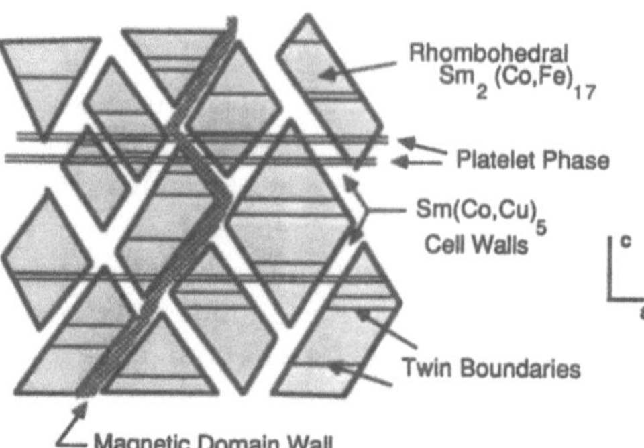

Figure 3. Schematic sketch of the microstructure of optimized Sm$_2$(Co,Fe,Cu,Zr)$_{17}$-type magnets.

transformation to its rhombohedral form and rejects excess Sm in the form of the 1:5 cell walls. The allotropic transformation nucleates homogeneously, creating a dense distribution of twins (interfaces between orientational variants) and antiphase domain boundaries (interfaces between translational variants). Since both forms of 2:17 can be created by ordered substitutions in the 1:5 phase, the 1:5 phase is the "disordered phase" in this system; it coats the antiphase boundaries and interfaces between orientational variants in the 2:17.[78] The topological constraints that apply to antiphase domain boundaries and interfaces between orientational variants are inherited by the 1:5 cell walls. The walls must exist as continuous, interconnected, films; if they were to be removed, high energy incoherent twin interfaces would remain. Elastic strain energy minimization[22] apparently also plays a role; it causes large segments of the 1:5 cell walls to be oriented parallel to the pyramid planes. The thin, Zr-rich plates tend to nucleate and grow throughout the cellular structure after it is well on its way to development;[78] their role in the development of coercivity is still not settled.

The cellular microstructure is a unique feature of the SmCo 2:17 system and is unlikely to be exactly reproduced in any new magnet, nevertheless, its discovery does point to the value of research into the metallurgy of these alloys. The creation of the cellular structure seems to have been the result of trial-and-error, coupled with repetitious use of practices that had been developed for other systems. Because attempts to make magnets using stoichiometric 2:17 had not been successful (for reasons that have never been fully delineated), and because it was assumed that free Co or Fe in the magnet would result in easy reverse domain nucleation, the composition was shifted toward SmCo$_5$. Cu was used as an alloying element because it had once been helpful for some SmCo$_5$ magnets.[69] Zr, and most of the other transition elements, were added in a purely trial-and-error mode;[21] Zr was later shown[79] to increase the anisotropy field of Sm$_2$Co$_{17}$. (To this date it is not known whether both Cu and Zr are necessary. or whether one or both may be eliminated.) Fe was added for its beneficial effects on magnetization. All these elements were combined, then subjected to a complex heat treatment that seems to have been modelled after the heat treatments used for AlNiCo alloys. Only after outstanding magnetic properties were discovered were the ultramicrostructures fully examined and the conditions for suitable microstructure delineated. Because this type of serendipity can not be expected in other systems, the best approach seems to be the systematic study of the metallurgy of the newer systems, coupled with an understanding of what makes a useful microstructure.

Sintered magnets based on $Nd_2Fe_{14}B$ moved very quickly into the marketplace after their initial discovery[1] because the technology that had been developed for $SmCo_5$ magnets was directly and immediately applicable. Following the same pattern as for $SmCo_5$, $Nd_2Fe_{14}B$ is melted together with small amounts of excess Nd and B, crushed, oriented in a magnetic field, pressed, and sintered. The excess Nd and B is added in order to avoid the formation of free Fe, which is known to be disastrous for coercivity (Fig. 2c). The excess B is present in finished magnets in the form of occasional particles of $Nd_{1+\epsilon}Fe_4B_4$; the excess Nd forms a non-ferromagnetic solid solution that continuously wets most of the grain boundaries.[34,80−82] Coercivity in this system depends on the resistance to nucleation of reversed domains within these magnetically isolated grains.[83]

$SmCo_5$ technology was applicable to $Nd_2Fe_{14}B$, in part, because the Nd-Fe-B system naturally provides a nonmagnetic phase that continuously wets the grain boundaries of $Nd_2Fe_{14}B$. The presence of the nonmagnetic phase in Nd-Fe-B magnets is especially critical because $Nd_2Fe_{14}B$ has a significantly higher magnetization and lower anisotropy field, than does $SmCo_5$. This has the effect that $Nd_2Fe_{14}B$ has a significantly smaller single domain particle size[52] than does $SmCo_5$; in contrast to $SmCo_5$, it normally exists in its thermally demagnetized state as an agglomerate of polydomain grains. Its high magnetization causes $Nd_2Fe_{14}B$ to tend to demagnetize itself more completely and increases the importance of magnetic isolation between grains. Thus, sintered $Nd_2Fe_{14}B$ would not have been such an immediately successful magnet if not for the the fact that Nd-rich compositions naturally result in a nearly continuous nonmagnetic grain boundary phase.

Alloying additions to the basic Nd-Fe-B system have resulted in improvements in both the intrinsic and the extrinsic properties of these magnets. Co is routinely substituted for Fe in order to increase the Curie temperature,[84−87] thus increasing the maximum use temperature for the magnet. Because it increases the magnetocrystalline anisotropy, Dy is often used to replace a small amount of the Nd.[84,86,88,89] Grain refinement has been accomplished following two different approaches. Nonboride-forming elements such as Ga, Al, Cu, and presumably Zn, lower the melting temperature of the grain boundary material, allowing for lower temperature processing and smaller grain sizes. There is some evidence that these elements improve the wetting of the grains by the Nd-rich phase when liquid, although they most probably cause the grain boundary material to pass through some form of eutectic reaction as it solidifies.[74,75,90] Al, and to a lesser extent Ga, seem to be helpful in that they form nonmagnetic $(Al,Ga)_3Fe_{11}Nd_6$ rather than soft-magnetic Nd_2Fe_{17}. Elements that are capable of forming two-metal borides with iron, such as V, Nb, Mo, or W may also be used as grain refiners; they can precipitate as $(Mo, Nb)FeB$, or $(V,Mo,Fe)_3Be_2$ and effectively pin the grain boundaries.[91,92] The reduced grain size does seem to result in higher coercivity at room temperature as well as reduced irreversible losses at elevated temperatures; apparently, by reducing the grain size, the negative effects of a single easy reverse domain nucleation site are significantly reduced.

Anisotropic NdFeB magnets can also be made by rapid solidification with consolidation by hot deformation.[48,49,93,94] In the best-known variant of the process, the elements are melted together, then melt spun to form microcrystalline ribbons. The ribbons are hot pressed to full density in a closed die, then upset $\approx$ 70% in an open die. The final magnets are composed of plate-shaped grains approximately 0.1 μm thick and 1.0 μm broad. Each such grain has its crystallographic c-axis normal to its broad dimension; the grains are stacked together with their c-axes in common as suggested by Fig. 4. These magnets also contain excess Nd; as in the case of the

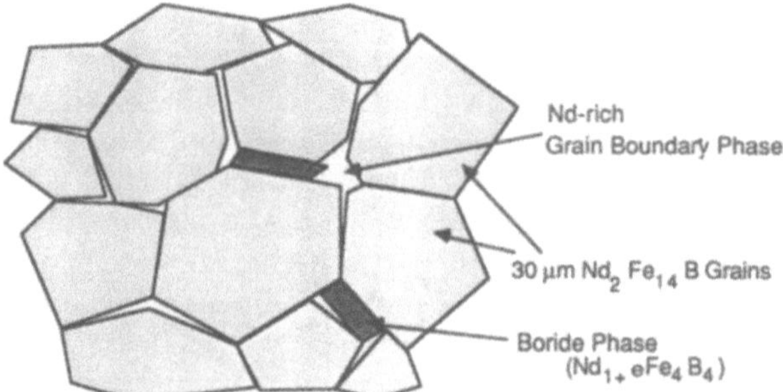

Figure 4a. Schematic sketch of the microstructure of a sintered $Nd_2Fe_{14}B$-based magnet. Most $Nd_2Fe_{14}B$ grains are isolated from each other by the presence of the Nd-rich grain boundary phase.

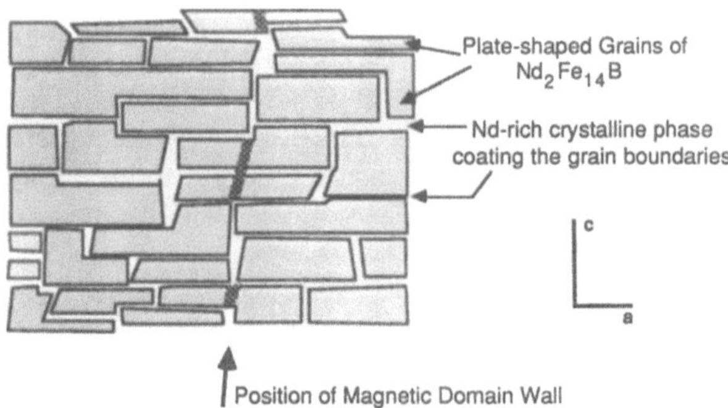

Figure 4b. Schematic sketch of the microstructure of a melt-spun, hot pressed, and die-upset $Nd_2Fe_{14}B$ magnet.

sintered magnets, the Nd coats the grain boundaries of the major phase. Excess B occurs as particles of $Nd_5Fe_2B_5$ that have essentially the same sizes and shapes as the $Nd_2Fe_{14}B$ grains.[95] In the thermally demagnetized state, "magnetic domain walls" tend to lie between the grains, occasionally traversing individual grains. It has been suggested that these are pinning magnets[94,96,97] because virgin magnetization curves show the effects of poor domain wall mobility, but it is also true that nucleation of a reversed domain within any one of these grains is difficult. Thus, once saturated, the magnetization reversal mechanism in these magnets is not easily distinguished from that of the conventional sintered magnets.

In addition to its role in isolating the grains, the Nd-rich grain boundary phase in these magnets is instrumental in the development of a strong crystallographic texture during hot deformation.[49] In the presence of the Nd-rich phase, the $Nd_2Fe_{14}B$ grain growth is anisotropic, causing the grains to become plate-shaped. During the hot deformation, the molten grain boundary material allows for easy grain boundary sliding. The plate-shaped grains slip back and forth until they are stacked with their broad flat faces normal to the principle compressive stress. The upshot is that a single deformation process results in a highly aligned, anisotropic magnet.

It is striking that in each of these systems, very favorable metallurgy has enabled easy access to relatively high coercivities such that $(BH)_{MAX}$ values greater than half of theoretical are reproducibly obtained.

V. The New Phases

Of the several new compounds that have been discovered and proposed as permanent magnet materials in recent years, the two most promising are $Sm(Fe,Ti)_{12}$ and $Sm_2Fe_{17}N_{3-\delta}$. Based on intrinsic properties (Table I), $Sm_2Fe_{17}N_{3-\delta}$ has the best chance of competing with $Nd_2Fe_{14}B$, especially when use temperatures are expected to be in the range 120 – 150°C. $Sm(Fe,Ti)_{12}$, with its high anisotropy field, might be able to replace $SmCo_5$ magnets in applications where high coercivity is more important than high magnetization. However, neither of these two has been successfully produced using the currently-available technologies.

The principal difficulty with $Sm_2Fe_{17}N_{3-\delta}$ is that it is not stable at the temperatures required for conventional sintering. $Sm_2Fe_{17}N_{3-\delta}$ is prepared by nitriding loose Sm_2Fe_{17} powders at 450°C < T < 550°C; at higher temperatures,[15] it tends to decompose into SmN and free Fe, precluding exposure to the usual sintering temperatures. This fact has focussed most current efforts at magnet fabrication on the use of some form of low melting point metallic or polymeric binder to hold particles together mechanically, but isolate them magnetically. It is not yet known whether this will prove to be a successful approach nor is it known whether the corrosion problem that haunts polymer-bonded $Nd_2Fe_{14}B$ magnets can be avoided in this system. In any case, the requirement that powders be nitrided at low temperatures severely limits the types of metallurgical processes that can be applied.

The SmFe binary alloy system provides few clues that might lead to alternative processes for the preparation of fully dense magnets based on $Sm_2Fe_{17}N_{3-\delta}$. The accepted SmFe phase diagram[98] is considerably simpler than the SmCo diagram.[99,100] Sm_2Fe_{17} remains rhombohedral at all temperatures; it equilibrates with the Fe-rich solid solution or with $SmFe_3$. There is no known $SmFe_5$ phase that might be exploited as a pinning phase as is done in the case of the Sm_2Co_{17} magnets. A 5:17 phase, analogous to the recently discovered Nd_5Fe_{17}, is known to exist in alloys containing some Ti,[6,18] but no evidence for a binary, or even a nitrogen-doped, Sm_5Fe_{17} phase is yet available. Fe, of course, is ferromagnetic at room temperature, it can serve as an easy reverse domain nucleation phase and can only be tolerated in $Sm_2Fe_{17}N_{3-\delta}$ - based magnets if it is isolated from the $Sm_2Fe_{17}N_{3-\delta}$ or divided into particles smaller than its critical domains size. $SmFe_3$ is also ferromagnetic; its high anisotropy field and small magnetization give it a relatively large critical particle size. $SmFe_3$ is, therefore, not so deleterious as is Fe; one might even imagine using it as a pinning phase if a process for the low-temperature consolidation of Sm-rich SmFeN magnets can be devised. This seems to be the only route worth pursuing within the very limited confines of the SmFe system.

Additions of ternary elements to the SmFeN system have not been investigated, and it is not clear what alloy modifications can be expected to allow $Sm_2Fe_{17}N_{3-\delta}$ to be developed into a fully dense anisotropic magnet. Since excess N, added to $Sm_2Fe_{17}N_{3-\delta}$, produces SmN, rather than any new ternary phases, addition of excess N does not seem to be a promising option to pursue. Strong nitride formers, such as Ti, Zr, or V will certainly extract the interstitial N from the Sm_2Fe_{17} lattice, resulting in soft magnetic Sm_2Fe_{17} and one or more nitride phases. Intermediate nitride formers, such as Cr or Mn, will tend to dissolve in the 2:17, but these two elements in particular are not generally beneficial for the intrinsic properties of these phases. Other than the first row transition metals and the rare earths, most elements are expected to be insoluble in $Sm_2Fe_{17}N_{3-\delta}$; it seems that the metallurgy of this system reduces to the search for an element or phase that will wet the $Sm_2Fe_{17}N_{3-\delta}$ grains at relatively low temperatures.

Although $Sm(Fe_{11}Ti)$ has a large anisotropy field, it has not shown high coercivities when prepared in bulk polycrystalline form.[6] In the ternary SmFeTi system, $Sm(Fe_{11}Ti)$ is known to equilibrate with the Fe-rich solid solution, with Fe_2Ti, and with Sm_2Fe_{17}, all of which are soft magnetic phases and should be avoided.[101] The details of the phase diagram have not been fully established, but there are indications[6] that $Sm(Fe_{11}Ti)$ may also equilibrate with a $Sm_5(FeTi)_{17}$ phase and/or the $SmFe_3$ phase, both of which are ferromagnetic with very high anisotropy fields. Since $Sm(Fe,X)_{12}$ phases melt incongruently, it has been necessary to resort to rapid solidification[102-104] or mechanical alloying techniques[105,106] in order to produce single phase material. Observations of the grain boundaries[107] in such samples indicate that they are free of secondary phases; thus, this system shows no natural proclivity for magnetic isolation of its grains. In contrast to the NdFeB system where the metallurgy is surprisingly favorable, the SmFeTi system simply does not cooperate to the extent that well-developed methodologies can be directly applicable.

Because of its inherent complexity, the SmFeTi system does present a variety of avenues to explore. It may be interesting to try to fabricate a pinning magnet based on $Sm(Fe_{11}Ti)$ and $Sm_5(FeTi)_{17}$ or $SmFe_3$. Very limited data[108] suggest that $Sm_5(FeTi)_{17}$ takes the form of needles having their long axes parallel to their easy axis of magnetization. An attractive microstructure would be a highly oriented $Sm(Fe_{11}Ti)$ matrix with needles of $Sm_5(FeTi)_{17}$ aligned parallel to the common c-axis. Alternatively, since $Sm(Fe_{11}Ti)$ is crystallographically very closely related to Sm_2Fe_{17}, it may be possible to precipitate Sm_2Fe_{17} from slightly supersaturated solution in $Sm(Fe_{11}Ti)$. The high temperature metallurgy of $Sm(Fe_{11}Ti)$ should be carefully investigated to see if there can be any deviation from stoichiometry in the direction of excess Sm that could be exploited to produce a uniform Sm_2Fe_{17} precipitate structure. Finally , additions of quaternary elements might be investigated in hopes of finding a nonferromagnetic phase that equilibrates with $Sm(Fe_{11}Ti)$ and wets its grain boundaries. Each of these has some chance of success; a systematic investigation of the metallurgy of this system surely will suggest further possibilities.

VI. Conclusions

Permanent magnet technology, like many other areas of Materials Science and Engineering, has seen dramatic progress in recent years and can hope to build on its successes for continued successes in the foreseeable future. The mid-1980's saw commercially available $(BH)_{MAX}$ values raised from 28 to 40 MGOe in two or three years by the discovery and rapid exploitation of the ternary intermetallic compound $Nd_2Fe_{14}B$. The end of the 1980's and beginning of the 1990's have seen the discovery of several potentially valuable permanent magnet phases, such that the variety of new materials being proposed as permanent magnets is larger than it ever before has been. Following a period of alloy development, some of these newer phases can be expected to find commercial applications based either on unique properties or on the fact that they avoid existing patents.

An analysis of the development of extrinsic properties in previous generations of permanent magnets gives some insight into the microstructural features that are necessary for outstanding performance in the new systems. It is seen that all successful permanent magnets can be classified as pinning magnets having (a) one ferromagnetic phase, or (b) two ferromagnetic phases, or (c) as reverse domain nucleation magnets. If a magnet does contain more than one ferromagnetic phase, permanent magnet behavior is only expected if the soft magnetic phase is present in particles smaller than the critical size for single domain behavior, or is completely isolated from the hard magnetic phase. The most recent magnets, based on Sm_2Co_{17} or on $Nd_2Fe_{14}B$, exhibit very favorable metallurgical behaviors such that these constraints have been very easily met. Since the metallurgical characteristics of some of the newer systems do not seem to be so favorable, systematic microstructural development efforts guided by the principles deduced from studies of existing magnets will be necessary for the continued expansion of this field.

References

1. M. Sagawa, S. Fujimura, N. Togawa, H. Yamamoto, and Y. Matsuura, *J. Appl. Phys.* **55**, 2083 (1984).
2. J.J. Croat, J.F. Herbst, R.W. Lee, and F.E. Pinkerton, *J. Appl. Phys.* **55**, 2078 (1984).
3. K.H.J. Buschow, *J. Magn. Magn. Mater.* **80**, 1 (1989).
4. J.M.D. Coey, *J. Magn. Magn. Mater.* **80**, 9 (1989).
5. M.Q. Huang, B.M. Ma, L.Y. Zhang, W.E. Wallace, and S.G. Sankar, *J. Appl. Phys.* **67**, 4981 (1990).
6. L. Schultz and M. Katter, in: *Supermagnets, Hard Magnetic Materials*, edited by G.J. Long and F. Grandjean, (Kluwer Academic Publishers, Netherlands, 1991), p. 227.
7. D. Givord, H.S. Li, and J.M. Moreau, *Solid State Commun.* **50**, 497 (1984).
8. J.F. Herbst, J.J. Croat, F.E. Pinkerton, and W.B. Yelon, *Phys. Rev. B* **29**, 4176 (1984).
9. M. McCaig, *Permanent Magnets in Theory and practice*, 2nd. ed. (J. Wiley and Sons, New York, 1977).
10. K.J. Strnat, in: *Ferromagnetic Materials*, edited by E.P. Wohlfarth and K.H.J. Buschow, (North-Holland, Amsterdam, 1988), Vol. 4, p. 131.
11. K.H.J. Buschow, in: *Ferromagnetic Materials*, E.P. Wohlfarth and K.H.J. Buschow, eds. (North-Holland, Amsterdam, 1988) Vol. 4, p. 1.
12. K. Ohashi, Y. Tawara, R. Osugi, and M. Shimao, *J. Appl. Phys.* **64**, 5714 (1988).
13. D.B. de Mooij and K.H.J. Buschow, *J. Less-Common Met.* **142**, 349 (1988).
14. J.M.D. Coey and H. Sun, *J. Magn. Magn. Mater.* **87**, L251 (1990).
15. Y. Otani, D.P.F. Hurley, H. Sun, and J.M.D. Coey, *J. Appl. Phys.* **69**, 5584 (1991).
16. D. Fruchart and S. Miraglia, *J. Appl. Phys.* **69**, 5578 (1991).
17. J.M. Moreau, L. Paccard, J.P. Nozières, F.P. Missell, G. Schneider, and V. Villas-Boas, *J. Less-Common Met.* **163**, 245 (1990).
18. M. Katter, J. Wecker, and L. Schultz, *IEEE Trans. Magn.* **26**, 1379 (1990).
19. H. Ido, K. Konno, and T. Ito, S.F. Cheng, S.G. Sankar, and W.E. Wallace, *J. Appl. Phys.* **69**, 5551 (1991).
20. F. Pourarian, S.K. Malik, E.B. Boltich, S.G. Sankar, and W.E. Wallace, *IEEE Trans. Magn.* **25**, 3315 (1989).

21. T. Ojima, S. Tomizawa, T. Yoneyama, and T. Hori, *IEEE Trans. Magn.* **MAG-13**, 1317 (1977).
22. J.D. Livingston and D.L. Martin, *J. Appl. Phys.* **48**, 1350 (1977).
23. R.K. Mishra, G. Thomas, T. Yoneyama, A. Fukuno, and T. Ojima, *J. Appl. Phys.* **52**, 2517 (1981).
24. L. Rabenberg, R.K. Mishra, and G. Thomas, *J. Appl. Phys.* **53**, 2389 (1982).
25. J. Fidler and P. Skalicky, *J. Magn. Magn. Mater.* **27**, 127 (1982).
26. G.C. Hadjipanayis, R.C. Hazelton, K.R. Lawless, and, L.S. Horton, *IEEE Trans. Magn.* **MAG-18**, 1460 (1982).
27. Y. Fukui, T. Nishio, and Y. Iwama, *IEEE Trans. Magn.* **MAG-23**, 2705 (1987)
28. H. Nagel, *J. Appl. Phys.* **50**, 1026 (1979).
29. K.-D. Durst, H. Kronmüller, and W. Ervens, *Phys. Stat. Sol.* **A108**, 403 (1988).
30. K. Morimoto, M. Watanabe, and T. Takeshita, *IEEE Trans. Magn.* **MAG-23**, 2708 (1989).
31. K. Hiraga, M. Hirabayashi, and N. Ishigaki, *J. Microscopy* **142**, 201 (1986).
32. B. Zhang, J.R. Blachère, W.A. Soffa, and A.E. Ray, *J. Appl. Phys,* **64**, 5729 (1988).
33. D.S. Tsai, K.S. Ho, T.S. Chin, Y.H. Chang, and W.T. Tsai, *J. Appl. Phys.* **61**, 3772 (1987).
34. J. Fidler, *IEEE Trans. Magn.* **MAG-21**, 1955 (1985).
35. N.A. El-Masry, and H.H. Stadelmaier, *Mater. Letters* **3**, 405 (1985).
36. K. Hiraga, M. Hirabayashi, M. Sagawa, and Y. Matsuura, *Jap. J. Appl. Phys.* **24**, L30 (1985).
37. K. Hiraga, M. Hirabayashi, M. Sagawa, and Y. Matsuura, *Jap. J. Appl. Phys.* **24**, 699 (1985)
38. G.C. Hadjipanayis, Y.F. Tao, and K.R. Lawless, *IEEE Trans. Magn.* **MAG-22**, 1845 (1986).
39. R.K. Mishra, J.K. Chen, and G. Thomas, *J. Appl. Phys.* **59**, 2244 (1986).
40. T. Mizoguchi, I. Sakai, H. Niu, and K. Inomata, *IEEE Trans. Magn.* **MAG-22**, 919 (1986).
41. R. Ramesh, K.M. Krishnan, E. Goo, G. Thomas, M. Okada, and M. Homma, *J. Magn. Magn. Mater.* **54–57**, 363, (1986).
42. P. Schrey, *IEEE Trans. Magn.* **MAG-22**, 913 (1986).
43. M. Tokunaga, M. Tobise, N. Meguro, and H. Harada, *IEEE Trans. Magn.* **MAG-22**, 904 (1986).
44. D. Givord, P. Tenaud, and T. Viadieu, *J. Appl. Phys.* **60**, 3263 (1986).
45. R. Ramesh, J.K. Chen, and G. Thomas, *J. Appl. Phys.* **61**, 2993 (1987).
46. T. Mizoguchi, I. Sakai, H. Niu, *IEEE Trans. Magn.* **MAG-23**, 2281 (1987).
47. R.J. Pollard, P.J. Grundy, S.F.H. Parker, and D.G. Lord, *IEEE Trans. Magn.* **MAG-24**, 1626 (1988).
48. R.K. Mishra, and R.W. Lee, *Appl. Phys. Lett.* **48**, 733 (1986).
49. T.Y. Chu, L. Rabenberg, and R.K. Mishra, *J. Magn. Magn. Mater.* **84**, 88 (1990).
50. A. Hütten and P. Haasen, *J. Appl. Phys.* **61**, 3769 (1987).
51. G.C. Hadjipanayis, R.C. Dickenson, and K.R. Lawless, *J. Magn. Magn. Mater.* **54–57**, 557 (1986).
52. J.D. Livingston, *J. Appl. Phys.* **57**, 4137, (1985).
53. W.F. Brown, *Rev. Mod. Phys.* **17**, 15, (1945).
54. W.F. Brown, *J. Appl. Phys.* **30**, 1305, (1959).
55. E.C. Stoner, and E.P. Wohlfarth, *Phil. Trans. Roy. Soc. (London)* **240**, 599 (1948).
56. E.A. Nesbitt and H.J. Williams, *Phys. Rev.* **80**, 112, (1950).

57. L.F. Bates, D.J. Craik, and E.D. Issac, *Proc. Phys. Soc.* **79**, 970, (1962).

58. H. Kaneko, M. Homma, K. Nakamura, and M. Miura, *IEEE Trans. Magn.* **MAG-8**, 347 (1972).

59. G.Y. Chin, J.T. Plewes, and B.C. Wonsiewicz, *J. Appl. Phys.* **49**, 2046 (1978).

60. Y. Belli, M. Okada, G. Thomas, M. Homma, and H. Kaneko, *J. Appl. Phys.* **49**, 2049 (1978).

61. K.G. Kubarych, M. Okada, and G. Thomas, *Met. Trans.* **9A**, 1265 (1978).

62. B. Zhang and W.A. Soffa, *IEEE Trans. Magn.* **26**, 1388 (1990).

63. G.S. Kandaurova, L.G. Onoprienko, and N.I. Sokolovskaya. *Phys. Stat. Sol.* **A73**, 351 (1982).

64. P. Gaunt, *Phil. Mag.* **13**, 579 (1966).

65. G.C. Hadjipanayis and P. Gaunt, *J. Appl. Phys.* **50**, 2358 (1978).

66. H. Stablein, in: *Ferromagnetic Materials*, edited by E.P. Wohlfarth, (North-Holland, Amsterdam, (1982), Vol. 3, Chapter 7.

67. F. Kools, *J. Phys.* **46**, C6 (1985).

68. E.A. Nesbitt, R.H. Willens, R.C. Sherwood, F. Buehler, and J.H. Wernick, *Appl. Phys. Lett.* **12**, 361 (1968).

69. E.A. Nesbitt and J.H. Wernick, *Rare Earth Permanent Magnets*, (Academic Press, New York, 1973).

70. J.D. Livingston, *Phys. Stat. Sol.* **A18**, 579 (1973).

71. K.J. Strnat, D. Li, and H.F. Mildrum, *J. Appl. Phys.* **55**, 2100, (1984).

72. L. Rabenberg, R.K. Mishra, and G. Thomas, *Proc. of the 6th Intl. Workshop on Rare Earth-Permanent Magnets and Their Applications*, (Technical University of Vienna, Vienna, Austria, 1982).

73. C.E. Maury, L. Rabenberg, and C.H. Allibert, In Preparation; to be submitted to *J. Mater. Research.*

74. K.G. Knoch, G. Schneider, J. Fidler, E. Th. Henig, and H. Kronmüller, *IEEE Trans. Magn.* **25**, 3426 (1989).

75. J. Fidler, *Proc. of Intl. Conf. on Magnetism, Magnetic Materials and Applications*, (Havana, Cuba, May 1991).

76. J. Fidler, P. Skalicky, and F. Rothwarf, *IEEE Trans. Magn.* **MAG-19**, 2041 (1983).

77. F. Delannay, S. Derkaoui, and C.H. Allibert, *J. Less-Common Metals* **134**, 249 (1987).

78. C.E. Maury, L. Rabenberg, and C.H. Allibert, In preparation.

79. M.V. Satyanarayana, H. Fujii, and W.E. Wallace, *J. Appl. Phys.* **53**, 2374 (1982).

80. M. Sagawa, S. Fujirama, M. Togawa, H. Yamamoto, and Y. Matsuura, *J. Appl. Phys.* **55**, 2083 (1984).

81. M. Sagawa, S. Fujirama, H. Yamamoto, Y. Matsuura, and K. Hiraga, *IEEE Trans. Magn.* **MAG-20**, 1584 (1984).

82. D. Givord, J.M. Moreau, and P. Tenaud, *J. Less-Common Metals* **115**, L7 (1986).

83. H. Kronmüller, K.D. Durst, and M. Sagawa, *J. Magn. Magn. Mater.* **74**, 291 (1988).

84. M. Sagawa, S. Fujimura, H. Yamamoto, Y. Matsuura, *IEEE Trans. Magn.* **MAG-20**, 1584 (1984).

85. C.D. Fuerst, J.F. Herbst, E.A. Alson, *J. Magn. Magn. Mater.* **54–57**, 567 (1986).

86. Y. Xiao, S. Liu, H.F. Mildrum, K.J. Strnat, and A.E. Ray, *J. Appl. Phys.* **63**, 3516 (1988).

87. H. Yamamoto, S. Hirosawa, S. Fujimura, K. Tokuhara, H. Nagata, and M. Sagawa, *IEEE Trans. Magn.* **MAG-23**, 2100 (1987).

88. S.Z. Zhou, C. Guo, and Q. Hu, *J. Appl. Phys.* **63**, 3327 (1988).

89. D.R. Gauder, M.H. Froning, and R.J. White, *J. Appl. Phys.* **63**, 3522 (1988).

90. J. Fidler, K.G. Knoch, H. Kronmüller, and G. Schneider, *J. Mater. Research* **4**, 806 (1989).

91. M. Sagawa, P. Tenaud, F. Vial, and K. Hiraga, *IEEE Trans. Magn.* **26**, 1957, (1990).

92. S. Hirosawa, A. Hanaki, H. Tomizawa, S. Mino, and A. Hamamura, *IEEE Trans. Magn.* **26**, 1960 (1990).

93. R.W. Lee, E.G. Brewer, and N.A. Schaffel, *IEEE Trans. Magn.* **MAG-21**, 1958 (1985).

94. R.K. Mishra, *J. Appl. Phys.* **62**, 967 (1987).

95. T.Y. Chu, L. Rabenberg, and R.K. Mishra, *J. Appl. Phys.* **69**, 6046 (1991).

96. F.E. Pinkerton, *J. Magn. Magn. Mater.* **54–57**, 579 (1986).

97. F.E. Pinkerton, and D.J. Van Wingerden *J. Appl. Phys.* **60**, 3685 (1986).

98. O. Kubaschewski, in: *Iron Binary Phase Diagrams*, (Springer-Verlag, 1982), p. 104.

99. Y. Khan, *Proc. 11th Rare Earth Res. Conf.*, Vol II, 652 (1974).

100. K.H.J. Buschow and A.S. Van der Goot, *L. Less-Common Metals* **14**, 323 (1968).

101. A. Müller, *J. Appl. Phys.* **64**, 249 (1988).

102. F.E. Pinkerton and D.J. Van Wingerden, *IEEE Trans. Magn.* **25**, 3306 (1989).

103. J. Strzeszewski, Y.Z. Wang, E.W. Singleton, and G.C. Hadjipanayis, *IEEE Trans. Magn.* **25**, 3309 (1989).

104. M. Katter, J. Wecker, L. Schultz, and R. Grössinger, *Appl. Phys. Lett.* **56**, 1377 (1990).

105. L. Schultz, K. Schnitzke, and J. Wecker, *J. Magn. Magn. Mater.* **80**, 115 (1989).

106. L. Schultz, K. Schnitzke, J. Wecker, and M. Katter, *IEEE Trans. Magn.* **26**, 1373 (1990).

107. C. Koestler, L. Schultz, and G. Thomas, *J. Appl. Phys.* **67**, 2532 (1990).

108. G. Schneider, F.J.C. Langraf, and F.P. Missell, J. *Less-Common Metals* **153**, 169 (1989).

Inductance Spectroscopy

R. Valenzuela

Instituto de Investigaciones en Materiales
Universidad Nacional Autónoma de Mexico
Apartado Postal 70-360
04510 Mexico, D.F.
MEXICO

Abstract

The basic principles and advantages of impedance spectroscopy are briefly described, and some recent results on dielectric and ferroelectric mateials are reviewed. The application of this analysis technique to magnetic materials is examined. A brief summary of recent results is also given. Impedance spectroscopy, or "Inductance spectroscopy" for magnetic materials is becoming a general characterization method for a wide range of materials.

I. Introduction

Measurement of electrical and magnetic properties of materials as a function of frequency has been performed since a very long time. The determination of the frequency behavior of materials is extremely important for applications and can be very useful in the basic investigation of polarization processes.

Recently, the availability of impedance analyzers controlled by microcomputers has allowed the development of very rapid measuring and analysing techniques for these studies. In turn, the basic significance of these measurements is being recognized.

The main advantage of measurements at many frequencies is that the several polarization mechanisms in the material can be resolved, since they possess different time-constants. At low frequencies, all the polarization mechanisms will be contributing to the global response of the sample, but as frequency increases, only the processes capable to follow the exciting field (the processes with smaller time-constant) will remain active.

The analysis of the results can be advantageously done by means of the complex impedance formalisms. The complex impedance plots and the related formalisms,

Advanced Topics in Materials Science and Engineering, Edited by
J.L. Morán-López and J.M. Sanchez, Plenum Press, New York, 1993

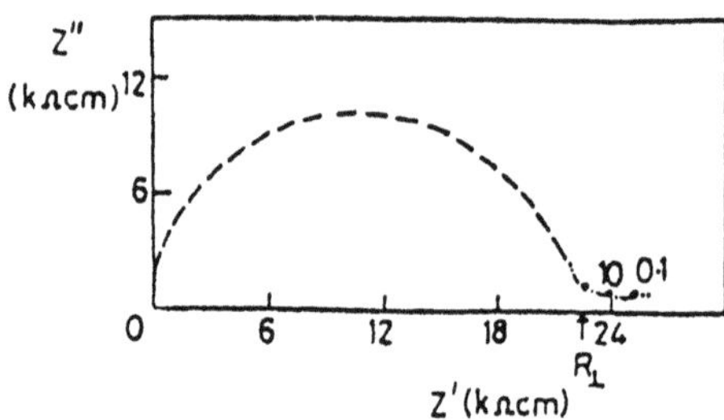

Figure 1. Complex impedance plot of the LiTaO$_3$ single crystal in the a direction.

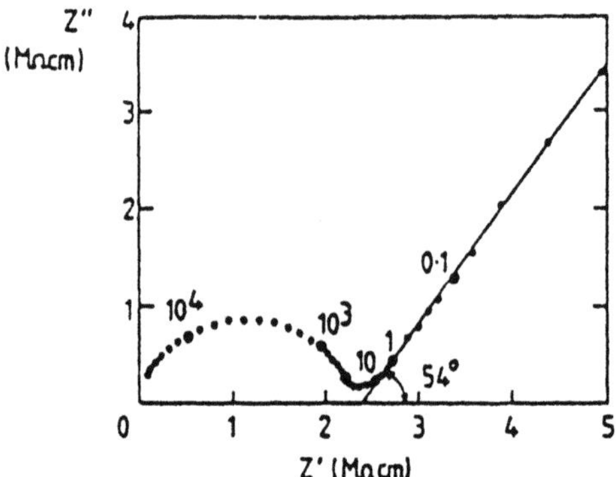

Figure 2. Complex impedance plot in the c direction.

admittance, permittivity and modulus,[1] as well as their spectroscopic representations can lead to a very clear picture of the frequency behaviour of the sample. The character of the dispersions (the manner in which a polarization mechanisms becomes unable to follow the excitation field) can be easily recognized for the relaxation and resonance cases. When experiments are performed as a function of temperature, an Arrhenius plot (logarithm of the conductivity as a function of the reciprocal temperature)can be constructed, from which the activation energy for the conductivity mechanism can be evaluated.

In many simple cases, the complex impedance analysis can lead to an equivalent circuit, which represents the sample behaviour. The significant point here is that the equivalent circuit elements can be associated with the actual physical parameters of the material, such as grain boundary resistance, ionic conductivity, grain (or bulk) ferroelectric capacitance, rotational permeability and many others. The material can be characterized.

II. Electrical Properties

In this brief review of the study of electrical properties by impedance spectroscopy, two examples wil be considered: the analysis of the conductivity of a single crystal of lithium tantalate, LiTaO$_3$,[2] and the separation of the contribution from grains and grain boundaries in polycrystalline barium titanate, BaTiO$_3$.[3,4]

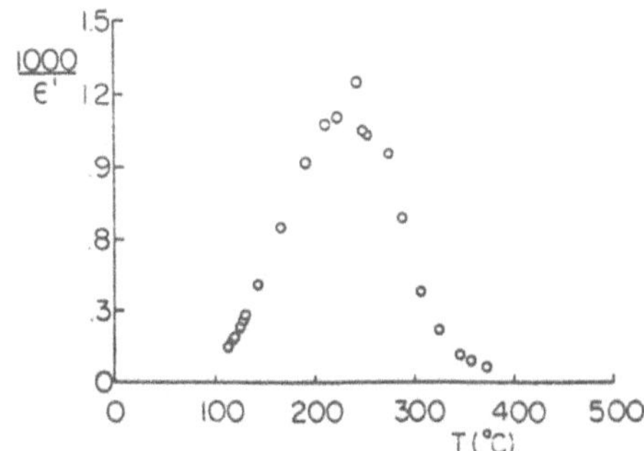

Figure 3. Electrical permittivity of polycrystalline barium titanate measured at 1 kHz, as a function of the inverse of temperature.

Lithium tantalate is a ferroelectric material with moderate ionic and electronic conductivity at high temperatures. The availability of a good quality, single crystal with considerable dimensions made possible the study of the conductivity and capacitance along the a and c crystallographic directions. The obtained complex impedance plots are shown in Figs. 1 and 2. The complex impedance is usually expressed as: $Z^* = Z' + jZ''$, where Z' is the real part (in phase) and Z'' the imaginary part (out of phase) of the total impedance, and j is $\sqrt{-1}$. A simple parallel RC circuit gives rise to a semicircle in the impedance plot, as shown in Fig. 1; C is the capacitance of the crystal in the a direction and R is the resistance to the electronic flux in that direction.

The complex impedance plot along the c direction showed other features, Fig. 2. In addition to the semicircle which represents the parallel resistance and the capacitance of the crystal, a "spike " appeared in the low frequency range. This spike is characteristic of a series capacitance, which occurs whenever a blocking mechanism prevents the flow of charges. It is commonly observed in the case of ionic carriers which cannot diffuse through the electrode material. This material exhibits thus a very anisotropic ionic conductivity, practically only along the c direction, while the conductivity in the a direction has an electronic character. the value of C, the parallel capacitance, revealed also that the ferroelectric properties are more intense in the c direction.

Polycrystalline materials are much more common than single crystals, particularly in the case of ceramics. During the last decade, the development of very sensitive characterization tools has revealed the importance of surfaces and interfaces, and the complexity of polycrystalline microstructures. Impedance spectroscopy can significantly contribute, since it allows a resolution of the various contributions to the total polarization. In particular, the contributions of the grains can be identified and separated from that of the grain boundaries.

The second example to be illustrated is based on electrical measurements of polycrystalline barium titanate,[3,4] which is a very widely used ferroelectric material with a Curie temperature about 120°C. Above this temperature, it is a paraelectric material which is expected to follow the Curie-Weiss law, exhibiting a linear relationship between the inverse of the electrical permittivity as a function of temperature. The electrical permittivity is usually determined from capacitance measurements at different temperatures, at 1 kHz. The experimentally observed relation, Fig. 3, is far from being a linear function.

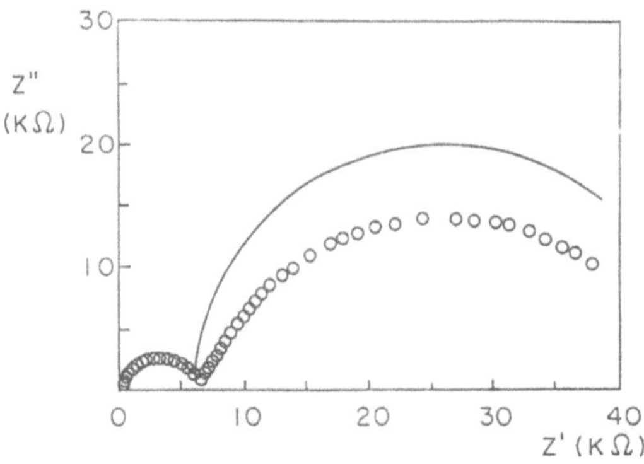

Figure 4. Complex impedance plot of barium titanate for frequencies in the range 5 Hz–13 MHz.

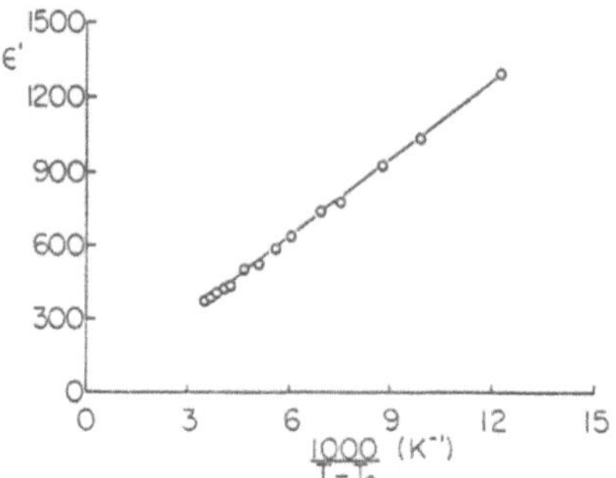

Figure 5. Inverse of the permittivity as calculated from the grain semicircle capacitance (Fig. 4), showing a Curie-Weiss relationship.

Measurements at different frequencies and an analysis of the complex impedance leads to a clear interpretation. When grain boundaries are more resistive than grains, both contributions can be easily resolved. The equivalent circuit is formed by two parallel RC branches in series, one for the resistance and capacitance of the grains, and the other for the corresponding elements of the grain boundaries, Fig. 4, where the larger semicircle corresponds to the grain boundaries. The resistances are simple the semicircle diameter in each case, and the capacitances can be extracted by the fact that for the highest value in each semicircle, the relation $R = \omega C$ holds. If now the permittivity calculated from the capacitance of the smaller semicircle, for several temperatures, is plotted in the form of the Curie-Weiss law, Fig. 5, it can be seen that a linear relationship is accurately obtained. The measurements at one fixed frequency can have no physical meaning since it is not possible to know if it is produced by the grains, the grain boundaries or the electrodes.

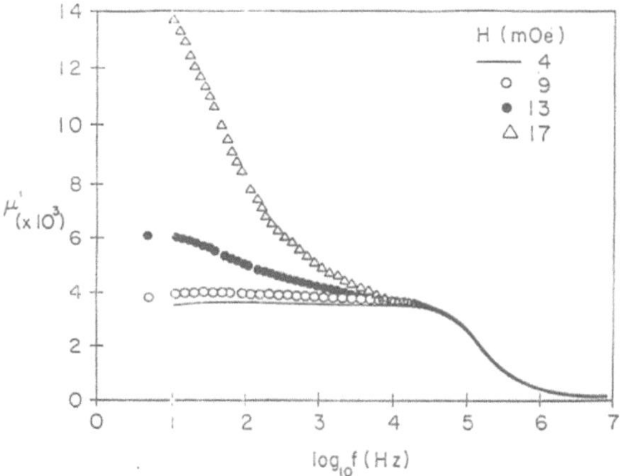

Figure 6. Real part of the magnetic permeability of an amorphous ribbon Vitrovac$^{®}$ 6025 as a function of the frequency. The amplitude of the applied is indicated for each curve.

III. Magnetic Properties

Magnetic properties as a function of frequency have been measured for many years. In some cases, complex impedance and permeability have been determined. However, to our knowledge, no attempt has been done to propose equivalent circuits, or associate specific magnetization mechanisms with complex formalism features.

In the case of ferro and ferrimagnetic materials, it is more clear to use the complex permeability instead of the complex impedance. They are related by:

$$\mu* = \mu' + j\mu" = (k/\omega)Z^*, \tag{1}$$

μ' and $\mu"$ are the real and imaginary permeabilities respectively, k is a geometrical factor and ω the angular frequency, equal to $2\pi f$. It is very instructive to measure the permeability as a function of frequency, Fig. 6, where results obtained[5] on an amorphous ribbon "Vitrovac$^{®}$ 6025" are shown for several applied fields. It is clear that the various contributions disappear as frequency increases. It is important to note several features. First, for low frequencies and fields above $\sim$ 4 mOe, the permeability is a function of the applied field, and decreases rapidly as the frequency increases. For fields lower than 4 mOe, the permeability is independent of the field; for frequencies above $\sim$ 10 kHz, the permeability is independent of the applied field, and finally, even at 10 MHz, a small value remains.

To account for these results, magnetic structure and basic magnetization mechanisms have to be examined. In these very soft ferromagnetic materials, a domain structure is formed to eliminate the magnetostatic contribution; three basic mechanisms[6] are available to change the magnetization state of the sample under the action of an external applied field: rotation of individual spins within each domain, reversible movement of domain walls and irreversible displacements of walls. The difference in their time-constants can be easily induced. Spin rotation is damped by spin-orbit coupling; reversible domain wall bowing involves the collective movement of pinned wall spins over small volume, and irreversible displacements of domain walls is the

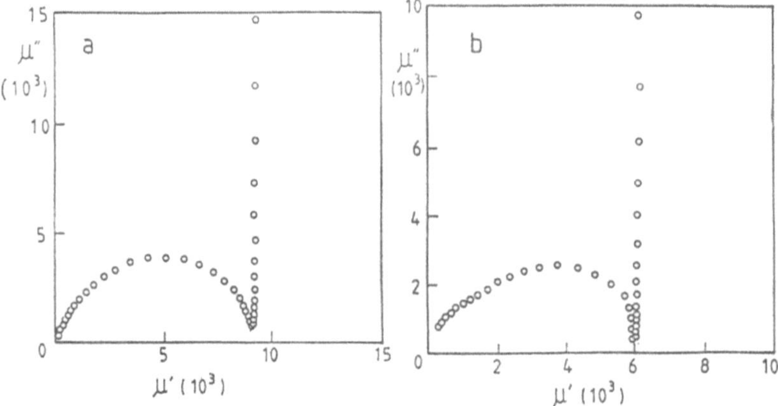

Figure 7. Complex permeability plot of the sample with an applied field below the critical value: a) Vitrovac amorphous ribbon, b) nickel-zinc ferrite.

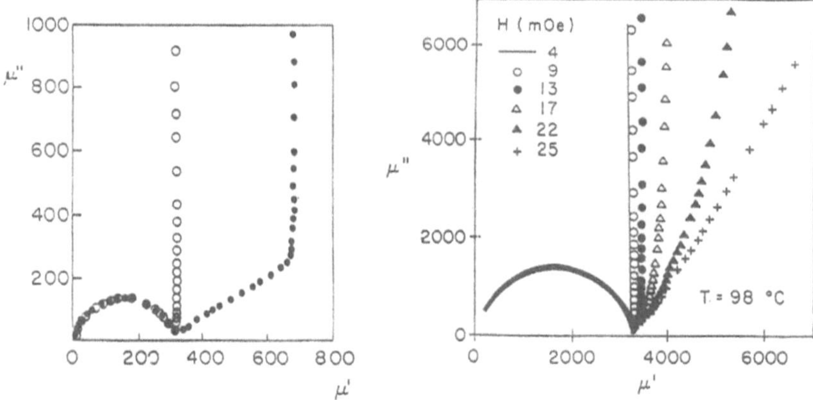

Figure 8. Complex permeability plots for applied fields higher than the critical value; a) Vitrovac and b) Metglas amorphous ribbons.

most complex mechanisms, involving bowing, depinning, displacement, pinning and flattening of the domain wall, four times per field period.[6] Obviously, the latter mechanism is the slowest, followed by wall bowing; spin rotation is expected to remain in operation even at very high frequencies, as observed in ferromagnetic resonance experiments.[7]

The observed dispersions can hence be associated with hysteretical, irreversible domain wall displacements, reversible bowing, and spin rotation, from the low to the high frequencies, respectively. These results agree also with the fact that domain wall reversible bowing and spin rotation are both independent of the applied field (for fields below the critical value), while magnetization produced by the hysteresis mechanism is larger and a function of the applied field.

The complex permeability representation provides a more clear image of magnetization phenomena than complex impedance.[8] For fields lower than the critical value, where only reversible bowing and spin rotation (which is also reversible as a function of the applied field) operate, the complex permeability plot ($\mu*$) shows a vertical spike and a semicircle, Fig. 7. An increase in the field above the critical value produces a change in the slope of the spike, as well as a displacement[8] toward higher real permeability values.

The investigation of the equivalent circuits which can represent the sample behaviour is straightforward. For the low applied field range, the equivalent circuit which models the observed behaviour is simply a parallel R_B L_B arrangement for the bowing mechanism, with a series inductance L_R representing the spin rotation magnetization. A small series resistance R_w has to be added to account for the coil resistance. The hysteresis mechanism is much more complex to model; the impedance element which seems to provide a convenient representation of this phenomena is a Warburg impedance.[9]

The following step in the process, which is to associate the equivalent circuit elements with the physical parameters of the sample, is direct for the inductances. The magnetization produced by spin rotation, wall bowing and wall displacement correspond to the series, parallel and Warburg inductances, respectively. The resistance elements are related to the losses in each of the processes; they are more difficult to associate with explicit parameters of the material. The domain wall bowing equivalent resistance has been associated[10] with the so-called viscous damping parameter and the effective mass[11] of the wall. The resistive elements in the Warburg impedance representing the hysteresis mechanism have not been related to the dissipation phenomena.

IV. Summary

Impedance spectroscopy provides a non destructive, rapid and precise method to analysis the electrical and magnetic properties of materials. This technique provides a clear resolution of the electrical and magnetic polarization phenomena in virtually any type of material, single crystal or polycrystalline aggregate. Some of the important surface-related properties whose significance has been recognized in the last few years, such as grain boundary properties, can conveniently be investigated. Recent developments in the dynamic magnetic properties of ceramic and amorphous materials have also been made.

Acknowledgements

The author thanks the European Commission for a Research Fellowship, as well as DGAPA-UNAM for a sabbatical leave.

References

1. J.R. Macdonald, *Impedance Spectroscopy*, (J. Wiley, N.Y., 1987).

2. D.C. Sinclair and A.R. West, *Phys. Rev. B* **39**, 486 (2989).

3. R. Flores-Ramírez, A. Huanosta, E. Amano, R. Valenzuela, and A.R. West, *Ferroelectrics* **99**, 195 (1989).

4. R. Flores-Ramírez, E. Amano, and R. Valenzuela, *J. Mat. Sci. Lett.* **10**, 951 (1991).

5. J.T.S. Irvine, E. Amano, and R. Valenzuela, *J. Magn. Magn. Mater.*, (to be published).
6. M.A. Escobar,. L.F. Magaña, and R. Valenzuela, *J. Appl. Phys.* **54**, 5935 (1983).
7. J. Smit and H.P.J. Wijn, *Ferrites*, (Wiley, N.Y., 1959) p. 158.
8. J.T.S. Irvine, E. Amano, and R. Valenzuela. *Mat. Sci. and Engin.* **A131**, 140 (1991).
9. J.T.S. Irvine, A.R. West, E. Amano, A. Huanosta, and R. Valenzuela, *Solid State Ionics* **40/41**, 220 (1990).
10. R. Valenzuela, J.T.S. Irvine, and A.R. West, *J. Magn. Magn. Mater*, (In press).
11. B.D. Cullity, *Magnetic Materials* (Addison-Wesley, 1972), p. 446.

Electron Tunneling in Superconducting Ba-K-Bi-O

R. Escudero

Instituto de Investigaciones en Materiales
Universidad Nacional Autónoma de México
Apartado Postal 70-360
04510 México, D.F.
MEXICO

Abstract

Some of the important aspects related to the superconducting state of $Ba_{1-x}K_xBiO_3$ are discussed. The main topic considered is electron tunneling and its implication in the understanding of the basic physical mechanisms of this system. Some recent experiments are discussed which emphasize the importance of electron-phonon interaction in these compounds, and may identify them as BCS superconductors between the intermediate to strong coupling limits. Nevertheless, some problems remain to be solved in these $BaBiO_3$-based compounds.

I. Introduction

The discovery of superconductivity in $Ba_{1-x}K_xBiO_3$ provides the possibility of comparing its characteristics to conventional phonon mediated superconductors and to the new Cu-based, high-T_c materials.[1] This compound has received much attention because it is the first oxide superconductor without Cu in which the transition temperature, T_c, is of the order of 30 K, well above the best *A-15* strong coupled superconductors. The crystal structure of this compound is a simple cubic perovskite formed from corner-shared BiO_6 octahedra with Ba and K on the cell origin.[2,3]

This three dimensional structure is isotropic with no magnetic effects, implying that the pairing mechanism might be of different nature to that of the *Cu-O* based compounds, where low dimensional behavior and/or magnetism could play a fundamental role in determining their abnormal high transition temperatures and for the

Advanced Topics in Materials Science and Engineering, Edited by
J.L. Morán-López and J.M. Sanchez, Plenum Press, New York, 1993

anomalous normal state properties. The absence of magnetic order in both $BaBiO_3$ and $Ba_{1-x}(K, Rb)_x BiO_3$ compounds is constrasting to the competition between antiferromagnetism and superconductivity in the cuprates.

From another point of view, Uemura[4] has found that it is possible to classify superconductors according to the value of the ratio T_c/T_F where T_F is the Fermi temperature, and T_c is the critical temperature. Uemura found that $T_c/T_F = 1/10$ to $\sim 1/100$ is high for materials such as cuprates, BKBO, organic superconductors including C_{60}'s, Chevrel phases and heavy-fermions, whereas ordinary BCS superconductors have ratios of the order of $T_c/T_F \leq 1/1000$. The point that Uemura suggests is the possibility that these systems may be classified as a kind of superconductors with exotic properties, different to the BCS systems. These kind of superconductors have features that differ from the BCS's systems, $e.g.$ high H_{c2}, small ξ, proximity to metal-insulator transition, spin and/or charge instabilities, highly correlated electronic behavior, and electrical resistivities close to the Mott limit. $Ba_{1-x}K_x BiO_3$ compounds together with $Ba_{1-x}Rb_x BiO_3$, belong to the family of compounds $BaBiO_3$, $BaPbO_3$, $BaPb_{1-x}Sb_x O_3$, and $BaPb_{1-x}Bi_x O_3$ all of which have a perovskite crystalline structure. All the above ceramics are superconductors, except for $BaBiO_3$, showing transition temperatures ranging from 0.5 K to 30 K, for $BaPbO_3$ to $Ba_{1-x}K_x BiO_3$ respectively.[5-7] An interesting feature of all these $BaBiO_3$-based superconductors, is that they have some peculiar characteristics which do not exactly fit the BCS simple theory to explain the transition temperatures, $i.e.$ their low electron density, Coulomb repulsion seems nearly absent, and the highest transition temperature at doping levels in the limit of the metal-insulator transition. Also the insulating state is intriguing and difficult to explain. It is improbable that a charge density wave induces the insulating state due to the nesting of the Fermi surface, because over a wide range of composition the crystal structure is nearly cubic, and implying a total nesting of the three dimensional Fermi surface to obtain the insulating behavior. In the case of the formation of charge or spin instability, both will compete to oppose the formation of the superconducting state. As we will see later in this chapter, the isotope effect has a considerable value close to the BCS prediction that again remarks the importance of the electron-phonon interaction. On the other hand the compound is diamagnetic over the entire range of compositions; no magnetism exist, nor Mott-type insulator behavior; then the following questions may arise: what is the origen of the insulating state?, why is the transition temperatures so high despite the low electronic density?, why is the superconducting state close to the boundary with the insulator state?, do both bismuthates and cuprates present a new state of matter? and lastly, are these ceramics in a way similar to the cuprates?. All these kind of physical properties deserve further investigation; in particular the study of the normal properties will bring light about the anomalous superconducting behavior in these interesting materials.

II. Gap Spectroscopies and Electron Tunneling

In a superconductor the energy gap Δ is one of the most characteristic features of the electronic condensate; it arises as a consequence of the many body interactions between quasiparticles that are scattered coherently in a region of the k space which is centered around the Fermi surface within a width $k_B T_c$. Due to its nature, the evolution with temperature, size, and value of the ratio $2\Delta/k_B T_c$ might give enough information that can be directly related to the microscopic processes that form the pairing condensate.

Several methods are presently used to measure and study the energy gap of a superconductor, *e.g.* infrared reflectivity, Raman scattering, ultraviolet angle-resolved photoemission spectroscopy, and electron tunneling. Each one is important by itself and may provide information related to the superconducting state as well as information concerning the normal state. For details related to those techniques, the reader is addressed to the vast literature in the field; see for example references (8–11) at the end of this chapter. In order to establish the historical background on this subject it is worth mentioning that the first experimental confirmation and measurement of the energy gap in a superconductor came from infrared spectroscopy. The experiment was realized by Glover and Tinkham[12] in the late 50's. More recently, Raman and photoemission spectroscopies have demonstrated that both are very powerful techniques to probe the nature of the interaction between electrons and phonons, or any other kind of elementary excitation in solids, as well as to study the basic mechanisms of the superconducting and the normal state.

On the other hand, electron tunneling is the only direct technique to study the superconducting energy gap, and it is perhaps the most sensitive probe for studying and analysing the superconducting state. This spectroscopic technique senses the microscopic processes which form the superconducting condensate, and can give information to completely characterize many of the microscopic processes that occur in the formation of a superconductor. The information that can be extracted from tunneling experiments is the temperature dependence of the energy gap, the phonon density of states, the coupling function $\alpha^2(\omega)F(\omega)$, where $\alpha^2(\omega)$ and $F(\omega)$ are, respectively, the functions that give the strength of the electron-electron interaction and the density of phonons in the material under study. It is worthwhile noticing that the coupling function or the weighted distribution function of phonons, $\alpha^2(\omega)F(\omega)$, plays a central role in the strong coupling superconductivity theory. Every piece of information related with the electrons, phonons, and the electron-phonon interaction, and thus important for superconductivity, is included here. A material differs from another according to the value of $\alpha^2(\omega)F(\omega)$. It is also important to remark that the distribution function $F(\omega)$ may have different physical meanings; in Eliashberg theory the only basic concept that needs to be introduced is that the distribution function must describe a bosonic kind of object, *e.g.* plasmons, polarons, bipolarons, exitons, holons, demons, etc.

Once the above information has been collected, it can be used to solve the two coupled Eliashberg equations. From there, all the thermodynamic information that gives the physical behavior of one particular material can be extracted, *e.g.* the critical magnetic field $H_c(T)$, the deviation function $D(t)$, the specific heat jump, the low temperature energy gap, the zero temperature energy gap, etc.

III. The Electron-Phonon Interaction

Early oxygen isotopic effect measurements in the *Ba-K-Bi-O* system have led to large values of $\alpha_{ox}{}^*$ implying a phonon-mediated pairing mechanism.[13] Recent work by Loong *et al.*[14,15] gives a clear indication that $Ba_{0.6}K_{0.4}BiO_3$ is indeed a weak-coupling

---

*   Note that α_{ox} arises from the BCS theory and the only permitted value must be $1/2$; thus $T_C = M^{-1/2}$. However, according to strong coupling theory $\alpha = 1/2\{1 - [\mu^*/(\lambda^* - \mu^*)]^2\}$, with μ^* the renormalized Coulomb interaction, given by the expression $\mu^* = \mu/[1 + \mu\ln(E_F/E_D)]$ and $\lambda^* = \lambda/(1 + \lambda)$.

superconductor with an isotope-effect exponent of 0.42 ± 0.05, with large matrix elements coupling the electrons to high-energy phonons involving oxygen vibrations.

Raman spectroscopy shows an optical phonon at 348 cm^{-1} (43 meV) with a Fano lineshape indicative of strong coupling between this phonon and the electron spectrum. For a non-superconducting sample of the material at a lower potassium content, the same phonon is not seen to be coupled to the electronic states.[16] Superconducting energy gap measurements, using infrared spectroscopy[17] and tunneling measurements,[18-24] are consistent with a coupling constant $2\Delta/k_B T_c$ in the range of 3.5 to 4.4. To finish this section, only rest to pointing out that all the above arguments imply, according to BCS theory, that electron-phonon interaction is one of the first contributing mechanisms in these $Ba_{1-x}(K, Rb)_x BiO_3$ compounds.

IV. Tunneling Measurements

In tunneling studies many of the effects observed in the current *vs.* voltage characteristics are so small that it is usually convenient to study the dynamic conductance $\sigma(V) = dI/dV$, the differential resistance $\sigma^{-1}(V)$ as well as its derivative $d\sigma/dV$. The wealth of information on the tunneling process itself, and on the microscopic excitation spectrum that can be obtained from these tunneling measurements, was in the past one of the most reliable pieces of information in the study of conventional superconductors.

With the discovery of new superconducting materials, workers in the field immediately tried to use this powerful tool to study the superconducting state. However, it was soon realized that reliable tunneling data in these new ceramic would be a very difficult task for various reasons, but mainly, due to the problem of making reproducible tunnel junctions. The particular difficulty attributable to the nature of the parameters involved in the high T_c superconductors is the small coherence length, ξ_0, which converts the technique from a bulk technique to a surface technique. The consequence of this change is that the experimental data depend on the behavior and characteristics of the surface. For example, changes on the surface due to possible variation of the oxygen stoichiometry may have effects on the superconducting characteristics of the surface that consequently will be reflected on the tunneling data. Degradation processes, defects, contamination, granularity, etc., will also bring changes on the superconducting properties, that again will modify the tunneling results. Nevertheless, the second and most serious problem concerns to the physical interpretation of the tunneling data, assuming of course that it is correct, due to the fact that actually there is no theoretical model to obtain or guide the interpretation of the experimental information. Furthermore, many experimentalist believe that this information must be understood in terms of the BCS model without thinking that, perhaps, nature is trying to indicate us a different kind of physical behavior for these exotic new materials. So, at this point, a warning should be given related to the interpretation that must be taken very seriously.

The tunneling techniques used in the past to fabricate tunneling devices with conventional materials were in general, made using evaporated thin films on a glass substrate. This consists of depositing a metallic film first, or electrode, typically aluminum metal, of a thickness of the order of 1000Å to 3000Å. The surface of this film is oxidized in a well controlled manner, in such a way that the thickness of the oxide is maintained in the range of 10Å to 50Å. The second step in the formation of the device consists of evaporating the second electrode on top of the first film, trying to maintain the junction area as small as possible. The purpose of the small area is to have minimum capacitance and to maintain the tunnel current distribution as homogeneous as possible.

This method gives the normal procedure to fabricate tunnel junctions using metals, *e.g.* lead, tin, indium, etc., and was used for the first time, many years ago, by I. Giaever. The implementation of this technique for alloys, or *A-15* superconductors, needs a little more care on the evaporating procedures but, however, gives in general good results. For materials such as single crystals or compounds which are difficult to evaporate, or where the stoichiometry and the characteristics of the compound can be changed with the evaporation, and therefore the properties, other type of tunneling junctions have been attempted. One type frequently used is the well known point contact tunnel junction. It consists of a metallic wire with a very fine tip in close proximity with the sample to be studied. The idea is to obtain enough tunneling current by regulating the distance between the tip and the sample. It is worth mentioning that this procedure is the basic idea of the tunneling microscope where one can control with exquisite precision the distance between the tip and the sample.

To end this description of the construction of tunneling devices, only rest to say that today many type of tunneling techniques have been implemented to study the new ceramic superconductors with relative success. An interesting description of the state of the art on tunnelig studies is the review by Hasegawa *et al.*[24]

In this chapter we will describe tunneling studies using a variation of the normal procedure. We used two different methods to fabricate the tunnel junctions; in both cases we use polycrystalline ceramic pellets. The first type of tunnel junction was prepared in the following way; we evaporated on one of the surfaces of the sample, which was previously cleaned with an etching solution, narrow strips films of Sn with thickness of the order of 2000Å–5000Å. The tunnel junction is formed by this electrode and the ceramic block, with the insulating layer being the native barrier formed naturally over the surface of the ceramic due to exposure to the atmosphere of the laboratory. The second type of tunnel junction was made using again a block of the ceramic sample. The insulating barrier was also the native oxide that grows on the surface, and the second electrode was formed using a very thin gold wire (the diameter of the wire was 5 μm) touching the surface of the sample. This last variation of the point contact method has proved to be good and reliable method to make tunneling junctions in these ceramics superconductors. The junctions were prepared using two different samples of bulk ceramic material with the optimal composition $Ba_{0.6}K_{0.4}BiO_3$. The transition temperatures were 21 K and 29.8 K. The melt processed samples were prepared as reported by Hinks *et al.*[3] The tunnel junctions were, as described before, metal-insulator-superconductor (MIS) junctions.[19−21] In one type, type (I), a pellet of high density material was encapsulated in high vacumm epoxy. Once the epoxy had cured, the surface of the pellet was polished and cleaned with an etching solution. The junctions were completed by evaporating a thin strip of Sn as the counterelectrode. The junctions dimensions were approximately 0.1×1.0 mm^2 with differential resistances (zero bias) between 20 to 100 Ohms at room temperature. Another type of junction, type (II), was made using a pellet of high density material. The surface was cleaned with an etching solution, left in the atmosphere of the laboratory for different periods of time and a 5 μm gold wire was stretched tight over the sample to touch the surface. The junctions dimensions were less than 5 μ $\times$ 0.5 mm. The differential resistances at zero bias were from 20 to 1000 Ohms at room temperature. The tunnel junctions were measured at different temperatures between 1.7 and 300 K. The measurements on the tunnel junctions were done using the conventional lock-in

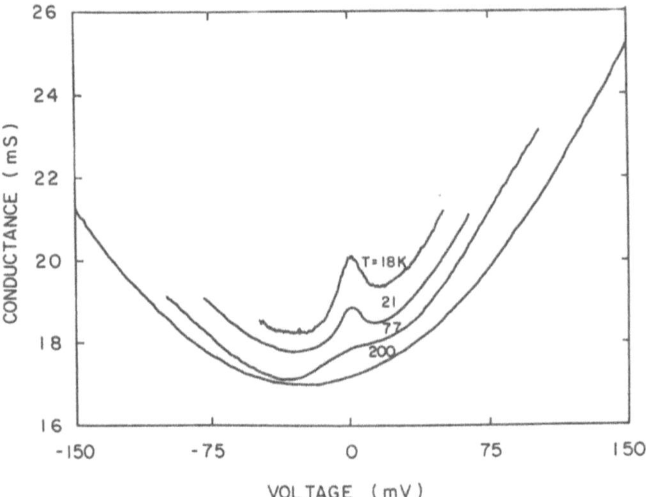

Figure 1. Conductance *vs.* voltage curves in the normal state for a tunnel junctions at different temperatures. The curves are characteristic of a type (I) junction. The curves are vertically shifted for clarity. T_c for this sample was 21 K.

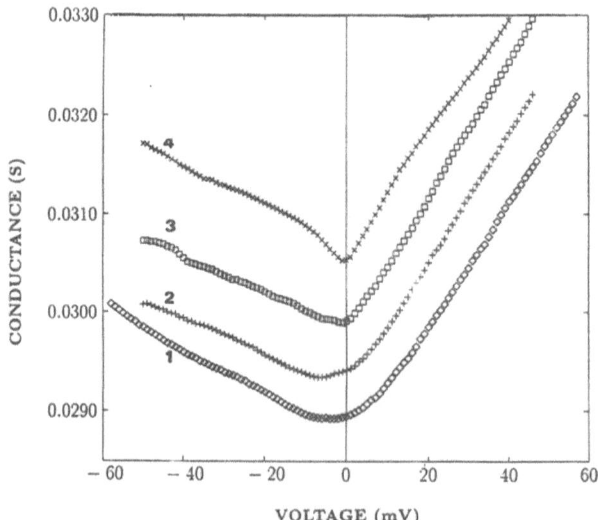

Figure 2. Dynamic conductance *vs.* voltage curves for a tunnel junction in the normal state at temperatures close to T_c. The curves show the characteristic behavior of a type (II) junction. The transition temperature of the ceramic pellet used to made the junction was 29.8 K. The curves are numbered indicating the temperature: 1, 2, 3, and 4 correspond to 35, 28, 27, and 26 K.

modulation technique. Figs. 1 and 2 show the conductance *vs.* voltage characteristics in the normal state, for the two type of tunnel junctions described before.

Fig. 1 shows the conductance *vs.* voltage characteristics of a typical tunnel junction of $Ba_{0.6}K_{0.4}BiO_3$-I-Sn taken at different temperatures above T_c. The curve at 200 K is the normal state. One notes a more or less parabolic shape for biases up to several hundred millivolts, and also a small negative offset which can be explained assuming a different average barrier height at each side of the insulating barrier, as proposed by Brinkman *et al.*[25] At 77 K and below, the conductance shows an extra feature that grows as the temperature decreases. The parabolic background remains but an increase at zero bias in the conductance curve becomes more and more noticeable as the temperature decreases. This zero bias anomaly (ZBA) may be due to different mechanisms some intrinsec to the device and others not. However, due to the type of physical processes that occur in the metals and/or the ceramic under study, these might be the following: Firstly, to a combination of tunneling and metallic contact. This effect may be due to the fact that at low temperatures the total scattering length l of the electrons is determined by the elastic scattering with the impurities l_{imp} and the inelastic scattering with the phonons l_{ep}. Then, according to Matthiessen's rule, we can write the total scattering length l as: $l^{-1} = l_{imp}^{-1} + l_{ep}^{-1}$ implying therefore, that as the inelastic scattering with phonons diminish, the conductivity tend to increase. A second type of Zero Bias Anomaly[26] could be related to different tunneling channels opening by inelastic processes very close to the Fermi surface and only existing for low energies. The last explanation might involve other type of physical processes, that could be related with an anomalous variations of the density of electrons close to the Fermi energy in systems with CDW or SDW.[27] Recent theoretical work by Machida and Nasu have modeled a similar behavior in systems with interacting charge density waves and superconductivity.[27] Going back to the description of Fig. 1, we must also note that when measuring the transition temperature by the resistivity method the transition temperature occurs at 21 K. However the tunneling conductance curve taken at 21 K does not show any feature that could be indicative of the opening of an energy gap (a small feature is observed at around and below 18 K).[19−21] One logical explanation for this behavior is the well known fact that tunneling senses regions in the sample to a depth related to the coherence length, in contrast to the resistances measurements, which senses the bulk percolative length.

Fig. 2 shows another set of data that is in clear contrast to the data of Fig. 1. Here the tunneling junction was made using the second type; the junction area is smaller than the first type and also, very probably, the current distribution is more homogeneous. The transition temperature of the pellet was higher than the first one, with $T_c = 29.8$ K. In particular we show in this figure the dynamic conductance *vs.* voltage curves above and below of T_c. The normal state presents a behavior rather different to the first case. The parabolic background does not exist, and a linear rise of the conductance with increased voltage is observed. It is also observed that the linear background is highly asymmetric for positive and negative signs of the bias voltage. Curve number 1 of this figure, shows the characteristic at 35 K. The parabolic background here is sligtlhy more definite that in the rest of the curves that compose Fig. 2. Curves 2, 3, and 4 show more clearly the linear background and, in particular, curve 4 shows a clear modification close to zero bias indicating that the energy gap is arising. The reason we are not observing the feature of the gap at the transition temperature ($T_c = 29.8$ K) is due, very probably, to thermal smearing and to the fact that the transition temperature on the surface is lower than in the bulk. Other physical explanation may be possible: thermal noise induces thermodynamic

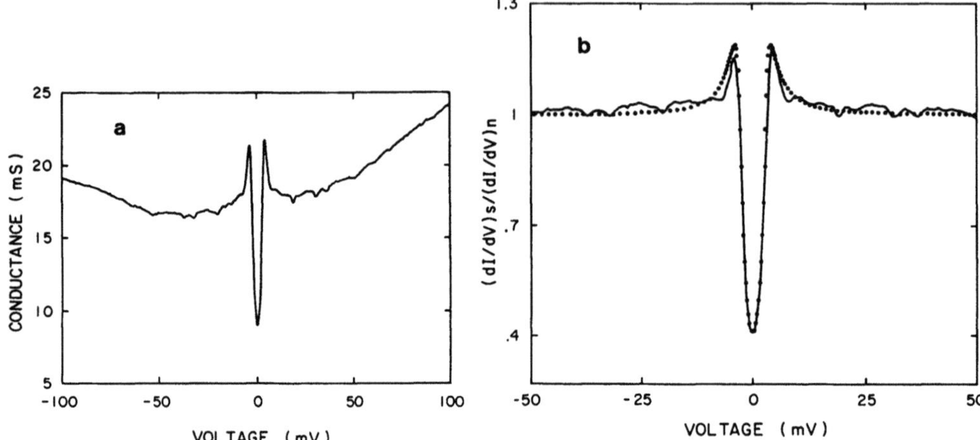

Figure 3. Conductance characteristic of a tunnel junctions of $Ba_{0.6}K_{0.4}BiO_3$-I-Sn. a) Measured at a temperature of 4 K. b) The curve is normalized with the curve at 18 K. The dotted curve is the BCS density of states smeared with a Γ parameter. The measured parameters are: $\Delta = 2.9$ meV and $\Gamma = 1.6$ meV.

fluctuations that break and create pairs at the surface, where the superconductivity is weak, and, therefore, reducing to a very small value the order parameter. Nevertheless, a feature that is a strong indication that exists an electronic modification at T_c is the change in shape of the curves when passing the transition.

As we explained before, the characteristics measured on superconducting tunnel junctions such as the current $vs.$ voltage, I $vs.$ V, the differential resistance, (dV/dI) $vs.$ V, its first derivative, d^2V/dI^2 $vs.$ V, at different temperatures, show many interesting features that can be related to the microscopic parameters of the superconducting state. In particular using the differential resistance we can extract the energy gap Δ. The measurements also display structure at higher voltages, which in conventional superconductors is the signature of phonon interactions with quasiparticles. The dynamic conductance (the inverse of the differential resistance) is proportional to the density of states in the superconducting state.[19] The second harmonic, d^2V/dI^2, gives information about the coupling function $\alpha^2(\omega)F(\omega)$. Dips in d^2V/dI^2 will correspond to peaks in the coupling function once the energy scale is corrected by ω-Δ. According to the above explanation, Fig. 3 shows the conductance $vs.$ voltage characteristic of a tunnel junction of $Ba_{0.6}K_{0.4}BiO_3$-I-Sn at 4 K. In this figure we observe the signature of the energy gap of the superconductor under study. T_c for this ceramic was 21 K.

It is worth mentioning that at temperatures below the transition temperature of Sn we observed the characteristic features of tunneling between two superconductors. In particular we observe the evolution of the sum of two gaps in both materials, $\Delta_{Sn} + \Delta_{BKBO}$. This is clear evidence that in fact the information we are observing is related with tunneling processes. The energy gap, Δ, was determined using a smeared BCS density of states, given by $N_s(E) = Re\{(E - i\Gamma)/[(E - i\Gamma)^2 - \Delta^2]^{1/2}\}$. In this equation Γ is a parameter which may be related to the density of states inside the gap, or to thermal smearing. The values for Δ and Γ that fit some of our experimental result are $\Delta = 2.9$ meV and the smeared parameter $\Gamma = 1.6$ meV. In Fig. 4 it is shown

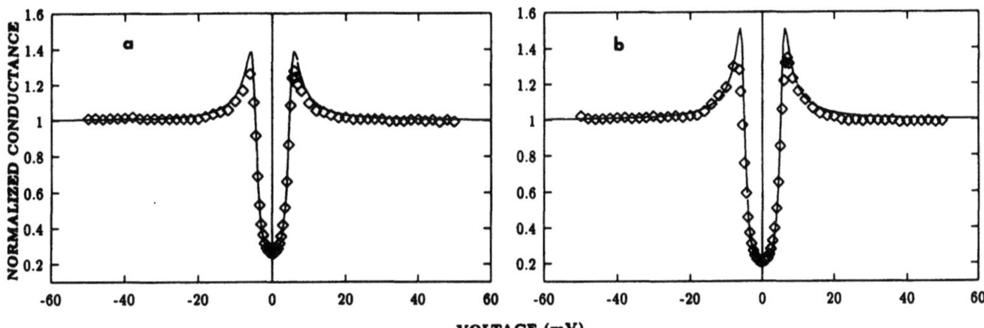

Figure 4. Conductance curves at 1.7 K normalized to 35 K for two $Ba_{0.6}K_{0.4}BiO_3 - I - Au$ contact junctions. The continuous curve is a smeared BCS density of states with a) $\Delta = 5.0$ meV and $\Gamma = 1.25$ meV, and $2\Delta/k_B T_c = 3.90$. b) $\Delta = 5.5$ meV, $\Gamma = 1.075$ meV and $2\Delta/k_B T_c = 4.29$.

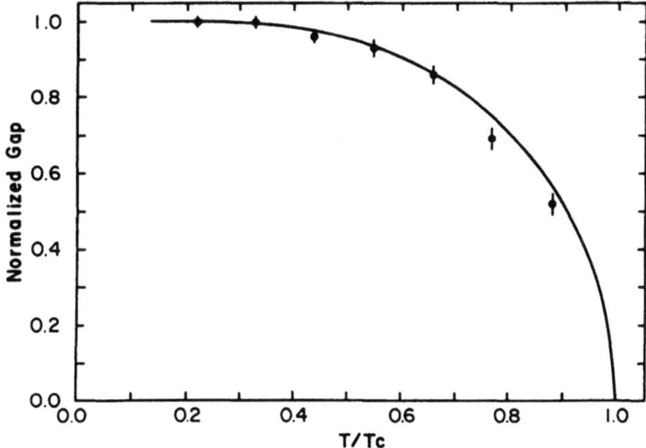

Figure 5. Temperature variation of the normalized energy gap $\Delta(T)/\Delta(4K)$ for $Ba_{0.6}K_{0.4}BiO_3$ with $T_c = 18$ K. The solid line represents the temperature dependence of the BCS theory.

the conductance of other tunnel junctions at 1.7 K, normalized with the conductance at the normal state. We also show the smeared BCS density of electronic states.

The experimental values of the energy gap from our different experiments show values that oscillate between 2.90 to 3.40 meV, using a smeared factor Γ from 1.6 to 0.58 meV. These data fit very well the experimental curves. One example of this fitting is illustrated in Fig. 3. According to this fitting we calculate the ratio $2\Delta/k_B T_c$, using a transition temperature of 18K, which changes from 3.70 to 4.39. Fig. 4 also shows data for the type (II) junction, here the maximum value of the ratio

$2\Delta/k_B T_c$ is equal to 4.29 using the values for the energy gap of $\Delta = 5.5$ meV and $\Gamma = 1.075$ meV. Using the same normalization routine for selected sets of data of the same tunnel junction, we find the development of the energy gap with temperature. The selection of the data was made considering the stability of the tunneling device with temperature. The evolution of the energy gap with temperature is shown in Fig. 5. In this figure we plot the energy gap normalized to the value of 4 K. The error bars mark the uncertainty in the measurement. The temperature scale also is normalized to 18 K (this value in the tunneling curves marks the onset where we start to see the feature of the gap). Normalizing to 18 K is not arbitrary and is justified by the fact that at 18 K the tunneling characteristic dI/dV vs. V shows the normal state behavior. We must remember that tunneling detects information in a superconducting material only up to the coherence length. The solid line in this plot is the expected BCS model. As was mentioned before, the information that can be extracted from the second harmonic of dV/dI vs. the energy ($eV\text{-}\Delta$) i.e. d^2V/dI^2, gives information about the coupling function $\alpha^2(\omega)F(\omega)$. Dips will correspond to peaks in the coupling function. This information already given in ref. 19 is well correlated with other experiments, e.g. neutron experiments reported by Loong et al.[14,15] and to optical phonons that were observed by Raman spectroscopy.[16] It is interesting to mention that theoretical calculation based on strong coupling theory of phonon mechanism for superconductivity, by Shirai et al.,[28,29] correlate also well with our experimental results.

Recently, point contact experiments that exhibit low leakage currents and sharp conductance peaks at the gap voltages, made by Huang et al.,[22,23] show high bias tunneling conductances which are characteristic of phonon effects as seen in conventional superconductors.

V. Conclusions

We summarize the results of this chapter with the following points:

Tunneling measurements indicate that $Ba_{1-x}K_xBiO_3$ compounds may be included in the class of materials for which superconducting properties are very probably framed within the BCS theory.

According to the ratio $2\Delta/k_B T_c$ the superconductor is between the intermediate to strong coupled regime, with the extreme values measured in our tunneling studies were $3.70 \leq 2\Delta/k_B T_c \leq 4.30$. For the maximum value of the ratio the smearing parameter Γ was 1.075 meV, whereas two times the energy gap 2Δ was 11 meV.

The energy gap dependence with temperature follows a normal BCS behavior.

The structure in the second derivative curve can be associated with phonon modes which have been observed in neutron experiments and in theoretical calculations using Eliashberg theory.

The normal state characteristics, as observed in our tunnel junction, present strong asymmetry between negative and positive bias voltage and linear tunneling conductance similar to that seen by other researches in the superconducting cuprates and in bismuthates. This behavior may have its explanation on the characteristics of the tunneling barrier, or in the material under study, and it is a topic that deserves further study.

Acknowledgements

I would like to thank F. Morales for valuable discussion and for the help on the experimental part. This work was supported by the Programa Universitario de Superconductores Cerámicos de Alta Temperatura Crítica, by the Dirección General de Apoyos al Personal Académico, of the Universidad Nacional Autónoma de México, by the Consejo Nacional de Ciencia y Tecnología and by the Organization of American States.

References

1. L.F. Mattheiss, E.M. Gyorgy, and D.M. Johnson, Jr., *Phys. Rev. B* **37**, 3745 (1988).
2. R.J. Cava, B. Batlogg, J.J. Krajewski, R. Farrow, L.W. Rupp Jr, A.E. White, K. Short, W.F. Peck, and T. Kometani, *Nature* **332**, 814 (1988).
3. D.G. Hinks, A.W. Mitchell, Y. Zheng, D.R. Richards, and B. Dabrowsky, *Appl. Phys. Lett.* **54**, 1585 (1989).
4. Y.J. Uemura, *Physica C* **185–189**, 733 (1991).
5. V.V. Bogatko and Yu.N. Venevtsev, *Sov. Phys. Solid State* **22**, 705 (1989).
6. A.W. Sleight, J.L. Gillson, and P.E. Bierstedt, *Solid State Commun.* **17**, 27 (1975).
7. R.J. Cava, B. Batlogg, G.P. Espinosa, A.P. Ramirez, J.J. Krajewski, W.F. Peck, Jr., L.W. Rupp, Jr., and A.S. Cooper, *Nature* **339**, 291 (1989).
8. R.R. Joyce and P.L. Richards, *Phys. Rev. Lett.* **24**, 1007 (1970); T. Timusk, D.B. Tanner, in: *Physical Properties of High Temperature Superconductors, Vol. I*, edited by Donald M. Ginsberg, (World Scientific, Singapore 1989), p. 339, and references therein.
9. C. Thompsen and M. Cardona, in: *Physical Properties of High Temperature Superconductors, Vol. II*, edited by Donald M. Ginsberg, (World Scientific, Singapore 1990), and references therein.
10. C.G. Olson, R. Liu, A.B. Yang, D.W. Lynch, A.J. Arko, R.S. List, B.W. Veal, Y.C. Chang, P.Z. Jiang, and A.P. Paulikas, *Science* **245**, 731 (1989); C.G. Olson, R. Liu, D.W. Lynch, R.S. List, A.J. Arko, B.W. Veal, Y.C. Chang, P.Z. Jiang, and A.P. Paulikas, *Phys. Rev. B* **42**, 381 (1990), and references therein; J.C. Campuzano, G. Jennings, M. Faiz, L. Beaulaigue, B.W. Veal, J.Z. Liu, A.P. Paulikas, K. Vandervoort, H. Claus, R.S. List, A.J. Arko, and R.J. Bartlett, *Phys. Rev. Lett.* **64**, 2308 (1990).
11. E.L. Wolf, *Principles of Electron Tunneling Spectroscopy,*Oxford University Press, New York, 1985.
12. R.E. Glover III and M. Tinkham, *Phys. Rev.* **108**, 243 (1957).
13. D.G. Hinks, D.R. Richards, B. Dabrowski, D.T. Marx, and A.W. Mitchell, *Nature* **335**, 419 (1988).
14. C.-K. Loong, D.G. Hinks, P. Vashishta, W. Jin, R.K. Kalia, M.H. Degani, D.L. Price, J.D. Jorgensen, B. Dabrowski, A.W. Mitchell, D.R. Richards, and Y. Zheng, *Phys. Rev. Lett.* **66**, 3217 (1991).
15. C.-K. Loong, P. Vashishta, M.H. Degani, D.L. Price, J.D. Jorgensen, D.G. Hinks, B. Dabrowski, A.W. Mitchell, J.D. Richards, and Y. Zheng, *Phys. Rev. Lett.* **62**, 2628 (1989).
16. K.F. McCarty, H.B. Radousky, D.G. Hinks, Y. Zheng, A.W. Mitchell, T.J. Folkerts, and R.N. Shelton, *Phys. Rev. B* **40**, 2662 (1989).

17. Z. Schlesinger, R.T. Collins, J.A. Calise, D.G. Hinks, A.W. Mitchell , Y. Zheng, B. Dabrowski, N.E. Bickers, and D.J. Scalapino, *Phys. Rev. B* **40**, 6862 (1989).
18. J.F. Zasadzinski, N. Tralshawala, D.G. Hinks, B. Dabrowski, A.W. Mitchell, and D.R. Richards, *Physica C* **158**, 519 (1989).
19. F. Morales, R. Escudero, D.G. Hinks, and Y. Zheng, *Physica C* **169**, 294 (1990).
20. F. Morales, R. Escudero, D.G. Hinks, and Y. Zheng, in: *Progress in High Temperature Physics*, Vol. **25**, edited by R. Nicolsky, (World Scientific, Singapore 1990), p. 366.
21. F. Morales, R. Escudero, D.G. Hinks, and Y. Zheng, *Physica B* **169**, 705 (1991).
22. Q. Huang, J.F. Zasadzinski, N. Tralshawala, K.E. Gray, D.G. Hinks, J.L. Peng, R.L. Greene, *Nature* **347**, 369 (1990).
23. Q. Huang, J.F. Zasadzinski, K.E. Gray, D.R. Richards, and D.G. Hinks, *Appl. Phys. Lett.* **57**, 2356 (1990).
24. T. Hasegawa, H. Ikuta, and K. Kitazawa in: *Physical Properties of High Temperature Superconductors III*, edited by D.M. Ginsberg, (World Scientific Publishing, 1992).
25. W.F. Brinkman, R.C. Dynes, and J.M. Rowell, *J. Appl. Phys.* **41**, 1915 (1970).
26. C.B. Duke, *Tunneling in Solids*, (Academic Press, 1969), p. 151, 301-306.
27. K. Nasu, *Phys. Rev. B* **35**, 1748 (1987). K. Machida, T. Koyama, and T. Matsubara, *Phys. Rev. B* **23**, 99 (1981).
28. M. Shirai, N. Suzuki, and K. Motizuki, *Solid State Commun.* **73**, 633 (1990).
29. K. Motizuki, M. Shirai, and N. Suzuki, *Research Report on Superconductivity II, Science Research on Priority Areas No. 031 Ministry of Education, Science and Culture.*, March 1991, page 481-490.

Vapor Deposition Processing

S. Purushothaman, C. Narayan and J.J. Cuomo

IBM T.J. Watson Research Center
P.O.Box 218, Yorktown Heights
New York 10598
U. S. A.

Abstract

Vapor phase deposition of materials is an area of long standing history and increasing diversity with significant practical importance to a wide variety of areas ranging from structural and automotive applications to optics and microelectronics. In this paper, we present one perspective view of this vast field beginning with an overview of the conventional deposition processes followed by some of the more novel techniques that have come to fore recently. We highlight the salient aspects of these techniques, and the key process parameters that can be controlled to achieve desired film properties. Examples of applications of these processes in industry are pointed out where possible. We conclude by describing some interesting process combinations that allow the deposition and synthesis of scientifically interesting and/or technologically useful materials with desired physical and chemical properties by synergistically combining the features of individual deposition processes.

I. Introduction

Vapor deposition of materials has been in use for a variety of applications from wear and corrosion protection coatings, decorative coatings, optical coatings and to a more active extent in the microelectronic and optoelectronic applications. The types of processes used vary over a wide spectrum comprising evaporation, laser ablative deposition, arc induced evaporation, plasma or ion beam sputter deposition and chemical vapor deposition. A wide range of materials including metals, alloys and compounds can be deposited by appropriately selecting source materials and/or deposition ambients. Given the diversity in the field, it is not surprising that a large number of reviews have been written on different aspects of the subject.[1-12] Hence, it will be redundant to attempt a regurgitation of the body of knowledge that is contained in

Advanced Topics in Materials Science and Engineering, Edited by
J.L. Morán-López and J.M. Sanchez, Plenum Press, New York, 1993

these articles in the present paper. Instead, we present a review with a perspective that will be most useful to applied scientists and technologists interested in a user oriented summary of the features and capabilities of different vapor deposition processes and their potential applications. Admittedly, we will not be elaborating on all the details of each specific process but instead will be referring the interested reader to the latest in depth reviews of the particular area. In what follows, we begin with a brief overview of the effects of species energetics (depositing or bombarding), on the general microstructure and physical properties of thin film deposits. We will then describe the salient features of the different deposition techniques and emphasize how the specific energetics of each of these techniques leads to the structure, property and morphology of the films deposited. Examples of applications where these processes are currently in use will be cited wherever possible. After this review, we will conclude with a description of some process variants and/or process combinations that can be used to achieve unique film properties or synthesize new materials.

II. Microstructure Evolution in Thin Films

Beginning with Movchan and Demchshin[13] several models have been proposed to describe the stages in the evolution of the micro and macrostructure of physical vapor deposited thin films.[14-16] The common underlying theme in all these models is the fact that the key factors that control the structure evolution are the mobility of the adatom on the substrate and the effects of energetic particle bombardment that occurs concomitantly with most of these deposition processes. Mobility of the adatoms arriving on a substrate is affected, to first order, by the kinetic energy acquired during the vapor generation process and once "bound" to the substrate, by the substrate temperature itself. In Table I we summarize the typical range of energies of the depositing and bombarding species associated with the more common deposition processes.

Clearly, the range spans several orders of magnitude and therefore the associated range of adatom mobility can be quite large. The substrate temperature employed in practical deposition processing (either by intentional substrate heating or self bias due to the processing conditions) is the other key factor and as a rule would be limited by the ability of the substrate to withstand such temperatures. In the first order models,[13] it is implicitly assumed that the arriving adatom kinetic energy and energetic bombardment are not significant. This may be applicable to evaporative processes which produce a thermalized atom flux at or close to the kinetic energy of the evaporant at its boiling point. Predictions are then made of possible sequences in structure evolution as a function of substrate temperature. At relatively low substrate temperatures, T_s, ($T_s < 0.3\ T_m$ where T_m is the melting point of the film material) the mobility of the adatoms is limited and they more or less get incorporated into the film as they land. The structure evolution is dominated by shadowing effects of pre-existing growth morphologies resulting in inclined columnar grains with a high level of porosity. As T_s is increased to between 0.3 and 0.5 T_m, surface diffusion of the adatoms on the substrate and on the pre-existing film surface begins to occur and dominate the structure evolution. This leads to reduced porosity films with columnar grains aligned more nearly parallel to the deposition flux. In the high temperature regime, when $T_s > 0.5\ T_m$, bulk diffusion effects become competitive and equiaxed and dense films result. This simple description has been further refined by detailing the morphological features such as the presence of fine equiaxed grains in the range

Table I

Typical species energies for some common deposition processes.
"N" indicates the energy range of neutrals
while "I" denotes the energies of ionized species.

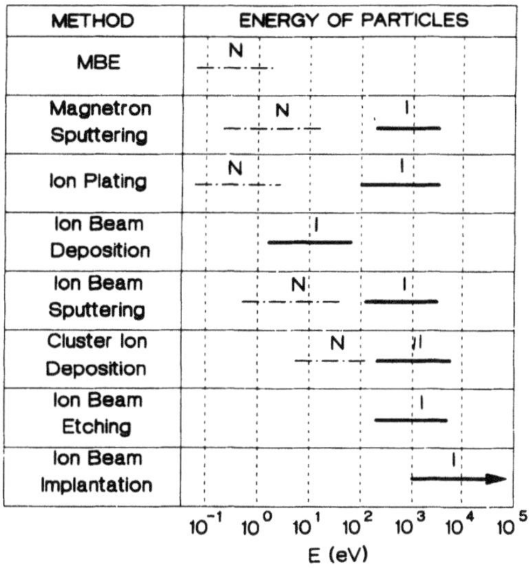

METHOD	ENERGY OF PARTICLES
MBE	N
Magnetron Sputtering	N I
Ion Plating	N I
Ion Beam Deposition	I
Ion Beam Sputtering	N I
Cluster Ion Deposition	N I
Ion Beam Etching	I
Ion Beam Implantation	I →

$$10^{-1} \quad 10^{0} \quad 10^{1} \quad 10^{2} \quad 10^{3} \quad 10^{4} \quad 10^{5}$$

E (eV)

of $T_s < 0.1\ T_m$ and bimodal grain sizes in the transition range bordering between 0.1 and 0.3 T_m.[16]

In the second set of models, attempt is made to incorporate the effects of energetic species bombardment on the structure evolution.[14,15] These models are more appropriate to describe structure evolution in processes where either the kinetic energy of the arriving atom flux is itself significantly higher than the thermalized levels noted above and/or where there is a secondary flux of bombarding species such as ions or neutrals from a plasma or a collimated ion source. In these instances, the structure and morphology of the films are affected strongly by these energetic species. The Thornton model[14] modifies the first order model and extends it to sputter deposition processes by considering the effects of sputtering process gas pressure used as a measure of the bombardment seen in these processes. Essentially, the model proposes that energetic species bombardment affects the transition from "lower" to "higher" reduced temperature structures described in the earlier model by suppressing the transitional regions dominated by lower mobility processes. A more recent analysis of the structure evolution[15] has focused on the species energy rather than on the sputtering pressure as the more fundamental parameter. These studies have shown that the effects of sputtering by arriving energetic species and the increased adatom mobility under energetic species flux are key in suppressing the porous structures associated with low mobility and promoting a more desirable and dense film structure.

In the following sections, we will draw upon the generic interplay of the species energy and substrate temperature effects on thin film structure and properties as outlined in the brief summary of the models, to emphasize ways in which desirable film structures and properties can be achieved in the different vapor deposition processes.

III. Evaporation Processes

Evaporation can imply a range of processes where a source material is heated and vaporized to produce a flux of species that can be condensed on a suitably placed substrate contained in a process chamber. The source of heating can be induction or resistive heating, or a scanned high energy electron or laser beam or a cathodically initiated arc. There are variants of evaporative deposition wherein either the depositing species are ionized and accelerated to the substrate or reacted with the process ambient to produce compounds on the substrate. In the following sub-sections some of the generic types of these evaporative processes will be described. The reader is referred to the excellent review by Deshpandey and Bunshah[17] for more detailed information on these processes.

III.1 Conventional Evaporation

Electron beam, resistive or induction heated melting and vaporization or sublimation of a source material is the most commonly practiced evaporation process in several coating applications. Typical energy of the species in the deposition flux is close to the thermal energy of atoms at the boiling point of the material and is usually in the range of 0.1 to 1 eV depending on the material being deposited. Little or no ionization of the evaporated species takes place prior to condensation on the substrate. Since evaporation is usually carried out in high vacuum chambers ($\sim 10^{-5}$ Pa) and the mean free path is long under these conditions, little or no collisional scattering of the species occurs. Deposition is therefore predominantly line of sight from the source. Hence, patterns can be generated easily by interposing either contact metal masks or conformally coated and patterned polymeric or inorganic stencils applied on to the substrates. Although the method is primarily used for the deposition of pure metals, deposition of alloys where the vapor pressure of the constituents are close, as well as compounds that congruently melt and evaporate (without decomposition) can be carried out by evaporation. Co-evaporation from multiple sources can be employed in cases where the components required have very different vapor pressures or if the compound to be evaporated decomposes before evaporation. Some interesting work on co-evaporated aluminum alloy nanostructural composites have shown that macroscopic thickness samples with unique properties can be fabricated by this method.[18,19]

Control of microstructure and film stresses in evaporation is mainly achieved by substrate heating to increase the adatom mobility. Generally, higher substrate temperatures result in denser films and lower intrinsic stresses in the deposits, as can be expected from the structure evolution process described in the last section.

Evaporation processes are primarily used in microelectronic device and packaging structures for deposition of contact and joining metallurgies making use of the line of sight deposition characteristic to generate pad or line patterns through masks. One concern associated with these applications is the requirement of reduced atom flux divergence as seen at the substrate surface to reduce penumbra formation under the mask openings or deposition of material on the stencil side walls. This requires that the throw distance between the source and the substrate be large enough so that the divergence across the substrate is reduced. This in turn implies larger and larger evaporation chambers required for larger substrate sizes and the associated higher equipment cost. Other uses of evaporation have been in depositing reflective optical coatings and decorative coatings.[20]

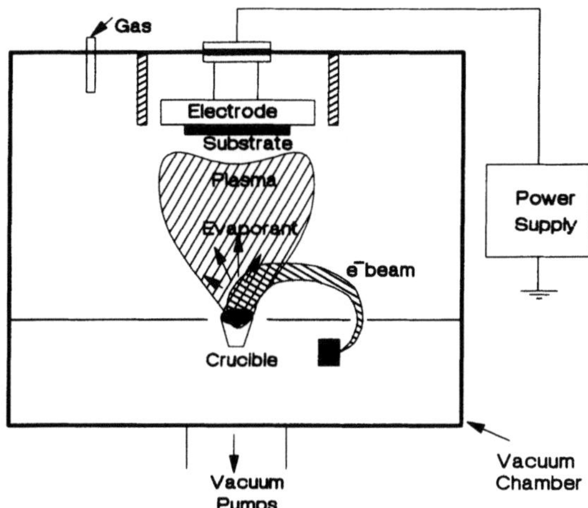

Figure 1. Schematic representation of ion plating.

III.2 Ion Plating

One of the drawbacks of conventional evaporation is the low species energy associated with a thermal source. This can be alleviated by the use of ion plating. In this process, a conventional evaporation source located in a differentially pumped source chamber is used to generate the atom flux which is then passed through a region of inert gas plasma generated in an intermediate chamber, see Fig. 1. Evaporant atoms collide with the plasma species and get ionized in the process. The substrate is held at a negative bias potential (100 to 5000 V) with respect to ground which results in the acceleration of the positively ionized species towards it.[21] Typical pressures in the plasma region would be in the 10^{-3} Pa range and hence significant ionization of the depositing species can occur due to collisions with the ions and electrons in the plasma. As a result, it is possible to deposit species at high incident energy levels and make use of the associated adatom mobility to achieve better microstructures. In addition, the bombardment of the growing films with the energetic ions (of inert gas as well as depositing material) can also lead to interface mixing and improved adhesion as well as lowering of film stresses. Additionally, a dilute mixture of a reactive and inert gases can be employed to achieve reactive ion plating of compounds such as nitrides or carbides.[6] One of the limitations of the process stems from the scattering of the depositing species in the plasma leading to reduced deposition rates. The primary use of ion plating has been in automotive and decorative coating applications.[22]

III.3 Activated Reactive Evaporation

As the name implies, activated reactive evaporation (ARE) involves the deposition of compound films by the reaction of evaporated species with activated gaseous species introduced into or generated in the process chamber. The most common configuration employed is shown schematically in Fig. 2. The metallic component is usually generated by evaporation from a source and the reactive species is introduced into the

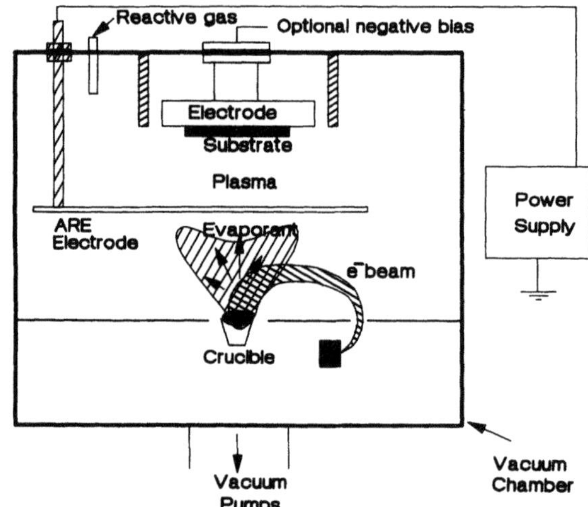

Figure 2. Schematic representation Activated Reactive Evaporation.

system separately. Several means could be used to generate the electrons required for the ionization of the species, for example, collection of secondary electrons generated in the evaporation sources due to e-beam bombardment, use of a separate thermionic source etc.; see the review by Deshpandey and Bunshah[20] for details. The electrons are accelerated towards the ARE electrode that is at a positive bias voltage and collide with the reactive gas and evaporant species to ionize them. The substrate is usually held at a negative bias voltage so that the ionic species are accelerated towards it. The process is very similar to ion plating described above except that the degree of ionization is lower since the chamber pressures are usually lower. Other variations can include the use of a supplementary supply of reactive species introduced into the chamber from a second source, such as a microwave down stream plasma (see later sections for details). ARE is used primarily for the deposition of compound materials such as oxides, nitrides, carbides and sulfides. One advantage of ARE would be the ability to maintain the film stoichiometry during the process made possible by independent control of the reactive plasma parameters and the deposition parameters (unlike in reactive sputtering to be discussed later). The primary applications have been in the deposition of oxides for optical and optoelectronic coatings,[23-27] wear resistant coatings (carbides and nitrides of Ti, Zr, Hf, V),[17,28-30] lubricating coatings (sulfides of Mo or Ti) and photovoltaic coatings.[17]

III.4 Cluster Ion Beam Deposition

This technique is a more recently explored variant of deposition from an ionized species flux and as a result still in early stages of development.[31-34] The source material is contained in a crucible that is indirectly heated by one of several means to melt and vaporize the contents. The crucible is provided with a special nozzle arrangement at the top with a length to diameter ratio of 0.5 to 2. This allows the species at high vapor pressure within the crucible to go through multiple collisions and interactions as it exits through the nozzle into the lower pressure region of the vacuum process chamber. This results in the formation of atom clusters which are claimed to consist of 100 to 1000 atoms which then proceed through an ionization region wherein they are bombarded by a cross flux of accelerated thermionic electrons. Depending on the electron flux, the cluster population can be ionized to varying degrees, ranging between a few percent to as high as 35% and acquire a charge. The ionized clusters are then accelerated by a grid biased at a negative potential, thus increasing their kinetic energy before they impinge on the substrate. Cluster energies of 100 to 1000 eV are claimed to be possible by this process. It is presumed that the impact of these energetic clusters results in higher mobility of adatoms on the substrate surface. Some amount of physical sputtering and implantation and adhesion enhancement can also be expected as a result. It has also been claimed that the physical sputtering combined with the enhanced mobility associated with the clusters leads to planarization and improved step coverage, features not possible with conventional evaporation. Since the cluster impact in general can provide more activated sites for nucleation, one can also expect a more refined film structure compared to a regular deposition from an atomic vapor flux. However, since the energy per atom is still small (<1 eV) the disruption of the substrate surface would not be as much as in processes with more energy per atom or ion and this could be important for sensitive applications where deposits with good adhesion are desired without significant disturbance of the substrate lattice near the surface. Because of these reasons, the primary area of potential application of this process may be in the electronic device fabrication. It must be pointed out however that the field of Cluster Ion Beam Deposition is one where several controversies still persist as to the exact energetics of the species and physics of the deposition process. Therefore, further studies are needed to better understand the technique before it can be utilized in practical applications.

III.5 Laser Ablative Deposition

In this process, as the name implies, a laser is used to locally heat, melt and vaporize or ablate the target material contained in a vacuum chamber and deposit it on to a substrate. In most applications, Nd:YAG or CO_2 lasers are used because of their high power capability, operational stability and reliability.[7] However, more recently, reports of the use of excimer lasers for ablative deposition have also been noted in the literature.[35-40] Both pulsed (1 ns at $0.5 GW/cm^2$ type power levels) and continuous wave (1 ms at $10 kW/cm^2$) operation are common depending on application. The power density employed plays a primary role in the nature of the species flux generated in the process.[7] At low energy densities, the laser energy leads to surface melting and thermal evaporation of the melt resulting in a flux of non-ionized species much as in conventional evaporation processes. However, as the power density is increased, due to the higher local vapor pressure in the near target region, a plasma is created which begins to modulate the flux of evaporating species. The net result is an increase in the ionized species and excited neutrals in the flux. Species energies on the order of few eV's can be easily generated under these conditions of strong plasma-surface interactions. However, as plasma density increases, its transparency to the laser energy is reduced thus reducing the efficiency of evaporation of the surface and mitigating the increase of plasma density. Thus a self limiting steady state is usually achieved at these high power density levels. Mixed mode behavior between these two extremes is usually seen in the intermediate range.

Although any material that can be deposited by a thermal evaporation process can also be deposited by the laser ablative deposition, the unique advantage of the laser process stems from the observation that congruent and stoichiometric deposition of compounds is possible in several cases. This has led to active investigation into the deposition of cuprate type 1-2-3 superconductor thin films and ferroelectrics by laser ablative deposition in the recent years.[35,36,38−40] Other advantages of this technique include the fact that the heat source used for evaporation is not internal to the process chamber which minimizes the outgasing from the chamber and fixtures; highly efficient heat transfer and target utilization with the use of a scanned focused laser beam which allows the use of small amounts of target material for deposition, an important factor in the case of targets of expensive and experimental materials; the potential for *in situ* cleaning of substrates using lower power laser heating and/or ablation without the need for other plasma type treatments; and extendibility to multi-target deposition as well as reactive deposition schemes if desired.

There are some drawbacks associated with laser deposition techniques. The major one is that of particulate ejection from the target surface, commonly referred to as "splashing". This is a result of the superheating of the under layer before the surface melt has an opportunity to evaporate. This can lead to sudden escape of dissolved gases from this sub-surface layer causing the melt to splatter. This situation, predictably, is worse under high power density pulsed operation mode. It is further accentuated by the wear of the target which can lead to a rough target texture. The usual way to reduce this effect would be to condition the target surface by heating and outgasing it at low power levels. There is also a critical maximum power density, characteristic of a target material, dependent on parameters such as its vaporization energy and thermal diffusivity, below which this phenomenon can be considerably reduced. This threshold can be determined experimentally and the splashing can be minimized by keeping the operating conditions safely below this level. Other extraneous methods such as the use of rotary vane filters to block out stray ejected particulates have also been reported to be useful in this regard.[41]

The use of laser ablative deposition is specially attractive for elemental or compound materials with high melting point. In particular it has been used for the deposition of inorganic dielectric materials (oxides), piezoelectric materials and refractory metals. Its ability to generate a flux of intact "molecular" species has made it very attractive for the deposition of superlattice materials (HgCdTe for example) and band gap engineered semiconductors and more recently the high T_c superconducting cuprates. Application of this technique to deposition of amorphous diamond films has also been recently reported.[42,43] The reader is referred to recent reviews[7,44] for a comprehensive listing of the applications of laser ablative deposition.

III.6 Cathodic Arc Deposition

This process is another variant of evaporative deposition in which the source of heat is a low voltage (20 to 40 V), high current (100 to 500 A) arc struck between the target surface and an anode.[45] Typical current densities in the arc spot can be on the order of $10^6 - 10^8$ A/cm^2. Systems with random arc spots on the target with an insulating confinement ring[46] as well as magnetically steered arc spots[47] are used in practice, as shown schematically in Fig. 3a and 3b. The arc is sustained by the locally high pressure created by the flux of evaporating species. The evaporated flux gets significantly ionized due to its interaction with the arc plasma and 30 to 100% ionization levels in the flux are possible with associated high kinetic energy in the 10 to 100 eV range. More recently, the use of a direct current magnetron (see later section for details) as an arc deposition source has also been reported.[48]

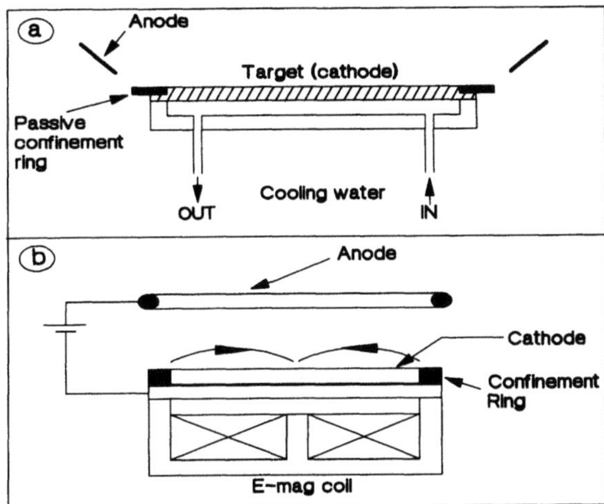

Figure 3. Schematic representation of cathodic arc source for deposition. (a) shows the arc source with passive arc confinement and (b) shows guided arc source.[45]

Inherently, cathodic arc deposition allows high rate deposition in the presence of a local plasma without the complications of a global process plasma as in sputter deposition processes. Because of the energetic nature of the species, films with good adhesion, microstructure integrity and dense structure can be deposited at relatively low substrate bias temperatures. Although pure metals can be easily deposited, it is also possible to deposit from alloy and compound targets and achieve good target composition replication. Also, using a suitable process gas ambient one can carry out reactive deposition of compounds.[49–51] Much like laser ablative deposition, this method suffers from ejection of particulates in the form of macro or microdroplets. Use of a diffused arc spot that is steered on the target surface and operating at a lower current level is usually very conducive to the reduction of this problem. Also, since these droplets generally appear to travel along a low angle trajectory from the target surface, use of low angle (0 to 30°) shielding has been employed as well with some degree of success to prevent the droplets from reaching the substrate surface.[45] Due to the highly ionized nature of the species flux, electrostatic and/or magnetic deflection of the ionized droplets using mass/charge discrimination has also been proposed and used in some critical applications.[52–54]

Primary application of the cathodic arc deposition process has been in the deposition of decorative coatings (nitrides), deposition of wear resistant coatings (carbides and nitrides) on intricate tool and die surfaces, deposition of metals and compounds (such as ZrN, ZrO) for microelectronic applications, deposition of coatings for optical applications (oxides of Zr and Ti, TiN)[45] and deposition of amorphous diamond films.[52,55,56]

IV. Sputter Deposition Processes

Sputter deposition is perhaps by far the most commonly used process for vapor deposition of a wide range of materials. The details of the sputter deposition schemes vary but the basic principle involves the bombardment of a negatively biased target by a flux of positively charged inert gas ions to sputter the target species from the surface and deposit the same onto a substrate held in the vicinity of the target. In what follows, we will quickly review the basic diode and triode sputter deposition schemes and then go on to describe enhanced schemes such as magnetron sputtering, hollow cathode enhanced magnetron sputtering and reactive sputter deposition. Readers interested in greater details of sputter deposition are referred to recent reviews on plasma assisted processing.[57−60]

IV.1 Diode/Triode Sputtering

These represent the simplest sputter deposition schemes used in practice. As the name implies, the diode system, is a two electrode system wherein the target to be sputtered is the cathode and the process chamber at ground potential acts as the anode. By means of a suitably throttled flow of inert gas (usually argon) through the system a plasma is started and the target gets bombarded by the positively charged argon ions. The collisions eject target atoms which traverse to the substrate with kinetic energies in the 10 eV to 2 keV range depending on the sputtering conditions (target bias, Ar gas pressure and associated thermalizing collisions). In a triode system, additional electron density and ionization is facilitated by the use of a third electrode which is at a slight positive voltage above ground potential and acts as a collector of secondary electrons as well as electrons supplied by a thermionic source. This has the advantage of increasing the degree of ionization at a given pressure and reducing undesirable substrate overheating due to secondary electron impact. The target is usually water cooled since the majority of the energy of the ion impact is converted to heat on the target surface. Both direct current and RF excitation can be used in the sputtering process, the latter being essential for sputtering insulating materials without the complications of space charge build up.[57]

IV.2 Magnetron Sputtering

In magnetron sputtering, the plasma density in the vicinity of the target is enhanced magnetically as shown schematically in Fig. 4. By placing an array of permanent magnets behind the target such that magnetic flux lines emerge from and close back into the target surface, the electrons in the region close to the target are forced to travel in helical paths around these field lines. This leads to increased electron path lengths in the plasma region near the target causing increased degree of collisions and hence ionizaion in that region. In effect, higher concentrations of inert gas ions at a given pressure are produced compared to a standard diode/triode system and higher deposition rates are possible. Most of the commercial sputter deposition tools in use today tend to be magnetron enhanced systems because of the associated throughput advantages. Magnetron sputter deposition, of course, retains all the benefits of the diode and triode sputtering described earlier. However, one disadvantage is that sputtering of highly magnetic materials is not very efficient and one may have to use other deposition techniques such as ion beam sputtering, described in a later section of this chapter.

One of the key attractive features of sputter deposition is the ability to uniformly coat large substrate areas using a very compact throw distance between the target and the substrate. In contrast, as described earlier, evaporation systems require larger and larger throw distances with increasing substrate sizes to assure uniformity and

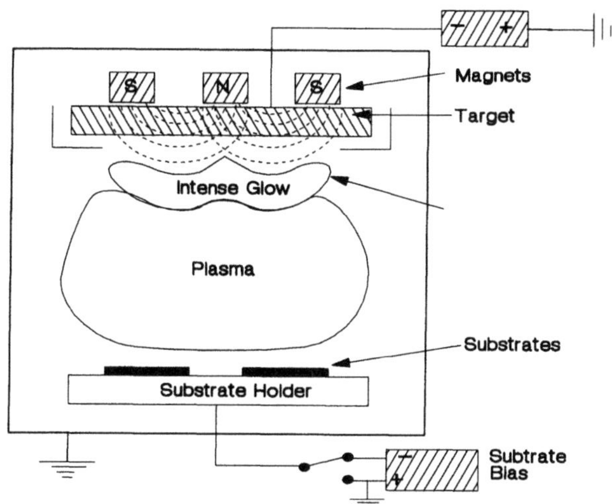

Figure 4. Schematic representation of magnetron sputtering.

divergence requirements. This in turn allows the vacuum chambers used for sputter deposition to be much more compact and hence requiring smaller pumping packages but more compatible with good pumping speed at moderate process gas flow rates. Sputtered atom flux from the target surface is usually non-directional and hence leads to a conformal coating and excellent step coverage of substrates with topography. However, one can bias the substrate negative with respect to ground and cause energetic bombardment with the inert gas ions to occur during deposition. This leads to simultaneous sputtering and deposition on the substrate surface. By varying the substrate bias level it is possible to "erode" the high spots on the deposit and achieve different degrees of planarization rather than conformal deposition.[61-63] In situ precleaning of the target and the substrate is normally practiced by negatively biasing the surface to be cleaned and protecting the other surface. This assures the deposition of cleaner films on to *in situ* precleaned substrate surfaces. Energetic bombardment by the depositing atom flux as well as the positive ions and energetic neutral species in the process gas facilitate good adhesion of the films to the substrate. The deposition pressure and the substrate bias can be used to vary this energetics and optimize the microstructure and intrinsic stress levels in sputtered films.[6,57,64-70]

Alloys can be deposited from a single target as long as the sputter yields of the constituent materials are comparable or by making calibration runs to determine a correlation between the target and the film composition. The restrictions on deposition from an alloy target are far less than in the case of evaporation from an alloy source. Deposition of alloys and compounds by reactive sputtering will be discussed later.

There are a vast majority of practical applications where sputter deposition is used. It is particularly suited for deposition of materials which are difficult to evaporate, such as refractory metals, oxides and other compounds.[57,70-74] A major user is the microelectronics industry where sputter deposited conductor metallizations (Al and its alloys), plating base layers (Cr or Ti and Cu) and contact pad metallurgies are common. Additionally, there have been reports of the use of sputter deposited films for magnetic recording (Co based alloys),[75,76] surface passivation coatings on base metals (stainless steel films on carbon steels for example),[77,78] magneto-optic coatings[79,80] and reflective coatings for optical and X-ray applications.[81]

IV.3 Reactive Sputter Deposition

As the name suggests, this is an enhancement to the conventional sputter deposition process by the use of a reactive ingredient in the process plasma. Typical examples are the use of oxygen, nitrogen and hydrocarbons with argon to deposit oxides, oxynitrides, carbides or carbonitrides.[6,59,60] The target is usually the pure metal that reacts with the reactive ingredient to form the compound. Depending on the processing conditions, the reaction can occur either in the plasma or more commonly on the substrate surface. One of the concerns is the ability to control these reactions to achieve the correct stoichiometry required in the film. More of a concern is the poisoning of the target surface by its reaction with the active gas ingredient, which in turn can result in drastic reductions in the sputter yield and potential changes in the film composition.[82] This is usually reduced by the use of localized delivery and distribution of the reactive species in close proximity to the substrate surface,[83] control of the degree of ionization and residence time of the reactive species in the chamber[60] and controlled gas flow patterns by baffling.[84-86] Typically reactive sputtering is most used for deposition of dielectric and optical coatings (oxides and nitrides), wear resistant or tribological coatings (carbides, nitrides or sulfides)[50,51,60,87-89] diffusion barrier and semiconductor coatings for microelectronics applications[90-94] and photovoltaic coatings.[6]

IV.4 Hollow Cathode Enhanced Magnetron Sputtering

A hollow cathode is a secondary cathode that is introduced into the magnetron system and comprises a refractory metal tube that is electrically isolated from the rest of the system and maintained at a slight cathodic potential and is used to deliver the inert gas into the chamber. It is heated to produce thermionically emitted electrons that create a plasma at the exit of the cathode tube. Configurations using the substrate itself as part of the hollow cathode instead have been used in semiconductor device etching and deposition.[95] In this case, apart from simple addition of electrons and ions into the plasma from the additional cathode, other effects such as mirroring and trapping of secondary electrons and hence their utilization in the ionization process can also contribute to the increased plasma density. In our discussions in this section, we will confine ourselves to the former type which is most commonly used to enhance magnetron sputter deposition. In this case, best results are accomplished by localizing the hollow cathode discharge to the region near the magnetron target, as shown in Fig. 5.

The intense plasma glow and the associated plasma and ion density can be used to significantly increase sputter rates at a given process gas pressure. Alternately, one can achieve high plasma densities and operating current levels at lower target voltage and lower pressure levels compared to a conventional magnetron process.[96-98] Operating pressures of 0.1 Pa are possible as opposed to .5 to 5 Pa normally used in conventional magnetron sputter deposition. The lower pressure operation results in less collisional scattering and hence a less divergent deposition flux compared to conventional sputtering. Further, it is possible to collimate the flux to different divergence levels by interposing collimators with different aspect ratio opening arrays.[99,100] Thus one can

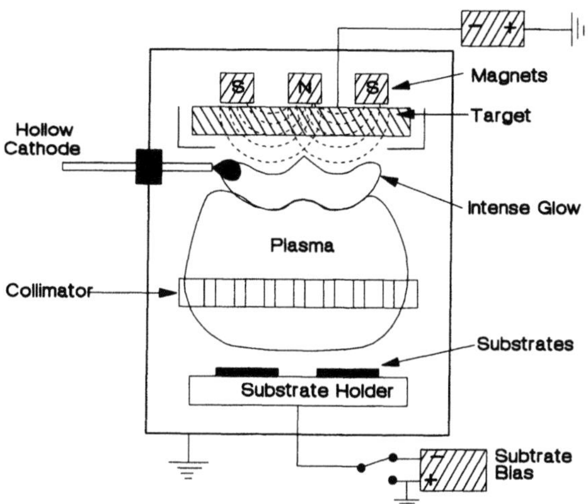

Figure 5. Schematic showing a hollow cathode enhancement to sputtering. Also shown is a collimator that can be used to achieve line of sight deposition made possible by the low pressure operation.

use the process for deposition through masks and stencils (very low divergence), and achieve line of sight deposition with a more energetic atom flux compared to thermal evaporation. Alternately, one can achieve via or trench fill (intermediate divergence) or conformal coating at high rates (high divergence). There will be, however, a trade off with lowered deposition rates with increasing degree of collimation. A further advantage could be derived by biasing the collimator block at a slight positive voltage with respect to ground enabling it to act as a secondary electron trap. This in turn reduces the substrate heating from their impact and this can be very important in depositing through polymer stencils or on to temperature sensitive substrates.

The main concern with this process is the durability of the hollow cathode tube since the intense heat and the high atom flux in its vicinity can lead to either melting or clogging of the tip. Uniformity of the deposition rate could be a problem but can be overcome by the use of proper hollow cathode geometry. Hollow cathode enhanced deposition has been used both to exploit the high rate capability as well as the collimated deposition process in microelectronic and magnetic thin film applications.[99,100]

V. ECR Enhanced Plasma Deposition

ECR or electron cyclotron resonance utilizes the well known phenomenon of helical motion of electrons in the presence of an axial d.c. magnetic field and an orthogonal a.c. electric field in the microwave frequency range. Essentially, the result is one of increased electron mean free path and hence increased ionization.[101−103] Greater than 10% ionization and high plasma density (as high as $5 \times 10^{11}/\text{cm}^3$) can be achieved at low pressures (0.1 Pa). In this respect, it could be viewed to be very similar to the hollow cathode enhanced plasma. However, additional features such as the potential use of the magnetic fields or electrostatic grids to confine and enhance the plasma and optional use of extraction grids to allow a down stream extraction of ions and

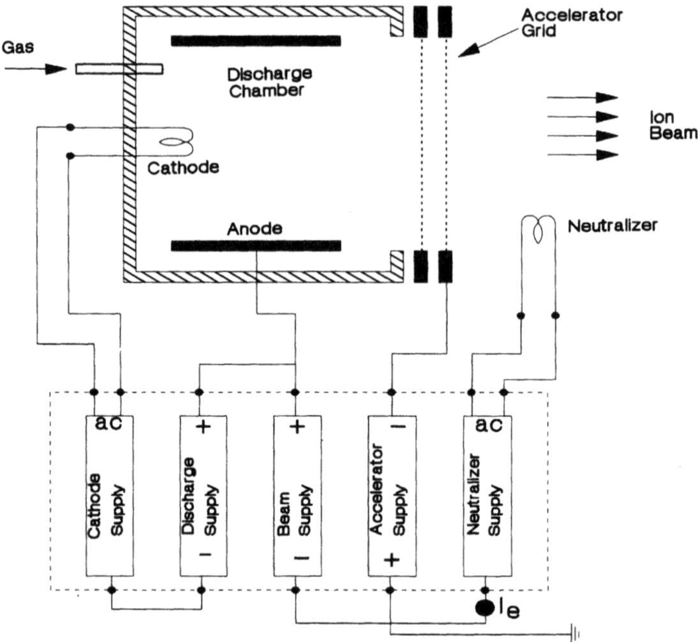

Figure 6. Schematic representation of a broad beam ion source.[111]

reactive species without the electrons provide significant process leverage with ECR enhancement.[104−107]

ECR enhancement is extensively used in the microelectronic industry for etching and ashing processes.[104] Other applications include its use as a means of providing reactive species for CVD (see later section on CVD) and reactive deposition processes.[104,108−110] The higher concentration of reactive species provided by ECR generally allows potentially lower substrate temperatures in CVD and it may be possible to use the plasma also to dope the deposited films. The use of ECR to enhance a physical sputtering plasma has not been reported in the literature to our knowledge.

VI. Ion Beam Sputter Deposition

As mentioned earlier, conventional evaporation and sputter deposition processes lack either the directionality of atom flux or the atom energetics for better adatom mobility. Ion beam sputter deposition overcomes these limitations and combines the best of both techniques. A broad beam (Kaufman type) ion gun is illustrated in Fig. 6. It consists of a thermionic cathode source and a suitably configured anode to generate energetic electrons which in turn collide with and ionize an inert gas atom flux maintained through the gun.[111] These ions are accelerated, focused and extracted out of the ion gun using a set of parallel grids (usually made of graphite or molybdenum) maintained at different negative potentials relative to ground. The net result is a stream of energetic inert gas ions (usually Ar) whose energy can be varied over a wide range (few tens to few thousands of eV) by adjusting the accelerator grid potential. The ion flux can be extracted with different levels of collimation by varying the focusing

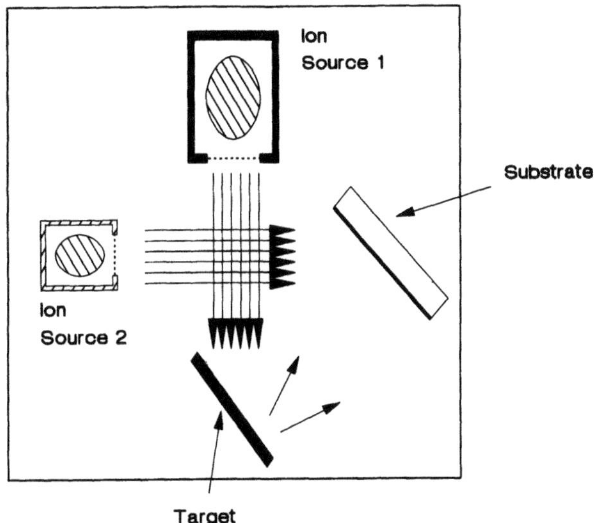

Figure 7. Schematic representation of ion beam sputer deposition. Also shown is a second ion source that can be used to modify the properties of the thin film deposit.[112,113]

grid potential. Since the generation of the ions and their extraction are independent of each other, the number density of ions and their directionality and energy can be independently controlled to meet specific application needs. An optional neutralizer (thermionic type or a plasma bridge type) can also be used to neutralize the net charge of the flux so as to minimize the complications due to charging when processing non-conductive substrates. In the ion beam sputter deposition application, the ion flux is then used to bombard a target surface and the sputtered material from it is deposited on to a substrate held at a suitable location as shown in Fig. 7. In a typical deposition system, the gas is fed through the ion gun and by the design of the gun and the pumping package, one can sustain a plasma in the gun even when the nominal deposition chamber pressure is less than a 0.1 Pa. Thus, the material ejected from the target has a long mean free path and one can achieve a line of sight deposition through masks or stencils and reduce back scattered impurities. But by virtue of the incident argon ion energy level, one can produce energetic sputtered species which will be conducive to high degrees of adatom mobility and microstructure development. Another advantage of the process is its ability to sputter magnetic materials at high rates which is difficult with conventional magnetron sputtering. Further, since the substrate to be coated is not immersed in a global plasma, one can eliminate negative ion and secondary electron bombardment and the associated overheating and substrate damage effects. Lastly, a second ion gun can be used to preclean the substrate surface *in situ* prior to film deposition to assure good adhesion. Ion beam sputtering of multicomponent systems is possible from targets that are composites of the components. By traversing the target and the beam relative to each other, alloy films can be deposited. The reader is referred to a concise review of these and other aspects provided by Rossnagel[112] for additional details.

More recently, ion guns have been made available that use an RF plasma in the gun chamber to produce the required ions. This eliminates the problem associated with cathode filament life and extends the use of ion guns to produce reactive ionic species.

Thus one can use these guns to achieve high ion currents and to modify substrate surfaces chemically prior to material deposition. These can also be used to tailor the chemistry of the growing films (reactive ion beam deposition) and some possible use of these guns in combination processes will be alluded to later in this paper. By far the major application of the non-reactive ion beam sputter deposition is in the areas of magnetic thin film deposition, contact metallization deposition for microelectronics[8] and hard wear resistant coatings of amorphous diamond films.[42,114–116]

VII. Chemical Vapor Deposition

All the deposition processes described earlier were characterized by the fact that the material deposited is either the same as the source material or an additive reaction product of the source and the controlled ambient of the chamber. In contrast, chemical vapor deposition or CVD involves the formation of the thin film deposit from a volatile precursor compound by decomposition or by reduction by the chamber gas ambient. Thus there is no solid target or source of the material internal to the deposition chamber unlike the physical vapor deposition processes. The precursor can be formed in the reactor chamber or more commonly introduced from an external source. The precursor has to contain the material species to be deposited and has to possess an adequate vapor pressure at or near room temperature so that it can be conveniently introduced into the process chamber. It should decompose or be reduced to the pure film deposit at a higher temperature at which the substrate to be coated is usually held within the chamber. Metalorganic coordination compounds, metal-halogen compounds or metalloid-hydrogen compounds meet these general requirements and are therefore commonly employed precursors in CVD. In many instances they are used in conjunction with oxygen, chlorine or hydrogen to achieve the reactions needed to form the film material. The above process is known as thermal CVD since the conversion is achieved using thermal activation. An exhaustive review of thermal CVD with a wealth of references is available in the literature[117] and we will only touch upon the very rudimentary aspects of this process in this section. Enhancement of CVD by a direct plasma (PECVD)[118] or a remote plasma,[119] laser photon energy (LCVD)[120] or ion beam flux (IBCVD)[121] could be used to increase the deposition rates by producing activated species or to achieve selective pattern deposition. In addition, selectivity can also be achieved in some systems by differences in the adsorption of the precursor material on different areas of the substrate. A typical example in microelectronic applications is the CVD of tungsten where the precursor adsorbs selectively on metals and silicon but not on silicon dioxide. This allows selective deposition of tungsten on all areas on the substrate except those covered with the oxide insulating layer.[117] Another variant of the thermal CVD is the Organometallic Vapor Phase Epitaxy (OMVPE) which is used for epitaxial growth of controlled chemistry, ultrathin compound semiconductors for optoelectronic and photovoltaic applications.[122]

Generally, the CVD processes provide faster throughput compared to other methods such as Molecular Beam Epitaxy (MBE) used in electronic device fabrication. One concern associated with CVD is the need for high substrate temperature required to obtain good film quality, without the incorporation of impurities from the precursor. This could be particularly important if the substrate material is sensitive to high temperature exposures. Enhanced CVD can be used to offset this to some degree. Another issue is the need for handling somewhat toxic precursor materials and the difficulty in scaling up the process to large scale manufacturing. At the present time, CVD processes are mostly used in microelectronic applications to selectively deposit conductors (Cu, Al, Au) or epitaxial semiconductor layers, for repairing photomasks (deposits of Cr, Au), or fabricating contacts and interconnects (W, silicides) and insulators (SiO_2). Exploratory work on PECVD of amorphous diamond films and other wear coatings[123−125] is another active area that is currently being pursued. The reader is referred to excellent reviews in the book edited by Vossen and Kern[3] for examples and further details.

VIII. Some Unique Process Combinations

It is apparent from the foregoing overview that each of the different deposition processes have advantages and limitations. Hence, a process development engineer can develop unique combination of processes that complement each other in a synergistic fashion. Some examples of these processes were briefly alluded to in the previous sections, as in the different enhanced CVD processes. In what follows, we will cite a few other examples wherein combinations of processes can be simultaneously active in a process chamber so as to improve the quality of the films produced or broaden the range of film types that could be deposited. As one would expect, in order for this to be accomplished, there has to be a regime of common operating conditions (chamber pressure, chamber gas chemistry, etc.) between the individual processes.

VIII.1 Ion Beam Assisted Processes

As reviewed recently by several authors[8,11,12,126−129] a large number of combination processes use ion beams to modify the physical and chemical nature of a film while it is being deposited. Typically, the primary deposition process can be e-beam evaporation (conventional or reactive), ion plating, hollow cathode enhanced magnetron sputtering or ion beam sputter deposition and a broad beam ion source can be used to produce and direct a flux of low energy bombarding ionic species on to the growing film substrate.[113,130] The bombarding species can be inert ions or reactive ions as needed. The effects of ion bombardment can be different at different stages of the film growth. If the ion bombardment is carried out when the film thickness is very small (say less than 10 nm), one can achieve mixing of the deposit into the substrate and promote better adhesion.[128,131−134] Secondary bombardment with high energy (tens of keV to MeV) ion beams typical of ion implantation has also been used to achieve interface mixing and adhesion improvements in systems where adhesion is otherwise very poor.[135] At higher film thicknesses one can achieve refinement of the thin film microstructure,[136,137] higher film densities[8,138,139] and lower intrinsic film stress levels.[126,137,140−143] Thus the purpose is mainly one of film modification.

Another application involves the use of a primary process to deposit a pure element and the lower energy ion source to provide species that can react with growing film on the substrate to produce a compound film.[11,12,124,144−148] Deposition of oxides and nitrides of refractory materials can be accomplished in this manner and a detailed table of references can be found in the review by Smidt.[8] This ion beam assisted deposition (IBAD) approach is better than sputtering from a compound target as well as deposition by reactive magnetron sputtering in that the rates are not limited by the sputter yield of the compound material which is usually much lower than the pure element involved. Additionally, the benefits of low energy ion bombardment described in the non-reactive case can also be realized. Low stress, high density films for optical, microelectronic and wear resistance applications are produced by reactive IBAD processes.[8,124,129]

VIII.2 Plasma Enhanced Processes

As described earlier, the ECR plasma process is a very convenient source of active species (ions and activated intermediates) which can be extracted and used downstream for participation in thin film processing. ECR plasma etching is widely practiced in microelectronics to achieve high throughput patterning and stripping of materials. But the same idea can also be used to enhance deposition processes, in particular CVD processes and reactive deposition techniques such as reactive sputtering and activated reactive evaporation. Injection of reactive species separately from an ECR plasma would allow them to be controlled independent of the main process chamber gas chemistry, thus increasing process flexibility and control of film chemistry and stoichiometry. ECR supplemented CVD has recently been reported for Cu deposition.[149]

Another area of recently rediscovered interest is one of inductively coupled plasma processes. This entails the use of an RF induction coil with a matching network to create an RF field in a process chamber. The field is used to generate a high density plasma localized close to the surface to be processed.[150-152] Depending on the type of process, the chamber gas pressure can be as high as 50 – 100 Pa (CVD, plasma oxidation)[150,152] or as low as 0.1 Pa (sputtering).[151] Plasma densities as high as 10^{12} cm^{-3} have been obtained in a sputter deposition application[151] thus opening the possibility of high rate deposition of materials with higher uniformity over larger substrate areas compared to other deposition techniques.[153,154] The re-emergence of inductively coupled plasma processing has spurred more interest in understanding the physics of these high density plasma discharges[155] to facilitate better harnessing of their benefits for processing applications in the microelectronic industry.

In conclusion, it is appropriate to note that vapor deposition processing is a field that appears to rejuvenate itself over the years, driven by the needs of specific application areas. There is an ongoing search for methods and processes where variants in process variables and unique process combinations are sought to accommodate ever changing demands of the applications, thus making it a dynamic and challenging field full of research and development opportunities.

References

1. L.I. Maissel and R. Glang (eds.), *Handbook of Thin Film Technology*, (McGraw-Hill, New York, 1970).
2. J.L. Vossen, G.L. Schnable, and W. Kern, *J. Vac Sci. Technol.* **11**, 60 (1974).
3. J.L. Vossen and W. Kern (eds.), *Thin Film Processes II*, (Academic Press, New York, 1991).
4. S.M. Rossnagel, J.J. Cuomo, and W.D. Westwood (eds.), *Handbook of Plasma Processing Technology*, (Noyes Publications, New Jersey, 1989).
5. J.J. Cuomo, S.M. Rossnagel, and H.R. Kaufman (eds.), *Handbook of Ion Beam Processing Technology*, (Noyes Publications, New Jersey, 1989).
6. K. Reichelt and X. Jiang, *Thin Solid Films* **191**, 91 (1990).

7. J.T. Cheung and H. Sankur, *CRC Critical Reviews in Solid State Materials Science* **15**, 63 (1988).

8. F.A. Smidt, *International Materials Reviews* **35**, 61 (1990).

9. P.B. Ghate, *Thin Solid Films* **93**, 359 (1982).

10. P.C. Johnson, *Metal Finishing*,61 (1991).

11. P.J. Martin and R.P. Netterfield, in: *Handbook of Ion Beam Processing Technology*, edited by J.J. Cuomo, S.M. Rossnagel, and H.R. Kaufman (Noyes Publications, New Jersey, 1989), p. 373.

12. G.K. Wolf, *Surf. and Coatings Technol.* **43/44**, 920 (1990).

13. B.A. Movchan and A.V. Demchishin, *Fiz. Metal. Metalloved.* **28**, 653 (1969).

14. J.A. Thornton, *Ann. Rev. Mat. Sci.* **7**, 239 (1977).

15. R. Messier, A. Giri, and R.A. Roy, *J. Vac. Sci. Technol.* **A 2**, 500 (1984).

16. C.R.M. Grovenor, H.T.G. Hentzell, and D.A. Smith, *Acta Metall.* **32**, 773 (1984).

17. C.V. Deshpandey and R.F. Bunshah, in: *Thin Film Processes II*, edited by J.L. Vossen and W. Kern (Academic Press, New York, 1991), p. 79.

18. R.L. Bickerdike, D. Clark, J.N. Eastabrook, G. Hughes, W.N. Mair, P.G. Partridge, and H.C. Ranson, *Intl. J. of Rapid Solidif.* **1**, 305 (1984–85).

19. R.L. Bickerdike, D. Clark, J.N. Eastabrook, G. Hughes, W.N. Mair, P.G. Partridge, and H.C. Ranson, *Intl. J. of Rapid Solidif.* **2**, 1 (1986).

20. C.V. Deshpande and R.F. Bunshah, in: *Handbook of Plasma Processing Technology*, edited by S.M. Rossnagel, J.J. Cuomo, and W.D. Westwood (Noyes Publications, New Jersey, 1989), p. 370.

21. V. Konig and H. Grewe, in: *Tribologie*,vol 1, edited by W. Bunk, J. Hansen, and M. Geyer (Springer, Cologne, Germany, 1981), p. 197.

22. D.M. Mattox, in: *Handbook of Plasma Processing Technology*, edited by S.M. Rossnagel, J.J. Cuomo, and W.D. Westwood (Noyes Publications, New Jersey, 1989), p. 338.

23. M. Aurwater, U.S. Patent 2,920,002, 1960.

24. W. Heitmann, *Appl. Opt.* **10**, 2414 (1971).

25. W. Heitmann, *Appl. Opt.* **10**, 2685 (1971).

26. H. Kuster and J. Ebert, *Thin Solid Films* **70**, 43 (1980).

27. J. Ebert, *SPIE* **325**, 29 (1982).

28. R.F. Bunshah and A.C. Raghuram, *J. Vac. Sci. Technol.* **9**, 1385 (1972).

29. P. Lin, C.V. Deshpandey, H.J. Doerr, R.F. Bunshah, K.L. Chopra, and V.D. Vankar, *Thin Solid Films* **153**, 487 (1987).

30. C,V. Deshpandey and R.F. Bunshah, *Thin Solid Films* **163**, 131 (1988).

31. T. Takagi, I. Yamada, M. Kumnori, and S. Kobiyama, in: *Proc. 2nd Intl. Conf. on Ion Sources*, (Osterreiciche Studiengeselshaft, Vienna, 1972), p. 790.

32. T. Takagi, I. Yamada, and A. Sasaki, *J. Vac. Sci. Technol.* **12**, 1128 (1975).

33. T. Takagi, I. Yamada, and K. Matsubara, *Thin Solid Films* **58**, 9 (1979).

34. I. Yamada and T. Takagi, in: *Handbook of Ion Beam Processing Technology*, edited by J.J. Cuomo, S.M. Rossnagel, and H.R. Kaufman (Noyes Publications, New Jersey, 1989), p. 58.

35. G.M. Davis and M.C. Gower, *Appl. Phys. Lett.* **55**, 112 (1989).

36. E.W. Chase, T. Venkatesan, C.C. Chang, B. Wilkens, W.L. Feldman, and P. Barboux *J. Mater. Res.* **4**, 1326 (1989).

37. A. Gupta and B. Hussey, RC 16334, 1990. IBM Watson Research Center, Yorktown Heights.

38. P.E. Dyer, A. Issa, and P.H. Key, *Appl. Phys. Lett.* **57**, 186 (1990).

39. M.G. Norton and C.B. Carter, *Physica C* **172**, 47 (1990).

40. R. De Reus, F. W. Saris, G.J. Van Der Kolk, C. Wittmer, B. Dam, D.H.A. Blank, D.J. Adelerhof, and J. Flokstra, *Mater. Sci. Eng.* **B7**, 135 (1990).
41. J. Dubowski, *Proc. Soc. Photo-Opt. Instrum. Eng.* **668**, 97 (1986).
42. J.J. Cuomo, J. Bruley, J.P. Doyle,D.L. Pappas, K. Saenger, J.C. Liu, and P.E. Batson, *Proc. Mat. Res. Symp.* **202**, 247 (1991).
43. D.L. Pappas, K.L. Saenger, W. Krakow, J.J. Cuomo, T. Gu, and R.W. Collins, *J. Appl. Phys.*, in press.
44. K. L. Saenger, *Processing of Advanced Materials*; In press.
45. P.C. Johnson, in: *Thin Film Processes II*, edited by J.L. Vossen and W. Kern (Academic Press, New York, 1991), p. 209.
46. W.M. Mularie, U.S. Patent 4,430,184, 1984.
47. S. Ramalingham, C.B. Qi, and K. Kim, U.S. Patent 4,673,477, 1987.
48. P. Robinson and A. Matthews, *Surface and Coatings Tech.* **43/44**,288 (1990).
49. P.C. Johnson and H. Randhawa, *Surface and Coatings Tech.* **33**, 53 (1987).
50. W.D. Munz, *J. Vac. Sci. Technol.* **A4**, 2717 (1986).
51. H. Randhawa, *Thin Solid Films* **153**, 209 (1987).
52. I.I. Aksenov, V.A. Belous, V.G. Padalka, and V.M. Khoroshikh, *Sov. J. Plasma Phys.* **4**, 425 (1978).
53. I.I. Aksenov, V.G. Padalka, N.S.Repalov, and V.M. Khoroshikh, *Sov. J. Plasma Phys.* **6**, 173 (1980).
54. V.A. Osipov, V.G. Padalka, L.P. Sablev, and R.I. Stupak, *Inst. and Exp. Techniques* **21**, 1650 (1978).
55. S.D. Berger, D.R. McKenzie, and P.J. Martin, *Phil. Mag.* **B 52**, 285 (1988).
56. C. Weissmantel, C. Schurer, F. Frohlich, P. Grau, and H. Lehmann *Thin Solid Films* **61**, L5 (1979).
57. J.S. Logan, in: *Handbook of Plasma Processing Technology* edited by S.M. Rossnagel, J.J. Cuomo, and W.D. Westwood (Noyes Publications, New Jersey, 1989), p. 140.
58. S.M. Rossnagel, in: *Handbook of Plasma Processing Technology*, edited by S.M. Rossnagel, J.J. Cuomo, and W.D. Westwood (Noyes Publications, New Jersey, 1989), p. 160.
59. W.D. Westwood, in: *Handbook of Plasma Processing Technology*, edited by S.M. Rossnagel, J.J. Cuomo, and W.D. Westwood (Noyes Publications, New Jersey, 1989), p. 233.
60. R. Parsons, in: *Thin Film Processes II*, edited by J.L. Vossen and W. Kern (Academic Press, New York, 1991), p. 79.
61. I.A. Blech, *Thin Solid Films* **6**, 113 (1970).
62. C.Y. Ting, V.J. Vivalda, and H. G. Schaeffer, *J. Vac. Sci. Tech* **15**, 1105 (1978).
63. Y. Homma and S. Tsunekawa, *J. Electrochem. Soc.* **132**, 1466 (1985).
64. D.M. Mattox and G.J. Kominiak, *J. Vac. Sci. and Technol.* **9**, 528 (1972).
65. D.W. Hoffman and R.C. McCune, in: *Handbook of Plasma Processing Technology*, edited by S.M. Rossnagel, J.J. Cuomo, and W.D. Westwood (Noyes Publications, New Jersey, 1989), p. 483.
66. J.A. Thornton and D.W. Hoffman, *Thin Solid Films* **171**, 5 (1989).
67. B. Window and K-H. Mueller, *Thin Solid Films* **171**, 183 (1989).
68. F.M. D'Heurle, *Int. Mater. Rev.* **34**, 53 (1989).
69. H. Windischmann, *J. Vac. Sci. Technol.* **A9**, 2431 (1991).
70. T.J. Vink, M.A.J. Somers, J.L.C. Daams, and A.G. Dirks, *J. Appl. Phys.* **70**, 4301 (1991).
71. G.H. Maher and R.J. Diefendorf, *IEEE Trans. PHP* **8**, 11 (1972).

72. S. Fukunishi, A. Kawana, N. Uchida, and J. Noda, *J. Appl. Phys. Suppl.* **2**, 749 (1974).

73. Y. Higuma, K. Tanaka, T. Nakagawa, T. Kariya, and Y. Hamakawa, *Japan J. Appl. Phys.* **16**, 1707 (1975).

74. K.K. Shih, D.A. Smith, and J.R. Krowe, *J. Vac. Sci. and Technol.* **A6**, 3 (1988).

75. Z.M. Li and R.R. Parasons, *J. Vac. Sci. Technol.* **A6**, 3062 (1988).

76. R. Ludwig, K. Kastner, R. Kukla, and M. Mayr, *IEE Trans. Magn.* **23**, 94 (1987).

77. M.J. Park, A. Leyland, and A. Matthews, *Surf. and Coatings. Tech.* **43/44**, 481 (1990).

78. A. Billard, M. Foos, C. Frantz, and M. Gantois, *Surf. and Coatings Tech.* **43/44**, 521 (1990).

79. P. Chaudhari, J.J. Cuomo, and R.J. Gambino, *Appl. Phys. Lett.* **22**, 337 (1972).

80. P. Chaudhari, J.J. Cuomo, and R.J. Gambino, *IBM J. Res. and Dev.* **17**, 66 (1973).

81. E. Spiller, *J. Vac. Sci. and Technol.* **A6**, 1709 (1990).

82. S. Schiller, U. Heisig, K. Goedicke, K. Schade, G. Teschner and J. Henneberger, *Thin Solid Films* **64**, 455 (1979).

83. P. Chang and D.M. McGarr, Diagnostic Techniques in VLSI Fabrication, (Semiconductor Equipment and Materials Inst, California, 1987), p. 189.

84. S. Maniv, C. Miner, and W.D. Westwood, *J. Vac. Sci. and Technol.* **18**, 195 (1981).

85. G. Este and W.D. Westwood, *J. Vac. Sci. and Technol.* **A6**, 1845 (1988).

86. S. Schiller, U. Heisig, C. Korndoefer, J. Strumpfel, and P. Frach, *Surf. and Coatings Technol.* **39/40**, 549 (1989).

87. J.E. Sundgren and H.T.G. Hentzell, *J. Vac. Sci. and Technol.* **A4**, 2259 (1986).

88. J.-PH. Nabot, A. Aubert, R. Gillet, and PH. Renaux, *Surf. and Coatings Technol.* **43/44**, 629 (1990).

89. U. Helmersson, J.-E. Sundgren, and J.E. Greene, *J. Vac. Sci. and Technol.* **A4**, 500 (1986).

90. A. Paccagnella, A. Callegari, N. Braslau, H. Hovel, and M. Murakami, *IEEE Trans. on Elec. Dev.* **36**, 2595 (1989).

91. A. Paccagnella, A. Callegari, A. Carnera, M. Gasser, E. Latta, M. Murakami, and M. Norcott, *J. Appl. Phys.* **69**, 2356 (1991).

92. J. Webb and C. Halpin, *Appl. Phys. Lett.* **47**, 831 (1985).

93. J.B. Webb, C. Halpin, and J.P. Noad, *J. Vac. Sci. and Technol.* **A4**, 379 (1986).

94. T.S. Rao, J.B. Webb, Y. Beaulieu, J.L. Brebner, J.P. Noad and J. Jackman, *J. Vac. Sci. and Technol.* **A7**, 1210 (1989).

95. C.M. Horwitz, in: *Handbook of Plasma Processing Technology*, edited by S.M. Rossnagel, J.J. Cuomo, and W.D. Westwood (Noyes Publications, New Jersey, 1989), p. 308.

96. J.J. Cuomo and S.M. Rossnagel, *J. Vac. Sci. Technol.* **A4**, 393 (1986).

97. J.J. Cuomo, H.R. Kaufman, and S.M. Rossnagel, U.S. Patent 4,588,490 1986.

98. Y.S. Kuo, R.F. Bunshah, and D. Okrent, *J. Vac. Sci. Technol.* **A4**, 397 (1986).

99. D. Mikalsen and S.M. Rossnagel, U.S. Patent No. 4,824,544, 1989.

100. S.M. Rossnagel, D. Mikalsen, H. Kinoshita, and J.J. Cuomo, *J. Vac. Sci. and Technol.* **A9**, 261 (1991).

101. K. Suzuki, S. Okudaira, N. Sakuda, and I. Kanamoto, *Japan J. Appl. Phys.* **16**, 1979 (1977).

102. N. Sakudo, K. Tokiguchi, H. Koiki, and I. Kanomoto, *Rev. Sci. Instrum.* **48**, 762 (1977).

103. N. Sakudo, K. Tokiguchi, H. Koike, and I. Kanomoto, *Rev. Sci. Instrum.* **49**, 940 (1978).

104. W.M. Holber, in: *Handbook of Ion Beam Processing Technology*, edited by J.J. Cuomo, S.M. Rossnagel, and H.R. Kaufman (Noyes Publications, New Jersey, 1989), p. 21.

105. M. Matsuoko and K. Ono, *J. Vac. Sci. and Technol.* **A6**, 25 (1988).

106. S. Matsuo and Y. Adachi, *Japan J. Appl. Phys.* **21**, 14 (1982).

107. B. Petit and J. Pelletier, *Japan J. Appl. Phys.* **26**, 825 (1987).

108. K. Machida and H. Oikawa, *J. Vac. Sci. Technol.* **B4**, 818 (1986).

109. S. Matsuo and M. Kiuchi, *Japan J. Appl. Phys.* **22**, L210 (1983).

110. K. Kobayashi, M. Hayama, S. Kawamoto, and H. Miki, *Japan J. Appl. Phys.* **26**, 202 (1987).

111. H.R. Kaufman and R. Robinson, in: *Handbook of Ion Beam Processing Technology*, edited by J.J. Cuomo, S.M. Rossnagel, and H.R. Kaufman (Noyes Publications, New Jersey, 1989), p. 8.

112. S. M. Rossnagel, in: *Handbook of Ion Beam Processing Technology*, edited by J.J. Cuomo, S.M. Rossnagel, and H.R. Kaufman (Noyes Publications, New Jersey, 1989), p. 362.

113. J.M.E. Harper, J.J. Cuomo, and H.T.G. Hentzell, *J. Appl. Phys.* **58**, 550 (1985).

114. J.J. Cuomo, J.P. Doyle, J. Bruley, and J.C. Liu, *Carbon* **28**, 761 (1990).

115. J.J. Cuomo, D.L. Pappas, J. Bruley, J.P. Doyle, and K.L. Saenger, *J. Appl. Phys.* **70**, 1706 (1991).

116. J.J. Cuomo, J.P. Doyle, J. Bruley, and J.C. Liu, *Appl. Phys. Lett.* **58**, 1 (1991).

117. K.F. Jensen and W. Kern, in: *Thin Film Processes II* edited by J.L. Vossen and W. Kern (Academic Press, New York, 1991), p. 283.

118. R. Reif and W. Kern, in: *Thin Film Processes II*, edited by J.L. Vossen and W. Kern (Academic Press, New York, 1991), p. 525.

119. G. Lucovsky, D.V. Tsu, and R.J. Markunas, in: *Handbook of Plasma Processing Technology*, edited by S.M. Rossnagel, J.J. Cuomo, and W.D. Westwood (Noyes Publications, New Jersey, 1989), p. 387.

120. J.G. Eden, in: *Thin Film Processes II*, edited by J.L. Vossen and W. Kern (Academic Press, New York, 1991), p. 443.

121. T.M. Mayer and S.D. Allen, in: *Thin Film Processes II* edited by J.L. Vossen and W. Kern (Academic Press, New York, 1991), p. 621.

122. T.F. Kuech and J.F. Jensen, in: *Thin Film Processes II* edited by J.L. Vossen and W. Kern (Academic Press, New York, 1991), p. 369.

123. L-P. Andersson, *Thin Solid Films* **86**, 193 (1981).

124. C. Weissmantel, R. Bewilogua, K. Breuer,D. Dietrich, H.J. Erler, B. Rau and G. Reisse, *Thin Solid Films* **96**, 31 (1982).

125. P.W. Carey and D.C. Cameron, *J. Mater. Process. Technol.* **26**, 117 (1991).

126. E. Kay and S. M. Rossnagel, in: *Handbook of Ion Beam Processing Technology*, edited by J.J. Cuomo, S.M. Rossnagel, and H.R. Kaufman (Noyes Publications, New Jersey, 1989), p. 170.

127. R. Roy and D.S. Yee, in: *Handbook of Ion Beam Processing Technology*, edited by J.J. Cuomo, S.M. Rossnagel, and H.R. Kaufman (Noyes Publications, New Jersey, 1989), p. 194.

128. J. Baglin, in: *Handbook of Ion Beam Processing Technology*, edited by J.J. Cuomo, S.M. Rossnagel, and H.R. Kaufman (Noyes Publications, New Jersey, 1989), p. 279.

129. J.K. Hirvonen, *Matl. Sci. Reports* **6**, 215 (1991).

130. J.M.E. Harper, J.J. Cuomo, and H.T.G. Hentzell, *Appl. Phys. Lett.* **43**, 547 (1983).

131. G.K. Wolf, *Vacuum* **39**, 1105 (1989).

132. G.K. Wolf, *Nucl. Instrum. Methods* **B 46**, 369 (1990).

133. E.H. Hirsch and I.K. Varga, *Thin Solid Films* **69**, 99 (1980).

134. J.J. Cuomo, J.P. Doyle, J. Bruley,, and J.C. Liu, *J. Vac. Sci. Tech.* **A9**, 2210 (1991).

135. J. Baglin, *Ion Beam Modification of Insulators*,eds. P. Mazzoldi and G.W. Arnold (1986), p. 585.

136. R.A. Roy, J..J. Cuomo, and D.S. Yee, *J. Vac. Sci. Technol.* **A6**, 1621 (1988).

137. R.A. Roy, D.S. Yee, and J.J. Cuomo, in: *Processing and Characterization of Materials by Ion Beams*, edited by L.E. Rehn *et al.*,(MRS, 1989), p. 23.

138. P.J. Martin, R.P. Netterfield, and W.G. Sainty, *J. Appl. Phys.* **55**, 235 (1984).

139. R.P. Netterfield, W.G. Sainty, P.J. Martin, and S.J. Sie, *Appl. Opt.* **24**, 2267 (1985).

140. J.J. Cuomo., J.M.E. Harper, C.R. Guarnieri, D.S. Yee, L.J. Attanasio, J. Angilello, C.T. Wu, and R. Hammond, *J. Vac. Sci. Tecnol.* **20**, 349 (1982).

141. D.W. Hoffman and M.R. Gaertner, *J. Vac. Sci. Technol.* **17**, 425 (1980).

142. D.S. Yee, J. Floro, D.J. Mikalsen, J.J. Cuomo, K.Y. Ahn, and D.A. Smith, *J. Vac. Sci. Technol.* **A3**, 2121 (1985).

143. W. Ensinger and G.K. Wolf, *Mater. Sci. Eng.* **A116**, 1 (1987).

144. J. Ebert, *Surf. and Coatings Tecnol.* **43/44**, 950 (1990).

145. F. Marchetti, M. Dapor, S. Girardi, F. Giacomozzi, and A. Cavalleri, *Mater. Sci. Eng.* **A 115**, 217 (1989).

146. S. Nakashima, M. Fukushima, M. Haginoya, K. Oohata, I. Hashimoto, and M. Terakado, *Mater. Sci. Eng* **A 115**, 197 (1989).

147. M. Barth, W. Ensinger, A. Schroer, and G.K. Wolf, in: *Proc. 3rd Intl. Conf. on Surf. Modif. Technol.*, (The Metallurgical Society, PA, 1990), p. 195.

148. E. McCafferty, G.K. Hubler, P.M. Natishan, P.G. Moore, R.A. Kant, and B.D. Sartwell *Mater. Sci. Eng.* **86**, 1 (1987).

149. J. Pelletier, R. Pantel, J.C. Oberlin, Y. Pauleu, and P. Gouy-Pailler, *J. Appl. Phys.* **70**, 3862 (1991).

150. V.Q. Ho and T. Sugano, *IEEE Trans. Elec. Dev.* **ED-27**, 1436 (1980).

151. M. Yamashita, *J. Vac. Sci. Technol.* **7**, 151 (1989).

152. R.A. Rudder, G.C. Hudson, J.B. Posthill, R.E. Thomas, R.C. Hendry, D.P. Malta, R.J. Markunas, T.P. Humphreys, and R.J. Nemanich *Appl. Phys. Lett.* **60**, 329 (1992).

153. P.H. Singer, *Semicond. Intl.*, 46 (1991).

154. C.R. Guarnieri, J.A. Hopwood, S.J. Whitehair, and J.J. Cuomo, *Proc. 38th Natl. Symp. of AVS*, 114 (1991).

155. J.A. Hopwood, C.R. Guarnieri, S.J. Whitehair, and J.J. Cuomo, Unpublished research, IBM Watson Research Center, 1992.

Thin Films for Photovoltaic Applications

R. Asomoza[1], A. Maldonado[1], D.R. Acosta[2] and J. Rickards[2]

[1] *Centro de Investigación y de Estudios Avanzados del IPN*
Apartado Postal 14-740
07000 México, D.F
MEXICO

[2] *Instituto de F!sica*
Universidad Nacional Autónoma de México
Apartado Postal 20364
01000 México, D.F.
MEXICO

Abstract

The interest in research and development activities focused in photovoltaic (PV) materials in Mexico is justified by the great potential this country has for photovoltaic applications. In fact, Mexico has one of the highest insolation rates in the World, in addition to more than 5 millions people who are potential users of PV systems.

A description of the research activities, at the Electrical Engineering Department of the Center of Research and Advanced Studies (CINVESTAV), concerning materials for photovoltaic applications, is given. The materials studied include semiconductor oxides, III-V compounds and ternary compounds. They were all prepared as thin films. In this paper we describe the preparation and characterization of these films.

I. Introduction

In Mexico there has been a great interest in developing technology for photovoltaic (PV) applications, mainly for two reasons: i) the Country has excellent insolation conditions in most of its territory and ii) there is a great number of potential users of photovoltaic systems.

There have been efforts at the CINVESTAV to develop technology for small scale production of solar cells and modules and to demonstrate the feasibility of PV systems. These efforts have resulted in the installation and operation of a pilot line for production of solar modules with a capacity of 25 kW/year. On the other hand,

Advanced Topics in Materials Science and Engineering, Edited by
J.L. Morán-López and J.M. Sanchez, Plenum Press, New York, 1993

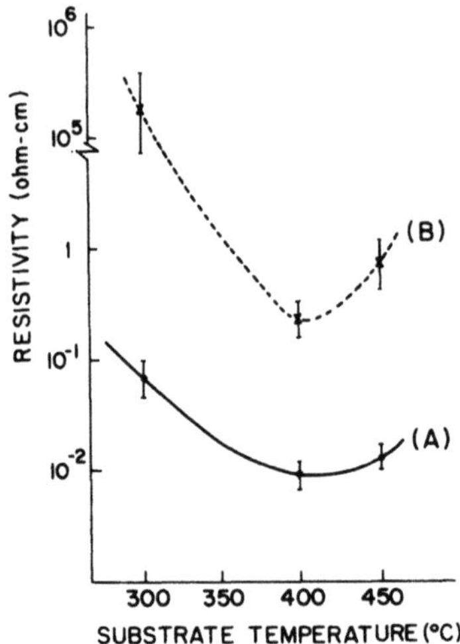

Figure 1. Resistivity as a function of substrate temperature for nondoped tin oxide samples. Curve (A) corresponds to a film deposited on glass and curve (B) to a sample deposited on KBr.

a great number of PV systems for specific applications have been installed in the field. The main applications, up to now, are: lighting systems, radio communications systems, educational TV and water pumping systems.

The technology developed so far has covered different aspects of those PV systems, such as: optimization of atomic diffusion processes in the fabrication of solar cells, improvement of the electrical characteristics of the cells and modules produced, and design and implementation of PV systems for specific applications, among others. Little emphasis was put, however, on materials research, mainly because the starting semiconductor material was commercially available single crystalline silicon wafers and high purity metals and chemicals.

Even though the goal mentioned before was the pilot production of PV systems, there have been, in parallel, research programs on high efficiency solar cells and thin film solar cells.

In what follows, a brief account of the results obtained so far, concerning the properties of the materials employed for thin film solar cells, is given. Different techniques for characterizing the films were employed. When they were not available at our laboratory, a collaborative research program was established, when possible.

II. Semiconductor Oxides

We have applied SnO_2 and ZnO as antireflecting coatings in the solar cells produced at the pilot line. The method used to prepare these films was Chemical Spray Pyrolysis. This method has the advantages of its simplicity and low cost. The films have reproducible and uniform properties over large areas, in our case three inches in diameter. The method is widely used and our particular setup has been described elsewhere.[1] The best deposition conditions, for the applications mentioned before,

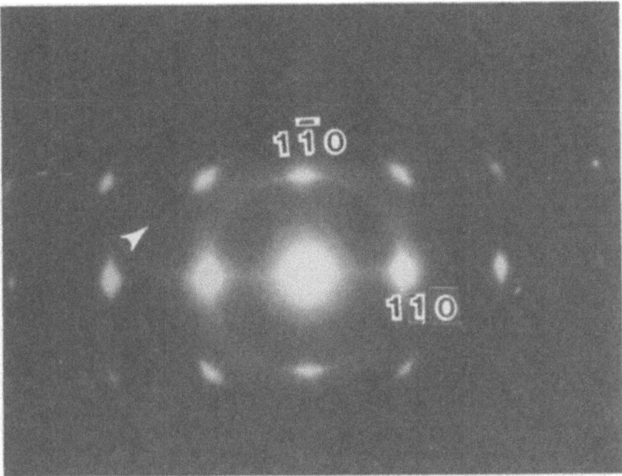

Figure 2. Transmission electron diffraction pattern of a tin oxide sample prepared at T_s =300°C. Several phases were deduced from this pattern (see text).

were determined by carefully exploring the influence that the deposition parameters have on the physical properties of the films, in particular, on the electrical and optical characteristics.

The techniques employed to characterize the films were: four point probe, Transmittance in the visible, Transmission Electron Microscopy (TEM), Auger Electron Spectroscopy (AES) and Resonant Nuclear Reactions (RNR).

II.1 Non-Doped SnO_2 Films

It was found that good quality films were those prepared at substrate temperatures in the range 420–450°C. The transmittance obtained was as high as 85% and the resistivity as low as 10^{-2} Ωcm, as shown in Fig. 1. Even though these values were acceptable for the applications mentioned before, we tried to correlate the results obtained to the preparation conditions of the films.

A TEM study indicated that in samples prepared at temperatures up to 400°C, a mixture of different phases was present. Fig. 2 shows the diffraction pattern of a sample prepared at 300°C. The crystalline structure deduced from this pattern was a tetragonal one with lattice parameters: a=0.472 and c=0.313 nm, corresponding to SnO_2. The extra spots in this figure indicated the presence of SnO and Sn_3O_4. On the contrary, Fig. 3 shows the diffraction pattern of a sample prepared at 400°C, the ring pattern corresponds to a single phase of SnO_2. The other phases mentioned before, if present, were negligible.

Fig. 4 shows the transmittance of SnO_2 samples prepared at different substrate temperatures, T_s. It is to be noted that the transmittance increases as T_s increases. According to the TEM results this is compatible with the fact that the samples prepared at 450°C are composed by nearly stoichiometric SnO_2, while the other samples are a mixture of other phases, usually less transparent.

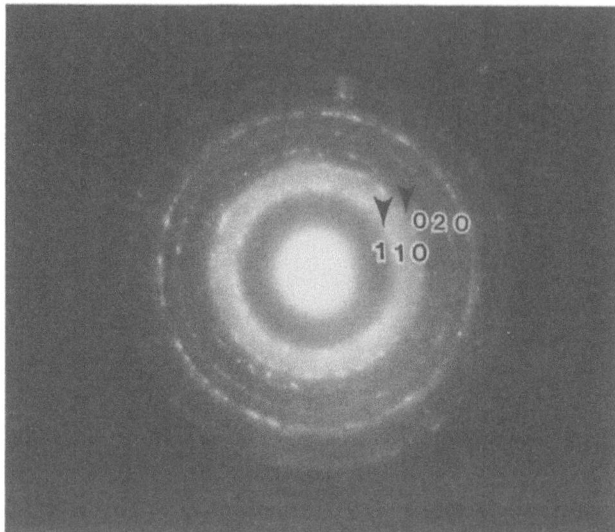

Figure 3. Transmission electron diffraction pattern of a tin oxide sample prepared at T_s =400°C. This pattern corresponds to a SnO_2 single phase.

II.2 Doped SnO_2 Films

In order to decrease the resistivity of SnO_2 films, without much decrease in transparency, fluorine was used as a dopant. The resistivity decreased to values in the range of 10^{-3} Ωcm. However, we could not get a clear picture of the incorporation of F into the films, in particular, because there was not a clear correlation between the resistivity of the films and the fluorine concentration in the departure solution.

In order to get more insight into the way the fluorine distributes into the films, two techniques were used to determine its concentration and depth profiles. These were AES and RNR. The AES technique is widely used to determine the composition of solid samples. However, in SnO_2:F, the fluorine desorption produced in the sample by the electron beam prevented a reliable determination of its concentration. On the contrary RNR proved to be a powerful technique not only to determine the fluorine concentration in the samples but also to give its depth profile.

The RNR experimental method has been described elsewhere,[2] but briefly it consists in the following: the sample is bombarded with nuclear projectiles having an energy in the vicinity of an isolated resonance in a reaction with the element to be detected. In our case we used a proton beam from the 700 kV Van der Graaff accelerator of the Instituto de Fisica (UNAM). The nuclear reaction was $^{19}F(p,\gamma)^{16}O$, with a resonance at 340 keV. Gamma rays at 6.14 MeV were detected. The resonance corresponds to the excitation of a discrete level in the compound nucleus ^{20}Ne; its width is 2.4 keV.

Fig. 5 shows the results obtained, using this technique for two SnO_2:F samples prepared with different concentrations in the departure solution. As can be seen, there is a good signal to background ratio which allows a precise determination of the gamma rays coming from the reaction studied. The gamma ray yield depends on the

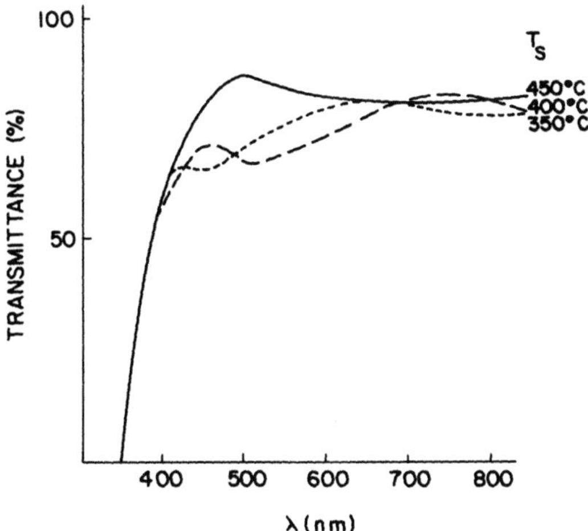

Figure 4. Transmittance *versus* wavelength for tin oxide samples prepared at different substrate temperatures.

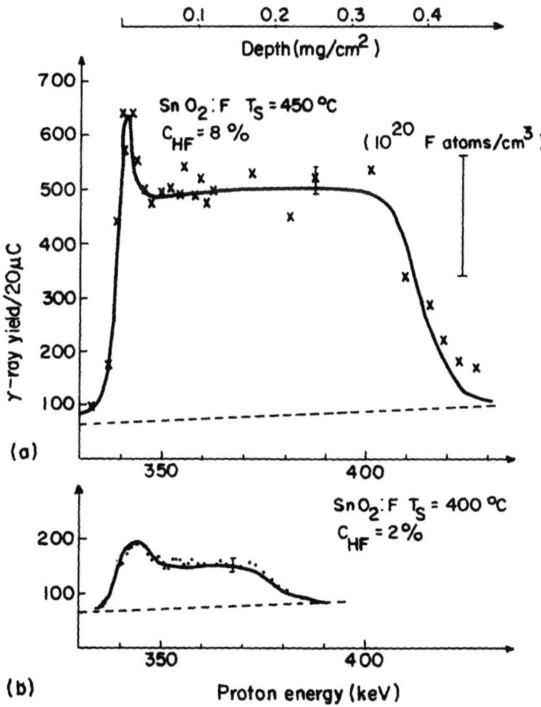

Figure 5. Excitation curves of the ^{19}F$(p,\alpha\gamma)^{16}$O reaction for two SnO$_2$:F samples. The preparation conditions were: (a) T$_s$ =450°C and 8% HF concentration in the departure solution, (b) T$_s$=400°C and 2% HF. The absolute vertical scale is indicated in (a).

number of fluorine atoms interacting with the impinging protons. By increasing the energy of these protons, fluorine atoms below the surface can be detected. In Fig. 5 a depth profile is also shown for the two samples mentioned before. In both cases there is evidence for an accumulation of fluorine atoms near the surface and an approximately constant concentration in the bulk of the film. The fall off at high energy corresponds to the interface SnO_2:F/substrate. By measuring the FWHM of the distribution curves, the thicknesses of the samples can be determined. In Fig. 5 a depth scale is also given. An absolute value of the fluorine concentration can be obtained if the system is calibrated by comparison to a target with a known concentration of fluorine. In our case we used a polycrystalline LiF target. It was then possible to assign a value to the vertical axis of Fig. 5, in units of fluorine atoms per cubic centimeter.

In summary, transmission electron microscopy results show that the films prepared below 400°C are composed of a mixture of several phases; small amounts of SnO and Sn_3O_4 embedded in SnO_2. On the other hand, films prepared above 400°C are almost stoichiometric tin oxide. The resonant nuclear reactions method was very sensitive to the presence of fluorine, allowing the determination of its absolute concentration, its depth profile as well as possible local segregations.

II.3 GaAs/GaAlAs Films.(*)

Solar cells based on GaAs are very attractive because they can attain high conversion efficiencies. Recently, efficiencies as high as 21% have been reported in AM 1.5 conditions.[3] Moreover, these cells can be used in concentration systems, lowering the costs of production, because less material is required.

Thin film solar cells have usually low conversion efficiency because a great number of defects are originated during growth. However, large areas can be covered with only small quantities of material, making them cheaper and more attractive than bulk solar cells. MOCVD deposition of GaAs has been proposed to produce terrestrial solar cells because with this technique is possible to cover large areas on different kinds of substrates, including Si, at lower cost than other deposition techniques. The quality of the films deposited is comparable to that of films obtained by using techniques like liquid phase epitaxy (LPE). Conversion efficiencies as high as 19% have been obtained by using MOCVD grown material.[4]

AsH_3 is currently used in MOCVD processes, however, this compound is highly toxic and dangerous. In order to avoid the use of AsH_3, and reduce the health risk, solid As was proposed as a substitution. Trimetilgallium (TMGa) and trimetilaluminium (TMAl) were used as precursors of Ga and Al respectively. There existed very few reports in the literature on this process[5] and it was then necessary to perform a complete study. For that purpose, a reactor was designed. It was composed of two hot zones, one for sublimating the As and the other to control the decomposition of the other gases. As is transported to the substrate by diffusion.[6]

The work was oriented in two directions:[7]

i) Growth and characterization of GaAs and GaAlAs to be used as the active layers of the solar cell, and

ii) Growth of GaAs on Si, in order to use low cost substrates.

Up to know, it has been possible to grow high quality GaAs on GaAs and GaAlAs on GaAs. In particular, it has been possible to obtain reproducible results concerning

---

* This work was performed by the MOCVD Group in the Solid State Electronics Section at CINVESTAV, Mexico.

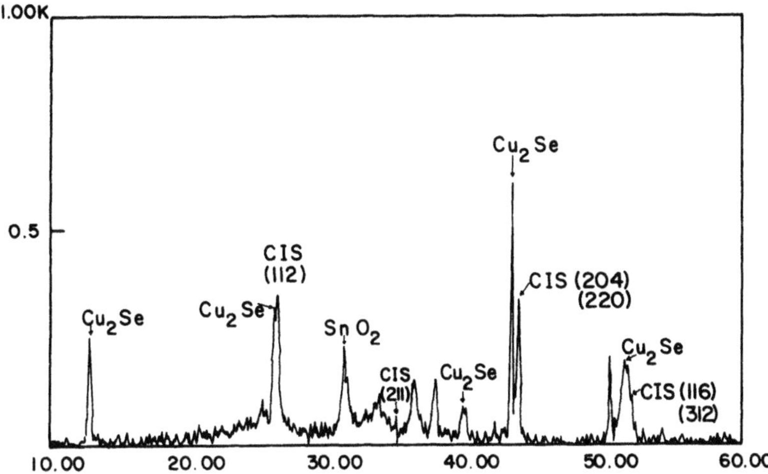

Figure 6. X-ray pattern of a CuInSe$_2$ (CIS) thin film prepared by electrodeposition. In the as prepared condition two phases were identified, Cu$_2$Se and CuInSe$_2$.

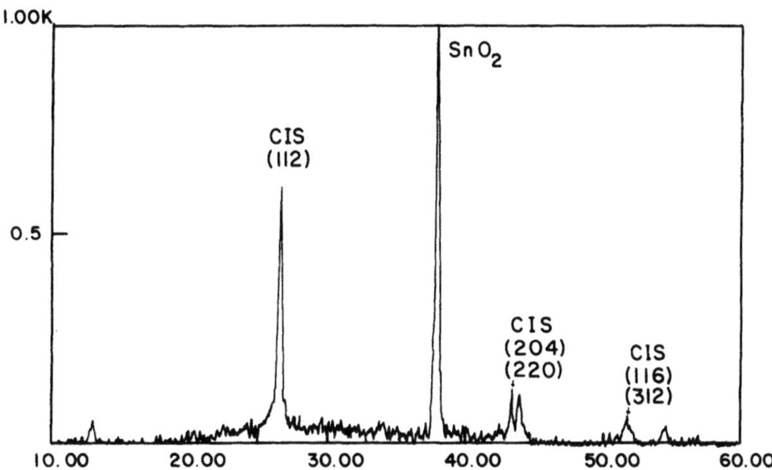

Figure 7. X-ray pattern of a CuInSe$_2$ (CIS) thin film prepared by electrodeposition and annealed at 350°C for two hours. The main compound present was CIS.

GaAs doped with C and Zn for p-type material and S for n-type. For GaAlAs it has been possible to obtain good surface morphology and crystallinity. The systematic study of doped films is in progress. As dopants Diethylzinc and Silane are to be used for p-type and n-type films, respectively. The properties of GaAs films grown on Si are still to be improved.

The Van der Paw technique was used to determine the resistivity, mobility and concentration of carriers. Typical results are: mobility at 77 K = 10000 cm^2/Vs; the best room temperature mobility was 5000 cm^2/Vs. The lowest carrier concentration was $2\times10^{15}cm^{-3}$ for n type and $3\times10^{15}cm^{-3}$ for p-type films. The surface of the films was mirrorlike.

III. Ternary Compounds

Thin film ternary compounds having the formula I-III-X_2 (X = S, Se Te) have shown good characteristics to be used as active absorbers in solar cells.[8] Efficiencies over 12% have been reported for $CuInSe_2$/CdS solar cells.[9] Different methods have been employed to prepare these films. In our laboratories, electrodeposition and chemical spray pyrolysis have been used to grow $CuInSe_2$[10] and $CuInS_2$[11] respectively. Up to now only partial results have been reached. In what follows a brief account of the results obtained in preparing $CuInSe_2$ is given.

III.1 $CuInSe_2$.(**)

As mentioned before, $CuInSe_2$ (CIS) films were obtained by electrodeposition. This technique has the following advantages:
- Large area deposition
- Low cost
- It can be easily adapted to large scale production

The starting solution had the following composition: $CuSO_4$, $5.2x10^{-3}$M; $In_2(SO_4)_3$, 10^{-2}M; SeO_2, 10^{-2}M; and $Na_3C_6H_5O_7$, $2x10^{-2}$M. The concentration of the sodium citrate, used as a complexing agent, was variable. The pH was kept constant at a value 1.87, the temperature of the solution was 38°C.

Tin oxide coated glasses were used as substrates. In some cases Ni coated glasses were used instead.

Fig. 6 shows an X-ray pattern of a sample prepared at a voltage v= 0.556V. It can be seen that in addition to CIS peaks, those corresponding to the phase Cu_2Se are present. Fig. 7 corresponds to a sample prepared under the same conditions but annealed at 350°C for two hours. In this case, the peaks characteristic of the phase Cu_2Se have almost disappeared, remaining only the peaks corresponding to $CuInSe_2$.

We can say that even if X ray analysis show that the main phase is $CuInSe_2$, the films obtained by this technique are not uniform at a microscopic scale. As a matter of fact they are quite porous making difficult their use in a device. Studies are under way to improve the adherence to the substrate as well as to reduce the porosity of the films.

References

1. A. Tiburcio-Silver, Ms. Sc. Thesis, Electrical Engineering Department, CINVES-TAV-IPN, Mexico (1985).
2. E.P. Zironi, J. Rickards, A. Maldonado, and R. Asomoza, *Nuclear Instruments and Methods in Physics Research* **B45**, 115 (1990).
3. R.P. Gale, R.W. McClelland, B.D. King, and J.C.C. Fan *Solar Cells* **27**, 99 (1989).

**   The preparation of the $CuInSe_2$ thin films was performed by Dr. Omar Solorza and Araceli Palafox, at the Chemistry Department of CINVESTAV, Mexico.

4. J.C. Chen, M. Ladle Ristow, J.I. Cubbage, and J.C. Werthen, *Appl. Phys. Lett.* **58**, 2282 (1991).

5. R. Bhat, *J. Electron. Mater.* **14**, 433 (1985).

6. R. Peña-Sierra, J.C. Castro-Zavala, and A. Escobosa, *J. Crystal Growth* **107**, 337 (1991).

7. R. Peña, V. Sánchez-Resendiz, and A. Escobosa. Private communication.

8. R.A. Mickelsen and W.S. Chen, *Proceedings of the 16th IEEE Photovoltaic Specialist Conference* (IEEE, New York, 1982).

9. K. Mitchel, C. Eberspacher, J. Ermer, and D. Pier, *Proceedings of the 20th IEEE Specialist Conference*, Las Vegas (IEEE, New York, 1988).

10. A. Palafox, O. Solorza, E. Arias, and R.Rivera., *XV National Congress of the Mexican Asociation of Engineers on Corrosion*, Toluca, Mexico, (1991).

11. A. Maldonado and R. Asomoza, *XI National Congress of the Mexican Society of Vacuum and Surface Science*, San Luis Potosi, Mexico, (1991).

Optical Properties of New Materials

Rubén G. Barrera

Instituto de Física
Universidad Nacional Autónoma de México
Apartado Postal 20-364
01000 México, D.F.
MEXICO

Abstract

An overall picture of the scientific activity of the mexican community in the area of optical properties of solids is presented. The latest work in new materials is emphasized and the main achievements of the theoretical group of the Institute of Physics at the National University of Mexico (UNAM) are reported. These achievements are in the field of optical properties of inhomogeneous media and they comprise: the determination of the surface impedance of the surface region, a new optical spectroscopy based on surface-induced optical anisotropies and a theory for the calculation of the effective dielectric response in composite materials.

I. Introduction

The main objective of the present paper is to give an overall picture of the scientific activity of the mexican community in the area of optical properties of solids emphasizing the latest work on new materials and the main achievements of our theoretical group from the Institute of Physics at the National University of Mexico (UNAM). In order to understand the capabilities and acomplishments of the mexican community in this field we display, in section 2, some figures about the size and productivity of the physics community as a whole, as well as its main fields of activity. This will give us an idea of the relative importance of the field of optical properties within Mexico and at the same time the proper scaling which has to be considered in the evaluation of its performance within the international community. In order to give an idea about the type of problems in this field that are actually being tackled, a short list of the titles of recently published articles is also included. In section 3 the main achievements of our theoretical group at UNAM are reported. This includes work on Surface Optics, on a new optical spectroscopy based on Surface Induced Optical Anisotropies and Optical Properties of Composite Materials.

Advanced Topics in Materials Science and Engineering, Edited by
J.L. Morán-López and J.M. Sanchez, Plenum Press, New York, 1993

Table I

Main groups	PhD's	T	E
Condensed Matter Physics	96	31	65
Statistical Physics and Thermodynamics	40	35	5
Astronomy and Astrophysics	40	2	38
Particle and Fields	27	24	3
Material Science	20	5	15
Relativity and Mathematical Physics	26	26	0
Atomic and Molecular Physics	23	15	8
Nuclear Physics	19	9	10
Optics	14	1	13

II. Mexican Physics in Figures

The oldest institution in Mexico devoted to basic research in physics is the Institute of Physics of the National University of Mexico which was founded in 1939. Although there were some other institutions dedicated to the study of areas related to physics like the School of Mines (Seminario Real de Minas), founded under the Spanish rule at the end of the 18th century, and the National Astronomical Observatory (Observatorio Astronómico Nacional) founded in the second half of last century by the republican government, their main objective was not to foster basic research in modern physics. In this sense we can say that the birth of basic research in physics in mexican institutions is just about 50 years old.

The latest studies[1,2] about the size of the mexican community dedicated to physics report 1086 full-time academic staff employed in 38 institutions. From these 497 have a Ph.D. degree (or equivalent), 308 a Master's degree and 261 a Physicist degree. This latest degree is obtained in Mexico after 9 semesters of courses in physics and mathematics and a thesis (not necessarily of original work) whose ellaboration usually takes between two to three semesters. Now, with respect to the location of the university where the highest (doctorate) degrees were obtained: 168 were obtained in Mexico and 285 abroad. On the other hand, from all the Master's degrees only 39 were obtained abroad and 212 in Mexico (these figures do not add up to the total reported above due to lack of information). If we now look to the countries where most mexicans obtained their doctorate, we find that 93 did it in the United States, 60 in Great Britain, 38 in France, 23 in Germany, 10 in the former USSR and the rest somewhere else.

Turning now to the fields of physics cultivated by the mexican community, we find that the fields of activity of the main groups are distributed as indicated in Table I. Here PhD's means the number of researchers with a PhD degree associated to the groups and T and E mean theorist and experimentalist, respectively.

If we now look at the productivity of the mexican physicists we find that they published 422 articles in 1989 and 503 in 1990. In this last year, the main journals where these articles appeared were: *Revista Mexicana de Física* (59), *Physical Review* (51) and *Journal of Physics* (19).

II.1 Optical Properties

The interaction between light and matter is studied in two major fields of physics: optics and optical properties of matter. In the first one the main objective is the light; matter is used in order to modify and study the properties of light. On the other hand, in the field of optical properties the aim is the study of matter, and light is used as the main tool in order to accomplish this aim. In this sense we could also refer to this field as: *optical spectroscopies.*

It is interesting to notice that in Mexico the field of optical spectroscopies has attracted the interest of a large (relatively speaking) number of scientists since more than fifteen years ago. Right now one can identify 8 active groups working in different institutions and involving at least 50 scientists with a PhD degree. From these, about 2/3 are experimentalist and a 1/3 are theoreticians. We want also to point out that one does not find such a large number of people working with electron spectroscopies. One reason could be the price of the equipment necessary to set up a laboratory.

In optical properties the main topics of research have been: (i) defects in solids, (ii) semiconductors and (iii) interfaces and disordered systems. In order to give an idea of the type of problems undertaken within these topics some titles of recent publications are listed below:

* "Non radiative energy transfer from Cu to Mo ions in monocrystaline NaCl "
Physical Review B **41**, 12270 (1990).
* "Self activated luminescence in lithium tantalate"
Solid State Communications **75**, 551 (1990).
* "Photoluminescence studies in $Zn_x Cd_{1-x} Te$ single crystals"
Journal of Vacuum Science and Technology **A 8**, 3255 (1990).
* "Collective modes in a multilayer p-n-p structure"
Solid State Communications **75**, 405, (1990).
* "Stark ladder resonances in the propagation of electromagnetic waves"
Physical Review Letters **64**, 1433 (1990).
* "Reflectance anisotropy of the (110) reconstructed surface of gold"
Physical Review B **42**, 9155 (1990).
* "Effective dielectric response of polydispersed composites"
Physical Review B **41**, 7370 (1990).

As it can be seen from these titles the work done by these groups deals more with the basic physics of the phenomena rather than with applications. Nevertheless, the concern in more applied problems is growing and the interest on problems of non-linear optics and opto-electronics devices is actually starting in most of these groups.

III. Main Achievements

In this section we report the main achievements of our theoretical group working at the Institute of Physics of the National University of Mexico (UNAM).

III.1 Surface Optics

One of the crucial problems in surface optics is to determine the influence of a surface perturbation on the optical properties of any given system. It is well known that when the surface of a material is modified, even slightly, there is an overall change in its optical response. Here we will be interested in modifications or perturbations whose extension in space is much less than the wavelength of light. These perturbations might be very different in nature. For example, the changes induced on a clean surface by the adsorption of foreign molecules or the changes in electron distribution in a metal-electrolyte interface after the application of an external field. One could also consider an actual surface as composed by a substrate and a perturbation. For example, the electronic structure of the surface region on a clean surface differs from the one in the bulk due to the breakdown of the bulk symmetry in the direction perpendicular to the surface; thus one could think of the surface region as a perturbation to the bulk-vacuum interface. In the same way a rough surface could be thought as a flat interface perturbed by the "rough" region.

From the optical point of view a flat interface is characterized by its surface impedance Z_0 which is defined as the ratio of the component of the electric field parallel to the surface to the corresponding component of the magnetic field, both measured right at the surface. Given the surface impedance of an interface, all its optical properties can be readily derived. The problem then is how to calculate the change in the surface impedance due to a perturbation of the actual interface.

We worked out,[3-6] in all its generality, the electromagnetic part of this problem and here we sketch a simple derivation[6] of the main formula. First, the z-axis is chosen perpendicular to the surface, the unperturbed interface is located at $z = 0$ and the system is considered to be isotropic in the xy-plane. Since we are assuming that the size of the surface region is small compared to the wavelength of the incident light, it is reasonable to expect that the final result is not sensitive to the actual distribution of the excess current density $\Delta\mathbf{j}(z)$ induced by the perturbation but rather that it depends only on the total surface current

$$\mathbf{i} = \int dz\,\Delta\mathbf{j}(z). \tag{1}$$

This suggests a very simple procedure for the calculation of the surface impedance of the system. In this procedure[6] one considers an unperturbed system characterized by its surface impedance Z_0, with an infinitesimal sheet on top of it carrying the total surface current given by Eq. (1). The great advantage of this procedure, as compared to others,[3-5] is that it does not require to specify any definite model for the unperturbed interface. On the other hand, since the parallel component of the electric field E_x and the perpendicular component of the displacement field D_z are slowly varying, the currents induced in the sheet are taken to be

$$i_x = \langle\langle\Delta\sigma_{xx}\rangle\rangle E_x(0), \tag{2a}$$

$$i_z = \langle\langle\Delta s_{zz}\rangle\rangle D_z(0), \tag{2b}$$

where the argument (0) means that the fields are calculated at $z = 0$ and $\langle\langle\Delta\sigma_{xx}\rangle\rangle$ and $\langle\langle\Delta s_{zz}\rangle\rangle$ play the role of the two independent components of a surface-response tensor which characterizes the optical response of the surface region. We obtained[6] for p-polarization, in cgs units,

$$Z = \frac{Z_0 + 4\pi Q^2 c \langle\langle \Delta s_{zz} \rangle\rangle / \omega^2}{1 + 4\pi Z_0 \langle\langle \Delta \sigma_{xx} \rangle\rangle / c}, \tag{3}$$

where ω is the frequency of the incident field, c is the speed of light and $Q = \omega \sin\theta_i / c$ is the parallel component of the incident wavevector. Here θ_i is the angle of incidence.

This very simple expression gives, in its most general form, the surface impedance of the system in terms of the surface impedance of the unperturbed interface Z_0 and two complex functions of frequency $\langle\langle \Delta\sigma_{xx} \rangle\rangle$ and $\langle\langle \Delta s_{zz} \rangle\rangle$ which characterize the response of the surface region. The calculation of $\langle\langle \Delta\sigma_{xx} \rangle\rangle$ and $\langle\langle \Delta s_{zz} \rangle\rangle$ require now the solution of the linear-response problem of the specific microscopic model used to describe the surface region.

The main difference between our results and previous calculations is that we do not need specific models for the unperturbed interface in order to apply our results. In case in which both the background and the perturbed system are real systems, the surface impedance of the background, a quantity that characterizes it completely for our theory, could actually be measured with optical methods. Thus, we believe that our theory will be useful to understand, for example, surface-sensitive experiments like electroreflectance, in which the response of a real metal (whose response is nonlocal, includes local-field effects, and changes continuously near the surface) is physically perturbed by a strong electric field in a small region measuring a few Angstroms in width.

III.2 Surface Induced Optical Anisotropy (SIOA)

One of the main beliefs in condensed matter physics is that in order to explore the atomic structure of a given system one requires an external radiation with wavelength of the order of inter-atomic distances; that is of the order of Angstroms. In this context, light had not been considered appropriate for the study of atomic structure. Nevertheless, we show here that one is able to study the atomic structure of monolayers adsorbed on different faces of cubic crystals using an optical spectroscopy. Although the incident beam has a wavelength much greater than the interatomic distances, the field scattered by the system has two contributions, both oscillating at the same frequency ω: one is the propagating coherent field which constitutes the reflected beam and the other one is a highly inhomogeneous non-propagating field which decays very rapidly away from the surface. It is this highly inhomogeneous non-propagating field the one which has Fourier components with wavelengths of the order of Angstroms. Thus the polarization of the adsorbed molecules by this non-propagating field conveys information to the reflected (coherent) beam about the local enviroment of the adsorbed molecules. Nevertheless, the contributions to the reflected beam come not only from the adsorbed monolayer but also from the bulk. Thus, the first question then is how to separate the contribution of the bulk from the one of the surface region, specially when the bulk's contribution is several orders of magnitude greater than the surface one. In order to circumvent this problem we have proposed[7] a differential procedure which applies to the case of cubic crystals.

From the optical point of view, the bulk of a cubic crystal is isotropic. Thus any change in the reflected beam for two perpendicular polarizations of the incident beam should arise necessarily from the surface region; it is in this region where the bulk symmetry breaks down. It is also in this sense that the measurement of anisotropies in the reflected beam of cubic crystals becomes a surface-sensitive spectroscopy. These anisotropies should exist even for clean surfaces of low symmetry faces due to the existence of a surface region that differs from the bulk. We estimated[7] that the relative size of these anisotropies should be of the order of 10^{-2} - 10^{-3} which is well within the range of experimental detection ($\sim 10^{-4}$ - 10^{-5}). These theoretical results stimulated the work of the experimentalists and the first measurements on air of optical anisotropies of oxidized faces of cubic semiconductors were reported by D. Aspnes and A. Studna[8-10] from Bellcore in 1985 and by A. Lastras and S. Acosta[11] from the University of San Luis Potosí (Mexico) in 1988.

A succesful interpretation of these optical-anisotropy spectra was made by us[12] using an extremely simple model for Si and Ge based in the surface local-field effect. The main idea was to consider that the crystal could be modeled as a semi-infinite *fcc* lattice of polarizable point dipoles. Each dipole represented a polarizable entity consisting of, say, a germanium atom plus its four tetrahedally-located electronic lobes. In this model each polarizable entity shared the 4 atoms located at the end of each lobe with another 4 polarizable entities. The lattice constructed this way has then an *fcc* structure instead of the diamond structure characteristic of germanium. The choice of these polarizable entities with two germanium atoms (the central one plus 4 at the end of each lobe contributing 1/4 each) instead of the germanium atoms themselves, arises from the fact that at optical frequencies the electronic transitions which contribute to the optical response involve only the valence electrons which are the ones located at the lobes; the electrons at the atomic cores are excited at much higher frequencies. On the other hand, the second-rank polarizability tensor of entities like these ones with tetrahedral symmetry is proportional to the unit tensor; therefore the polarizability of each entity is a scalar. If we now assume that all the polarizable entities interact with each other only through the dipolar interaction, the induced dipole moment $\mathbf{p}_i$ at the entity located at $\mathbf{R}_i$ is given through the solution of the following system of equations:

$$\mathbf{p}_i = \alpha(\mathbf{E}_i^{ex} + \sum_j \overleftrightarrow{\mathbf{t}}_{ij} \cdot \mathbf{p}_j), \tag{4}$$

where α is the polarizability of the polarizable entities, $\mathbf{E}_i^{ex}$ is the external field at $\mathbf{R}_i$,

$$\overleftrightarrow{\mathbf{t}}_{ij} = (1 - \delta_{ij})\vec{\nabla}_i\vec{\nabla}_j(1/R_{ij}), \tag{5}$$

is the dipole-dipole interaction tensor and the sum over j runs over all the sites of the semi-infinite lattice. Here δ_{ij} is the Kronecker delta which appears in order to avoid self-interaction. Notice that we have assumed that all the entities have the same polarizability α. This hypothesis is not justified for the entities that lie close to the surface where one expects a polarizability different than the one in the bulk due to changes in the electronic structure as one approaches the surface. What we are actually doing here is to freeze the electronic structure up to the surface and to explore only the effect due to the difference in the local field at the surface region as compared with the one in the bulk. This is what we call: *the surface local-field effect.*

The polarizability α is obtained through the Clausius-Mossotti relation[13] which relates the dielectric function ϵ of an infinite cubic crystal of point dipoles with its polarizability α. We took the experimentally-determined dielectric function ϵ and solve for α. Since the Clausius-Mossotti relation is exact for cubic crystals this procedure yields the effective bulk polarizability of the chosen entities.

From the symmetry of the system one concludes that the dipoles p_n induced at a lattice plane n parallel to the surface are all alike. Eq. (4) was then solved numerically

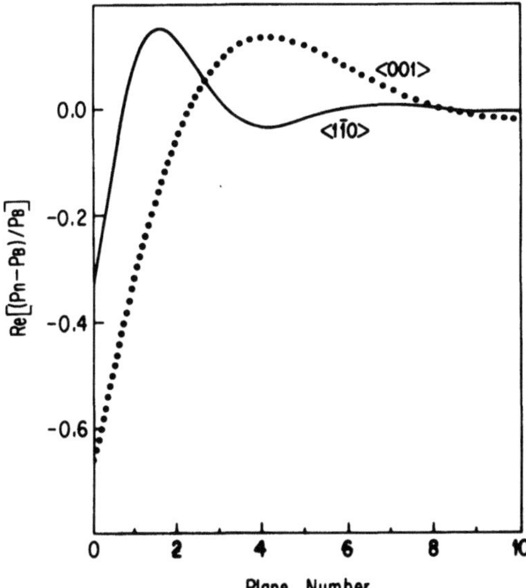

Figure 1. The real part of $(p_n - p_B)/p_B$ as a function of the plane number for an external field polarized along the $\langle 1\bar{1}0 \rangle$ (solid line) and the $\langle 001 \rangle$ (dotted-line) directions. Here p_n is the dipole moment of a polarizable entity on the n-th plane and the outermost plane is labeled $n = 0$.

for p_n. In Fig. 1 we show the real part of $(p_n - p_B)/p_B$ as a function of n. Here p_B is the dipole moment induced in the bulk and $n = 0$ is the outermost plane. The calculation was done for the (110) face of germaniun with the external field polarized along the $\langle 001 \rangle$ and $\langle 110 \rangle$ directions. It is clear that the induced dipole moment differs from the one in the bulk within the first ~ 10 layers where it finally heals to the bulk value. In the first layer the difference can be as big as 60%. It is also seen that the size of the induced dipole moments in the surface region (~ 10 layers) is very sensitive to the polarization of the external field: this is the source of the reflectance anisotropy. We are showing in Fig. 1 only the real part of $(p_n - p_B)/p_B$; the imaginary part is related to a difference in phase between the dipoles and the external field which yields energy absorption.

The question now is how these changes in polarization within the first ~ 10 layers of the sample affect the overall reflectance and how big is the difference between two different polarizations of the external field. A quantitative measurement of this effect is given through the differential reflectance

$$\frac{\Delta R}{R} = \frac{R_\gamma - R_\beta}{\langle R \rangle}, \tag{6}$$

where R_γ is the reflectance when the external field is polarized along the γ direction and $\langle R \rangle$ is the arithmetic average of the reflectance in the two directions (γ and β); $\Delta R = 0$ means that there is no anisotropy. The substraction $R_\gamma - R_\beta$ cancells all the isotropic contribution coming from the bulk and leaves out only the difference in contributions coming from the surface region.

The solution of Eq. (4) for the p_n's allows a straightforward calculation of $\Delta R/R$.[12] In Fig. 2 we show the calculated $\Delta R/R$ as a function of frequency (in eV) for the

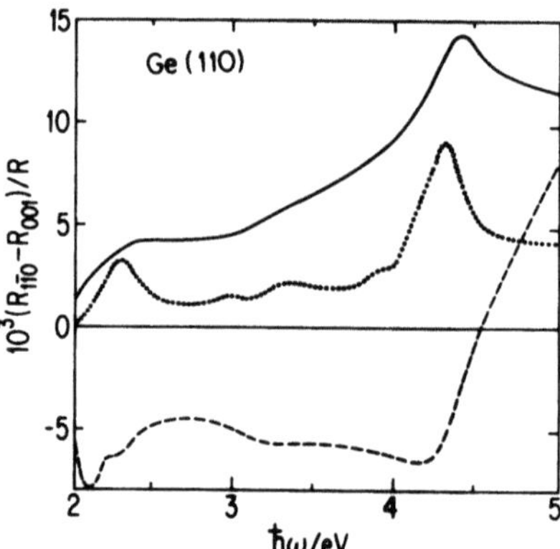

Figure 2. The differential reflectance $\Delta R/R$ as a function of frequency (in eV) for the (110) face of germanium. Here $\Delta R = R(1\bar{1}0) - R(001)$ is multiplied by 10^3 The dotted line is the experiment, the solid line is theory and the dashed line is the theoretical result obtained using a diamond structure instead of an FCC structure.

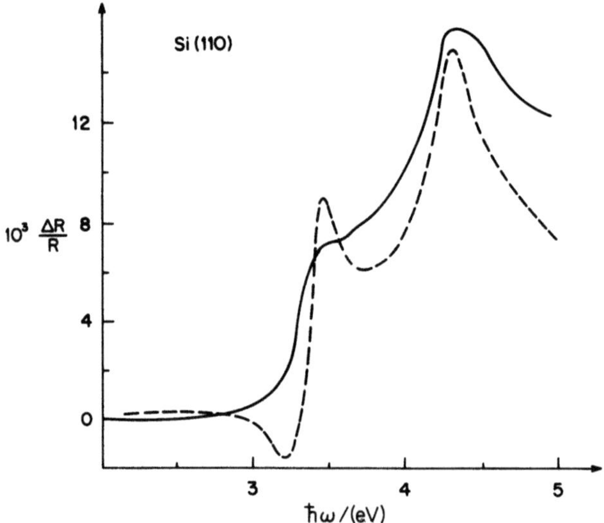

Figure 3. The same as in Fig. 2 but for the (110) face of silicon. Here the dashed line is experiment and the solid line is theory.

(110) face of germanium along with the experimental results. Again the external field was polarized along $\gamma = \langle 110 \rangle$ and $\beta = \langle 001 \rangle$. One can see that there is a very good agreement between theory and experiment specially when one recalls that the theoretical calculations have no adjustable parameters. In Fig. 3 we show the same results as the ones in Fig. 2 but now corresponding to the (110) face of silicon. Here the agreement is even better. We remark that the experiments were done in air, and thus, one expects the presence of an oxide layer at the interface. The fact that the agreement is so good might indicate that (i) the presence of the oxide emulates well

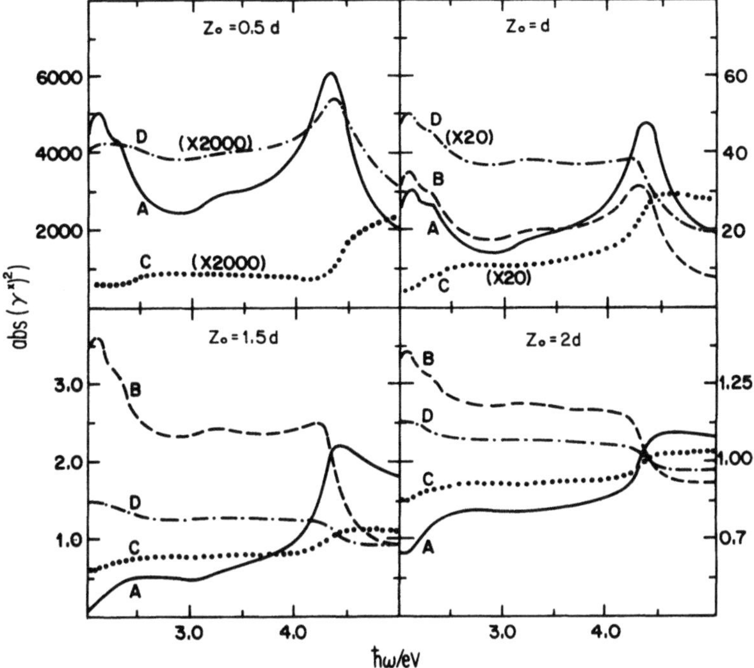

Figure 4. The absolute value of the parameter $(\gamma^x)^2$ as a function of frequency (in eV) for the (110) face of germanium. The four panels correspond to four different heights z_0 of the adsorbed molecules above the nominal plane ($z = 0$) in units of the interplanar distance d. The different curves in each panel labeled A, B, C, D correspond to four different positions of the adsorbed molecules on top of the outermost layer as described in the text.

a semi-infinite crystal with no change in electronic structure close to its surface and (ii) that the choice of the polarizable entities was adequate.

These ideas were further extended[12] in order to calculate the changes in the SIOA spectra due to the presence of an adsorbed monolayer. Assuming that the principal axis $\mu(= x, y, z)$ of the polarizability tensor of the adsorbed molecules coincide with the corresponding ones of the substrate, we showed that the μ-components of the surface conductivity of the adsorbate $\langle\langle\Delta\sigma^\mu\rangle\rangle$ are proportional to the square of the μ-components of the local field $(E^\circ_{loc})^\mu$ at the adsorbate position but in the absence of the adsorbed monolayer, that is[14]

$$\frac{\langle\langle\Delta\sigma\rangle\rangle^\mu}{\langle\langle\Delta\sigma_0\rangle\rangle^\mu} \sim \left[\frac{(E^0_{loc})^\mu}{E^\mu_M}\right]^2 \equiv (\gamma^\mu)^2, \tag{7}$$

where $\langle\langle\Delta\sigma_0\rangle\rangle^\mu$ and E^μ_M are the μ-components of the surface conductivity of the adsorbed monolayer in the absence of the substrate and the macroscopic electric field at the surface, respectively. This expression shows explicitly that the surface conductivity of the adsorbed overlayer is very sensitive to the adsorbates position because the local electric field $\mathbf{E}^0_{loc}$ is an extremely sensitive function of that position.

In Fig. 4 we show the absolute value of $(\gamma^x)^2$ as a function of ω for four different positions A, B, C, D of the adsorbates on the 2D unit cell of the monolayer and four different distances from the nominal surface plane of the substrate. The calculation was performed for Br_2 molecules adsorbed on the (110) face of germanium using for the substrate the same model mentioned above. The positions of the adsorbate correspond to: (A) directly above a lattice point, (B) halfway between two lattice points along the x, $\langle 110 \rangle$, direction, (C) halfway along the y, $\langle 001 \rangle$ direction and (D) above the middle of the 2D unit cell. The distances from the surface are measured in units of $d = 5.658/2\sqrt{2}$Å, the interplanar distance. This figure shows the extreme sensitivity of $(\gamma^x)^2$ as a function of both the lateral and vertical position of the adsorbed overlayer. Having obtained the surface conductivity of the adsorbed overlayer, it is straightforward to calculate the SIOA spectra $\Delta R/R$ as defined by Eq. (6) with R along the x and y directions. We assumed that the Br_2 molecule polarizes along its molecular axis with a polarizability obtained from measurements in the gas phase.[15] Then we compare our calculated spectra with the experimental one as measured by Aspnes and Studna.[8] The best agreement with our calculations corresponds to Br_2 oriented along the $\langle 001 \rangle$ direction; the best positions are A and B at a distance from the surface $z_0 = 1.17d = 2.34$Å and $z_0 = 0.89d = 1.78$Å, respectively. The theoretical spectra for the two positions chosen above are almost identical and the calculated line shape resembles the experiment quite well. In order to decide between these two positions we require, probably, additional information coming from an anlysis of the most favorable geometry for chemical bonding. Nevertheless, what this example shows is the extreme surface sensitivity of the SIOA spectra. This also explains the fact why SIOA spectra are actually being used to monitor the perfection of epitaxial growth of cubic semiconductors and the MBE growth of semiconductor superlattices.

III.3 Optical Properties of Composites

A composite is a system made of a mixture of two or more materials. The difference between a composite and an alloy is that while in the alloy the mixture is done at the molecular level, in the composite is at the macroscopic level; that is, each of its constituents allows a macroscopic description. That is why composites have also been called "granular" matter. There are also different ways of mixing two materials and according to its topology they have been classified as: (i) laminar, (ii) granular and (iii) agglomerates. Laminar composites are built as sequence of sheets and a typical example are the recently fabricated superlattices. In granular composites, grains of one material are always completely surrounded by the other one, while in agglomerates there are regions in which one material is embedded in the other one and in other regions is the other way around.

The study of the physical properties of composites poses different types of problems. For the theoretician the problem is to explain or predict the properties of the mixture in terms of the properties of their isolated constituents and the statistical information of their distribution. For the material scientist the problem is that of actual fabrication and tayloring of new types of materials through blending. The properties of the mixture might become unexpectedly different from the ones of its isolated constituents. For example, while the low frequency dielectric response of rocks and brine are around 10 and 80, respectively, when porous rocks are filled with brine the real part of their dielectric constant can be as great as 10^6 ; four to five orders of magnitude different from the one corresponding to either of its constituents.[15]

A quite general problem in the physics of composites is the determination of the *effective* linear response of a system consisting of the blend of two materials each of which respond linearly. Here we report our work on the calculation of the *effective* dielectric response of a composite with a simple geometry: a collection of identical spheres within an otherwise homogeneous and isotropic host matrix . Although the problem is very old[16] and a vast amount of effort has been devoted to its analysis,[17] there is still no generally accepted solution.

One of the first and most fruitful approaches to this problem was given by J.C. Maxwell Garnett[18] and his theory is known today as the Maxwell Garnett theory (MGT). It is completely equivalent to the Clausius-Mossotti relation developed in the molecular theory of the dielectric properties of liquids and dense gases.[19] The approximation involved in MGT corresponds to a self-consistent mean-field approximation (MFA) which neglects the fluctuations of the local field felt by each inclusion. Thus most of the recent theoretical work in this problem has been aimed to finding ways of dealing with the local field fluctuations. Some of these approaches have been: decoupling procedures of the integral equation obeyed by the electric field, substitution of the spatial disorder by a substitutional disorder in a fictitious cubic lattice, multiple-scattering techniques, phenomenological ansatz, etc.[20]

The main problem in deciding which of the theoretical approaches or calculations is the correct one has been the comparison with experiment. This has been a painful process because the models used in the theoretical work are far from being adequate to the structure found in actual samples. This is due to the fact that in actual samples there is a distribution of sizes and non-sphericity of the inclusions as well as clustering effects. Another problem has been that the only statistical parameter reported by most experimentalists, during decades, was the volume fraction of the inclusions. It was not until recently that it has been recognized that in order to have a proper comparison between theory and experiment more microstructural information about the system is required. Part of our work has been directed precisely towards this aim of vindicating and disseminating this message.

In order to illustrate this point we present here the results of a simple and intuitive approach[21] that we have developed for the system of identical spheres embedded in a matrix. In the usual mean-field approach, the effective (or macroscopic) dielectric response ϵ_M can be written as

$$\epsilon_M = \epsilon_h \frac{1 + 2f\alpha}{1 - f\alpha}, \tag{8}$$

where ϵ_h is the dielectric function of the host matrix, f is the volume filling fraction of the inclusions and

$$\alpha = \frac{\epsilon_s - \epsilon_h}{\epsilon_s + 2\epsilon_h}, \tag{9}$$

is its polarizability in units of a^3 in the cgs system. Here a is the radius of the spheres.

In our approach[21] we take account of the local-field fluctuations through a renormalized polarizability α^* which replaces the bare polarizability α in Eq. (8) and obeys a simple second-order algebraic equation

$$\frac{\alpha^*}{\alpha} = 1 + (\alpha^*)^2 \Delta T^2, \tag{10}$$

where ΔT^2 is related to the variance of the dipole-dipole interaction field which describes the interaction among the inclusions. When one neglects the fluctuautions of this field $\Delta T = 0$, then $\alpha^* = \alpha$ and one recovers the mean-field approach.

Now we present the results of our theory, which from now on will be referred as the renormalized polarizability theory (RPT), for a system of metallic inclusions

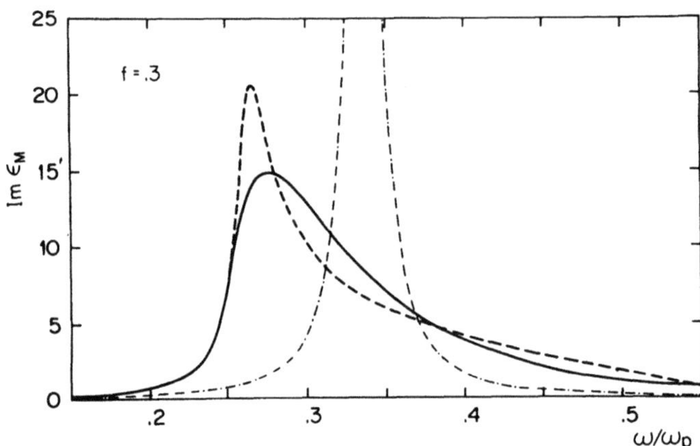

Figure 5. The imaginary part of the effective dielectric response ϵ_M as a function of frequency (in units of the plasma frequency ω_p) The dot-dashed line corresponds to the Maxwell Garnett theory (MGT) and the dashed line to the renormalized polarizability theory (RPT). The solid line corresponds to a more sophisticated calculation based in a diagramatic approach and reported in Ref. 23.

embedded in gelatin. For the dielectric function of the metal, ϵ_s, we chose a simple free electron model (Drude model) and we assume that the gelatin is transparent and dispersionless with a dielectric function $\epsilon_h = 2.37$. In Fig. 5 we show the imaginary part of the effective dielectric function ϵ_M as a function of frequency for a filling fraction $f = 0.3$. The frequency is measured in units of ω_p, the plasma frequency of the electrons, and the relaxation time τ of their non-radiative dissipation was assumed to be equal to $\tau = 46/\omega_p$. Typically, for a free-electron like metal such as silver, $\hbar\omega_p$ is taken around 10 eV. The imaginary part of ϵ_M can be determined from the absorption spectra and the frequency region where it peaks indicates the region where the system favors energy absorption through the coupling of the external probe with the normal modes of the system. We show the results of mean-field theory (MFT) along with the ones of RPT. One can clearly see that the effect of local field fluctuations is to reduce dramatically the height of the absorption peak and to make it asymmetrically wider. If one would have taken τ as infinitely large (no dissipation) the MFT peak would have become a delta function, which means that MFT accounts only for a single optically-active mode. This is understandable because in the absence of fluctuations the system behaves like an ordered one where all the sites are equivalent. In these systems and in the long-wavelength limit there is only one optically-active mode and the MFT is exact.[22] The width of the MFT peak comes from the non-radiative dissipation given by the finite value of τ and is independent of the filling fraction.

On the other hand the width of the absorption peak in RPT is due, besides the contribution coming from dissipation, to the excitation of a collection of optically-active electromagnetic modes whose proper frequencies are spanned over a continuos range. The physical origin of these modes is due precisely to the presence of disorder. In this same figure we present also the results of a more sophisticated calculation which was done using a different formalism based on a diagrammatic technique.[23]

One of the main advantages of RPT when compared with other approaches is that it incorporates in an extremely simple way the effects of the field fluctuations. This allows extensions of the theory to more complicated and/or more realistic systems;

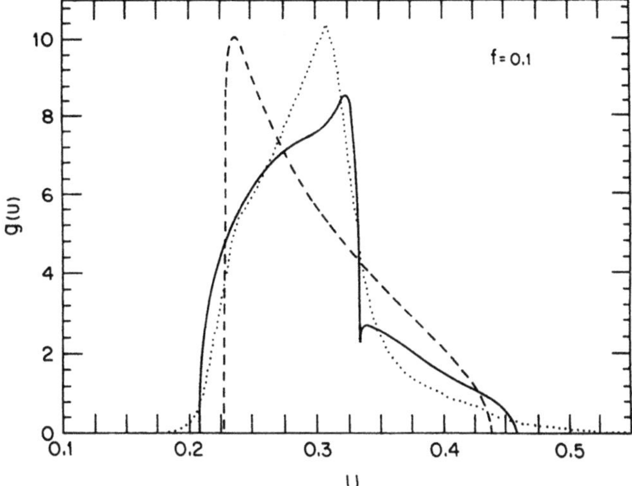

Figure 6. The spectral function $g(u)$ as a function of u. The dotted line refers to the results of the numerical simulations reported in Ref. 27 and the dashed line corresponds to the results of the renormalized polarizability theory (RPT). The solid line corresponds to a more sophisticated calculation based in a summation of low-density diagrams and reported in Ref. 30.

with other approaches these extensions would have required a superlative effort. For example, the extension of RPT to a polydispersed composites has render a valuable insight on the effects of a distribution of sizes on the width of the absorption peak.[24,25] RPT has been also extended to anisotropic systems such as composites with aligned spheroidal inclusions[26] and its results are found to be also relevant to the problem of dichroism in liquid crystals.

Nevertheless the most fundamental problem that still remains open is to determine how good the theory actually is. As it was commented before, the comparison between theory and experiment has been a difficult endeavor specially due to the lack of information on the statistical distribution of the inclusions in the samples. Fortunately there are now a couple of "numerical experiments"[27,28] which have smooth out this difficulty. In both of these numerical simulations[27,28] the model is precisely a collection of identical spheres interacting only through the dipole-dipole interaction. Here we will only refer to the work of Cichocki and Felderhof[27] who present their results in terms of the spectral function $g(u)$ defined through[29]

$$\frac{\epsilon_M}{\epsilon_h} = 1 - f \int_0^1 \frac{g(u)}{t-u} du, \tag{11}$$

where the variable t is given by

$$t = \frac{1}{1 - \epsilon_s/\epsilon_h}. \tag{12}$$

The main advantage of this spectral representation of the effective dielectric function ϵ_M is that $g(u)$ does not depend on the dielectric properties of the materials which compose the system. It is essentially a property of the geometry and the spatial distribution of the inclusions and it measures the weight of the normal electromagnetic modes of the system. For example, in our system of spherical inclusions, the mean-field approximation yields $g(u) = \delta[u - (1 - f)/3]$ because in the absence of fluctuations and in the long-wavelength limit, only one mode can be excited. Here $\delta[x]$ is Dirac's delta function.

Cichocki and Felderhof[27] performed their computer simulation at six different volume fractions. At each volume fraction the average was taken over N_x independent configurations generated earlier in a Monte Carlo simulation of the hard-sphere system in equilibrium. The configuration considered N_c particles within a cube which was then repeated periodically. Since our theory (RPT) is valid only in the low-density regime here we compare our results only for $f = 0.1$, the lowest filling fraction considered in Ref. 26. For this filling fraction N_c was taken equal to 300 and $N_x = 500$.

In Fig. 6 we plot $g(u)$ as a function of u for $f = 0.1$. The dotted lines are the results of the computer simulation (CS) and the dashed lines are the results of RPT. As it can be seen RPT is quite succesful in determining the region in which $g(u)$ is different from zero but with respect to the line-shape the agreement is not so good. Nevertheless, in order to obtain a better agreement between theory and "experiment", we have performed, within the diagrammatic approach,[23] new summations of different classes of diagrams which yield the results shown by the solid lines.[30] It is clear that in this latter case the agreement is excellent. For larger filling fractions ($f \simeq 0.3$) the problem becomes more complicated because then one has to take into account the effects of, at least, three-particle correlations. Efforts along this direction are actually under way.

Finally we want to add that we have also started a collaboration with the polymer industry (*Industrias Resistol SA*) in order to study optical properties of composite materials made of inclusions of transparent rubber within a transparent polymer matrix. The problem here is to determine the relative importance of roughness scattering *versus* the scattering due to the inclusions and its relation with gloss and haze.

Acknowledgements

This work was partially supported by Dirección General de Asuntos del Personal Académico of Universidad Nacional Autónoma de México (Mexico) under contracts IN-104689 and IN-102592.

References

1. *Catálogo de Programas y Recursos Humanos en Física (Catalogue of Programs and Human Resources in Physics)*. Published by the Mexican Physical Society, 1991.
2. M.A. Pérez-Angón, *Bull. of the Mex. Phys. Soc.* **51**, 67 (1991).
3. A. Bagchi, R.G. Barrera, and B.B. Dasgupta, *Phys. Rev. Letters* **44**, 1475 (1980).
4. R.G. Barrera and A. Bagchi, *Phys. Rev. B* **14**, 1612 (1981).
5. A. Bagchi, R.G. Barrera, and R. Fuchs, *Phys. Rev. B* **25**, 7086 (1982).

6. W.L. Mochán, R. Fuchs, and R.G. Barrera, *Phys. Rev. B* **27**, 771 (1983).

7. W.L. Mochán and R.G. Barrera, *J. de Physique*, **45** Colloque C5, 207 (1984).

8. D.E. Aspnes and A.A. Studna, *Phys. Rev. Letters* **54**, 1956 (1988).

9. D.E. Aspnes, *J. Vac. Sci. and Technol. B* **3**, 1138 (1985).

10. D.E. Aspnes, *J. Vac. Sci. and Technol. B* **3**, 1498 (1985).

11. S. Acosta-Ortiz and A. Lastras, *Solid State Commun.* **64**, 809 (1988).

12. W.L. Mochán and R.G. Barrera, *Phys. Rev Letters* **55**, 1192 (1985).

13. See for example *Classical Electromagnetic Radiation* by J.B. Marion (Academic Press, New York, 1965) p. 283.

14. W.L. Mochán and R.G. Barrera, *Phys. Rev. Lett.* **56**, 2221 (1986).

15. F. Browers, A. Ramsamugh, and V.V. Dixit, *J. Mat. Sci.* **22**, 2759 (1987).

16. J.C. Maxwell, *A Treatise on Electricity and Magnetism.* Vol 1 (Reprint: Dover, New York, 1954) Sec. 314, p. 440.

17. See for example the historical review of Rolf Landauer on *Electrical Transport and Optical Properties of Inhomogeneous Media.* edited by J.C. Garland and D.B. Tanner (American Institute of Physics, New York, 1978), p. 2.

18. J.C. Maxwell Garnett, *Philos. Trans. R. Soc. London* **203**, 385 (1904).

19. See for example J.D. Jackson, *Classical Electrodynamics*, Second Edition (J. Wiley and Sons, New York, 1975), Chapter 4.

20. For a very complete account of the different procedures proposed, see for example: *Electrical Transport and Optical Properties of Inhomogeneous Media.* (AIP Conference Proceedings. Number 40). edited by J.C. Garland and D.B. Tanner (American Institute of Physics, New York, 1978); *Electrodynamics of Interfaces and Composite Systems*, Advanced Series in Surface Science. Vol 4. edited by R.G. Barrera and W.L. Mochán (World Scientific, Singapore, 1988); *ETOPIM 2.* Proceedings of the Second International Conference on Electrical Transport and Optical Properties of Inhomogeneous Media. edited by J. Lafait and D.B. Tanner (North Holland, Amsterdam, 1989).

21. R.G. Barrera, G. Monsivais, and W.L. Mochán, *Phys. Rev. B* **38**, 5371 (1988).

22. See for example J.D. Jackson, *Classical Electrodynamics*, Second Edition (J. Wiley, New York, 1975), Sec. 4.5.

23. R.G. Barrera, G. Monsivais, W.L. Mochán, and E. Anda, *Phys. Rev. B* **39**, 9998 (1989).

24. R. G. Barrera, P. Villaseñor-González, W.L. Mochán, M. del Castillo-Mussot, and G. Monsivais, *Phys. Rev B* **39**, 3522 (1989).

25. R.G. Barrera, P. Villaseñor-González, W.L. Mochán, and G. Monsivais, *Phys. Rev. B* **41**, 7370 (1990).

26. R.G. Barrera, J. Giraldo, and W.L. Mochán, *MRS Proceedings,* **253** Boston (1991); *Phys. Rev B*, submitted (1992).

27. B. Cichocki and B.U. Felderhof, *J. Chem. Phys.* **90**, 4960 (1989).

28. S. Kumar and R.I. Cukier, *J. Phys. Chem.* **93**, 4334 (1989).

29. D.J. Bergman, *Phys. Rep.* **43**, 377 (1978).

30. R.G. Barrera, C. Noguez, and E.V. Anda, *J. Chem. Phys.* **96**, 1574 (1992).

Recent Advances in Computational Materials Science and Engineering

J.M. Sanchez

Center for Materials Science and Engineering
The University of Texas
Austin, Texas 78712
USA

Abstract

Theories of phase equilibrium have been successfully used to characterize the main contributions to alloy phase stability, the interpretation of complex and extensive experimental data and, in some instances, the prediction of metastable phases. At present, there is increased interest in the application of microscopic quantum theory in order to produce a reliable description of phase equilibrium and, in particular, phase diagrams. Here we discuss a first-principles statistical mechanics theory of alloy phase stability which incorporates the calculation of electronic total energies in the local density approximation, configurational entropies and vibrational modes into the total free energy of disordered systems and intermetallic compounds. Applications of the theory to the binary Ni-Al and ternary Ni-Al-Ti systems are given using the Linear Muffin-Tin Orbital method for the total energy calculations, the Cluster Variation method for the description of the configurational entropy, and the Debye-Grüneisen approximation for the vibrational modes.

I. Introduction

Although theory and computation in materials science and engineering have traditionally played a role only in the *post-mortem* analysis of most developments of materials properties, recent advances clearly indicate that a predictive computational component in the complex process of alloy design is within reach. At present, a number of physical properties such as, to name a few, cohesive energies, elastic moduli and expansion coefficients of elemental solids and intermetallic compounds can be rou-

Advanced Topics in Materials Science and Engineering, Edited by
J.L. Morán-López and J.M. Sanchez, Plenum Press, New York, 1993

tinely calculated from first principles using, as input, only the atomic numbers of the constituent elements and the crystal structure of the solid. These achievements are a direct consequence of a mature theoretical and computational framework in solid state physics which, to be sure, has been in place for some time. More relevant to the current status is the ever increasing availability and capability of computer hardware, and a genuine interest in the development of efficient and computationally tractable algorithms. The later is primarily driven by the perceived needs of the materials science and engineering community and by the recognition that breakthroughs in this area are likely to result in a sizable pay-off in science and technology.

Most of the first principles computational capabilities currently available apply to solids that can be idealized as having a perfect crystal structure, void of grain boundaries, surfaces and other unpleasant imperfections. The realm of engineering materials, be it for structural, electronics or other applications, is, however, that of "defective" solids. Defects and their control dictate the properties of real materials. There is, at present, an impressive body of work in materials simulation which is aimed at understanding properties of real materials. These simulations rely heavily on either a phenomenological or semiempirical description of atomic interactions which, in general, tend to severely limit the predictive capability of the approach.

Here we will focus on recent theoretical and computational developments in alloy theory which make possible the non-empirical calculation of phase diagrams and, in general, the study of phase stability strictly from first principles. This particular approach to the computation of phase diagrams requires the application of electronic structure theory, which as pointed out is well developed for perfect crystals, to systems with substitutional disorder induced by temperature and, in compounds, off-stoichiometric effects. Although this type of defects represent a relatively simple deviation from a perfect crystal, as defined by the existence of a finite unite cell, their presence translates into a real challenge for electronic structure calculations. Despite the difficulties, a predictive phase diagram methodology that does not rely on empirical data shows considerable promise at this time.

The importance of phase diagrams in materials science and engineering is, of course, widely recognized. In fact, the limited information usually available for equilibrium and metastable phase diagrams represents a major hurdle in the design and development of new materials systems. In engineering alloy systems, especially for structural applications, a number of equilibrium and metastable phases compete closely for stability within given temperature and composition ranges. As such, the accurate experimental determination of the equilibrium phase diagram, particularly for multicomponent alloys, is generally a time consuming undertaking capable of significantly slowing down any alloy development program.

Early studies of alloy phase equilibrium have shown that simple models were capable of reproducing quite well the most important features of alloy phase diagrams.[1-5] The essence of these models was the description of the energy in terms of pair and many-body interactions. In these developments, the Cluster Variation Method (CVM) of Statistical Mechanics[6] played key role since it provided an efficient and computationally economical way of describing the configurational thermodynamic of alloys.

The early success of the CVM calculations revived the old dream of computing alloy phase diagrams from first principles; *i.e.* from the knowledge of the electronic structure of the alloy. Indeed, one of the most significant recent developments in alloy theory, density functional theory and its computational version, the local-density approximation (LDA),[7] was fully developed and ready to be applied together with the statistical models to tackle the delicate problem of alloy stability at finite temperatures. This local-density approximation has been used repeatedly to calculate the total energy at zero temperature of pure solids,[8−10] relatively simple compounds[11,12] and disordered alloys.[13−15] In general, results of the calculations reproduce physical properties within a few percent of the experimental values.

Other developments in the statistical thermodynamics of alloys led to the description of the energy in terms of multisite correlation functions.[1,16] In this cluster expansion, the energy is given by pair and many-body interactions. An attractive feature of this cluster representation for the configurational energy is that the alloy problem becomes essentially isomorphic to a generalized Ising model where the chemical interactions are relatively short ranged. Although these interactions include, in general, many-body terms as well as temperature and volume dependence, they can be easily treated using the CVM to calculate configurational free energies and, from them, the solid state portion of phase diagrams.

The success of this cluster representation for the energy of alloys led Connolly and Williams[17] to propose the use *ab-initio* total energy calculations of ordered compounds in order to obtain the set of effective pair and multisite chemical interactions. The proposal of Connolly and Williams clearly reflected the degree of confidence with which total energies could be calculated using the local-density approximation, and opened the door to first principles calculation of phase diagrams.

Among the first applications of this first-principles approach were studies of temperature-composition binary phase diagrams of noble-metal alloys[18] and semiconductor alloys.[19] Subsequently, numerous other cases have been investigated with relatively good results.[20−27] In general, the cluster expansion of the configurational energy converges relatively fast, thus offering a practical method for the determination of the chemical interactions in a disordered alloy system from total energy band calculations for a relatively small set of ordered compounds.

In the next section we present a cluster theory of the configurational thermodynamic of alloys. The description provides the formal framework for the treatment of short-range order (SRO) effects in the configurational energy and in the configurational entropy. The contribution to the free energy due to vibrational modes is also discussed. We make contact with microscopic electronic theories via the Linear Muffin-Tin Orbital (LMTO) approximation, which is used to calculate the total energies of selected compounds in the Ni-Al-Ti systems. The first principles statistical thermodynamic theory, which includes configurational, vibrational and local volume relaxation effects, is applied to the computation of the equilibrium solid state phase diagrams for the Ni-Al system and for the ternary Ni-Al-Ti system. The Ni-Al system is chosen as an interesting example in which *bcc*- and *fcc*-based phases coexist giving rise to a relatively complex binary phase diagram. The Ni-Al-Ti system, on the other hand, exemplifies the application of the method to ternary alloys.

II. Configurational Thermodynamics

In this section we review briefly the general formalism used in the description of the configurational thermodynamics of alloys. The main result is the cluster expansion which provides the basis for the description of disordered alloys from the knowledge of the energy (binding curves) of ordered compounds. For the sake of simplicity we consider only binary systems although the theory can be easily extended to multicomponent alloys.[16]

II.1 Cluster Expansion

The configuration of a crystalline binary alloy is described in terms of spin or occupation numbers σ_i at each lattice site i which take, respectively, values $+1$ and -1 for components A and B. Any configuration of the system is then fully specified by the N-dimensional vector $\vec{\sigma} = \sigma_1, \sigma_2, ..., \sigma_N$, where N is the number of lattice points. In general, one is confronted with the problem of describing functions that depend explicitly on the occupation variables σ_i, such as the energy of formation of the alloy. In order to provide an unambiguous description of such functions, it is convenient to introduce an orthogonal functional basis in configurational space. Although in the thermodynamic limit the dimension of the complete orthogonal basis is infinite, judicious choice of the basis functions allow us to obtain accurate approximations to the thermodynamic potentials in terms of subset of finite dimension.

For a single site i, the set of two polynomials in the discrete variable σ_i, namely the polynomial of order 0, $\phi_0(\sigma_i) = 1$, and the polynomial of order 1, $\phi_1(\sigma_i) = \sigma_i$, form a complete and orthonormal set, with the inner product between two functions of configuration $f(\sigma_i)$ and $g(\sigma_i)$ in the one-dimensional discrete space spanned by σ_i defined as:

$$\langle f(\sigma_i) \bullet g(\sigma_i) \rangle = \frac{1}{2} \sum_{\sigma_i = \pm 1} f(\sigma_i) g(\sigma_i). \tag{1}$$

The set of orthonormal characteristic functions in the N-dimensional discrete space spanned by the vector σ is obtained from the direct product of the $\{\phi_0(\sigma_i), \phi_1(\sigma_i)\}$, where i spans all crystal sites ($i = 1, 2, ... N$). For a binary system, the resulting characteristic functions, $\Phi_\alpha(\vec{\sigma})$, are given by products of the spin operator si over the sites of all possible clusters $\alpha = \{i_1, i_2, ... i_n\}$ in the crystal:[16]

$$\Phi_\alpha(\vec{\sigma}) = \prod_{i \in \alpha} \sigma_i = \sigma_{i_1} \sigma_{i_2} ... \sigma_{i_n}. \tag{2}$$

Accordingly, there is a one to one correspondence between the set of orthogonal functions $\Phi_\alpha(\vec{\sigma})$ and the set of all clusters α in the crystal, including the empty cluster for which $\Phi_0(\vec{\sigma}) = 1$.

The orthogonality of the characteristic functions $\Phi_\alpha(\vec{\sigma})$ is expressed by:[16]

$$\frac{1}{2^N} \sum_\sigma \Phi_\alpha(\vec{\sigma}) \Phi_\beta(\vec{\sigma}) = \delta_{\alpha,\beta}. \tag{3}$$

In view of Eq. (3), any function of configuration, $F(\vec{\sigma})$, may be written as,

$$F(\vec{\sigma}) = \sum_\alpha F_\alpha \Phi_\alpha(\vec{\sigma}), \tag{4}$$

where the sum extends over all clusters in the crystal, including the empty cluster, and where F_α is given by,

$$F_\alpha = \langle F(\vec{\sigma}) \bullet \Phi_\alpha(\vec{\sigma}) \rangle = \frac{1}{2^N} \sum_{\vec{\sigma}} F(\vec{\sigma}) \Phi_\alpha(\vec{\sigma}). \tag{5}$$

Thus, the terms F_α are the projections of $F(\vec{\sigma})$ on the orthogonal cluster basis.

It should be noted that the space group symmetry of the crystal requires that the cluster projections F_α of the function $F(\vec\sigma)$ be the same for all clusters α which are related by a symmetry operation (translation or point group). Accordingly, the cluster expansion in Eq. (4) becomes:

$$F(\vec\sigma) = \sum_{n=0}^{N} F_n \Theta_n(\vec\sigma), \tag{6}$$

where n labels the set of inequivalent clusters in the crystal. In the case of a disordered lattice, these cluster are only distinguished by their number of points and their geometry. In Eq. (6), the $\Theta_n(\vec\sigma)$ are given by:

$$\Theta_n(\vec\sigma) = \sum_{\alpha \in n} \Phi_\alpha(\vec\sigma). \tag{7}$$

In view of the orthogonality of the $\Phi_\alpha(\vec\sigma)$, we also have:

$$\frac{1}{2^N} \sum_\sigma \Theta_n(\vec\sigma)\Theta_m(\vec\sigma) = z_n N \delta_{n,m}, \tag{8}$$

where $z_n N$ is the total number of n-type clusters in the crystal.

The most common applications of Eq. (6) are for the cluster expansion of expectation values of functions of configurations, such as the average of the configurational energy. With the notation $\xi_n = \langle \Phi_\alpha(\vec\sigma) \rangle$ for the expectation value of the characteristic functions, where α is any cluster belonging to the equivalent set n, we obtain:

$$\bar{F} = \langle F(\vec\sigma) \rangle = N \sum_{n=0}^{N} z_n F_n \xi_n. \tag{9}$$

The usefulness of this cluster expansion rests on the fast convergence of the projections F_n. In Section III, the cluster expansion given by Eq. (9) is used to obtain the renormalized contributions to the configurational energy arising from chemical interactions, local volume relaxations and vibrational modes.

II.2 Electronic Structure Calculation

As mentioned, the development of local-density functional theory[7] has been instrumental in our ability to calculate the total energy of ordered compounds from the knowledge of their electronic structure. These calculations, which use only atomic numbers as input, correctly reproduce 0 K ground state properties of the elements and of ordered compounds.

Here, the electronic structure results for the binary Ni-Al system[20,28] were obtained using the augmented spherical wave (ASW) approximation.[8,11,12] For the ternary Ni-Al-Ti system, the calculations were carried out using the linear muffin-tin orbital (LMTO) method,[9] with exchange and correlation treated in the local-density approximation. Total energy calculations for the pure elements and for each compound are performed in the atomic sphere approximation, including corrections due to the overlap of atomic spheres, for approximately 15 values of the average Wigner-Seitz radius centered around the minimum of the electronic binding curve. In all cases, the energy is minimized with respect to the ratio of atomic radii for each element in the compound.

A convenient representation of the calculated total energy curves for a given compound is provided by a Morse function of the form:[29]

$$E(r) = A - 2Ce^{-\lambda(r-r_0)} + Ce^{-2\lambda(r-r_0)}, \tag{10}$$

where $E(r)$ is the calculated electronic total energy of the rigid lattice and A, C, λ, and r_0 are fitting parameters. Here, the variable r, is the Wigner-Seitz atomic radius related to the volume per atom by the relation $\Omega = (4\pi/3)r^3$. For the compounds, r is the "effective" Wigner-Seitz radius obtained from the average of the constituent atomic volumes. It follows from Eq. (1) that r_0 is the Wigner-Seitz radius corresponding to the minimum in the binding curve and that C is the cohesive energy of the rigid lattice.

The results of the electronic structure calculations can be extended to include the vibrational modes of configurationally ordered systems, *i.e.* pure elements and ordered compounds, by means of a Debye-Grüneisen analysis[29] of the calculated binding energies.[27] In particular, the analysis yields theoretical bulk moduli, Debye temperatures, and Grüneisen constants from which the vibrational free energies of pure metals and hypothetical, chemically ordered compounds, are obtained. For simple elemental systems[29] and alloys,[27] the theoretical thermal properties are in reasonable agreement with experiment.

In terms of properties of the rigid lattice, the vibrational free energy, $F(r,T)$, is given by,

$$F(r,T) = \frac{9}{8}k_B\Theta + E(r) - k_B T \left[D\left(\Theta/T\right) - 3ln\left(1 - e^{-\Theta/T}\right) \right], \tag{11}$$

where k_B is Boltzmann's constant, $D(x)$ is the Debye function, $E(r)$ is the electronic binding energy, and where the volume dependence of the Debye temperature, Θ, is given by:

$$\Theta = \Theta_0(r_0/r)^{3\gamma}, \tag{12}$$

with Θ_0 the Debye temperature corresponding to r_0, and with γ the Grüneisen constant.

For the ordered compounds, the free energy given by Eq. (11) represents the volume and temperature dependent binding energy in the absence of configurational disorder.

Thus, the results of the electronic structure calculations are cast in the form of a set of Morse parameters (see Eq. (10)), the Debye temperature Θ_0 and the Grüneisen constant γ for several compounds and for the pure elements. For the purpose of proceeding with the computation of equilibrium phase diagrams, we present next the determination of the effective interactions, using the cluster expansion developed in Section II.1, followed by the calculation of the configurational entropy in the CVM approximation. We defer until Section III further discussion of the electronic structure parameters obtained for the binary Ni-Al and ternary Al-Ni-Ti systems.

II.3 The Effective Interactions

The cluster expansion developed in Section II.1, and in particular Eq. (9), may be applied to the energy of a set of ordered compounds, for which the correlation functions are known *a-priori*, in order to obtain effective chemical interactions applicable to the disordered system. Here we apply the procedure to the vibrational free energies of the ordered compounds calculated in the Debye-Grüneisen approximation. The resulting temperature and volume dependent effective interactions are then used in a CVM treatment of the configurational entropy in order to include contributions due to configurational disorder into the total free energy. Phase diagrams calculations that include vibrational entropies have been carried out previously for Cu-Ag[27] and Ru-Nb[26] alloys.

In practice, the applicability of the method depends upon the convergence of the cluster expansion for relatively small clusters, and on the availability of a set of total energies for which the required inversion of Eq. (9) is defined. At present, selection criteria for clusters giving a converged cluster expansion are not available. Thus, a maximum interaction range is assumed *a-priori*, much as is done with the correlation range in the CVM. In some instances, attempts to ascertain the accuracy of the approximation have been made by comparing the total energy of compounds not included in the inversion procedure, with the values obtained using the assumed cluster expansion.[23,24]

For the *fcc* lattice in the tetrahedron approximation, the inversion of Eq. (9) is straightforward, requiring the calculation of total energies for only five high symmetry structures. For larger cluster approximations, this inversion is not immediately apparent and *ad-hoc* approaches, such as least-square fitting of the calculated total energies, have been proposed. It can be shown, however, that within a given maximum cluster approximation, there is always a natural set of relevant structures, given by the vertices of a convex configurational polyhedron, for which the inversion of Eq. (9) is unique.[30]

Defining the correlation functions $\xi_{k,n}$, where k labels a set of ordered compounds and n labels the interactions included in the cluster expansion, the vibrational free energy (per atom) for each of the ordered structures, $F_k(r,T)$, takes the form:

$$F_k(r,T) = \sum_{n=0}^{m} z_n V_n(r,T)\xi_{k,n}, \tag{13}$$

where $V_n(r,T)$ are the volume and temperature dependent effective interactions.

By properly choosing the set of ordered compounds ($k = 0,1,...m$) and of the cluster interactions ($n = 0,1,...m$), Eq. (13) can be inverted. At a fixed Wigner-Seitz radius and temperature, we have:

$$V_n(r,T) = \sum_{k=0}^{m} \omega_{n,k} F_k(r,T), \tag{14}$$

where $\omega_{n,k}$ are obtained by inversion of the matrix with elements $z_n\xi_{k,n}$.

If interactions are obtained using the same volume for all five compounds, the approach yields, once configurational effects are included, a total free energy functional that depends on the configurational variables ξ_n and volume (or r). Volume relaxation can then incorporated globally by minimizing this functional with respect to r, in addition to the usual minimization with respect to the correlation functions ξ_n. This global volume relaxation is based on the assumption that the effective local volumes occupied by tetrahedron clusters in the disordered alloy are independent of their configuration. Although this assumption has been applied to most of the first-principles calculations done to date, it appears physically implausible in cases where there is a significant difference between the atomic volumes of the constituents elements.

Other schemes aimed to account for the effect of local volume relaxations, have also been proposed. For example, the cluster interactions may be obtained from the total energies $F_k(r_k, T)$ of each compound calculated at different Wigner-Seitz radii r_k. The Wigner-Seitz radius, r_k, corresponds to the equivalent volume per atom in the disordered alloy occupied by clusters with configurations characteristic of compound k. In this local volume relaxation scheme, Eq. (14) becomes:

$$V_n(\vec{r}, T) = \sum_{k=0}^{m} \omega_{n,k} F_k(r_k, T), \tag{15}$$

where $\vec{r}$ stands for the set of Wigner-Seitz radii.

The set $\vec{r}$ is chosen by minimizing the total free energy with respect to each of the r_k subject to appropriate constraints. In the absence of any constraint, the approach is equivalent to the minimization of the vibrational free energy for each of the five compounds independently, and it is commonly referred to as total volume relaxation. Thus, in this scheme, the local volume of each tetrahedron cluster in the alloy is allowed to relax fully to the value found in the ordered state implying the existence of well define interatomic distance which are approximately independent of the surrounding chemical environment.

An approach intermediate to global and total volume relaxations, both of which appear physically implausible, is to define atomic radii for each constituent, r_A and r_B, such that the Wigner-Seitz radius r_k for each ordered structure k is given by:

$$r_k^3 = (1 - c_k)r_A^3 + c_k r_B^3, \tag{16}$$

where c_k is the concentration of component B in the ordered compound k. The local atomic radii, which are clearly dependent on the state of ordered and temperature, are then obtained variationally by minimization of the total free energy with respect to both r_A and r_B. This approach, which results in partial relaxation of the local volumes in the alloy, has been used to calculate the phase diagram of the Ru-Nb[26] and Ag-Cu systems.[27]

II.4 Configurational Entropy

The CVM, originally proposed by Kikuchi[2] in 1951, is formulated here in terms of a cluster expansion of the configurational entropy.[16] For a given probability distribution $X(\vec{\sigma})$, the configurational entropy is given exactly by:

$$S = -k_B \sum_{\vec{\sigma}} X(\vec{\sigma}) ln X(\vec{\sigma}), \tag{17}$$

where the sum is carried over all 2^N configurations in the crystal.

A tractable representation of Eq. (17) may be obtained by considering an infinite series of clusters with entropies defined by:

$$S_\alpha = -k_B \sum_{\vec{\sigma}_\alpha} X_\alpha(\vec{\sigma}_\alpha) ln X_\alpha(\vec{\sigma}_\alpha), \tag{18}$$

where the cluster probability distribution $X_\alpha(\vec{\sigma}_\alpha)$ is given by the sum of the $X(\vec{\sigma})$ over all configurational variables σ_i outside cluster α. This series of cluster entropies trivially converges to the exact configurational entropy as the size of the cluster α increases to include all points in the crystal. Using an exact Möbius transformation, we may also write the cluster entropies, S_α, in terms of a set of irreducible cluster contributions, $\hat{S}_\alpha$, as:[16]

$$S_\alpha = \sum_{\beta \in \alpha} \hat{S}_\beta \tag{19}$$

where the sum runs over all the subclusters of α, including α, and excludes the empty cluster.

The key approximation made in the CVM consists of neglecting the irreducible entropy contributions $\hat{S}_\alpha$ for clusters larger than a given maximum cluster. This closure condition allow us to express the total configurational entropy, S, in terms of a finite sum of irreducible contributions. Using the space group symmetry of the crystal, Eq. (19) for $\alpha \to N$ becomes:[16]

$$S = N \sum_{n=1}^{m} z_n \hat{S}_n, \tag{20}$$

or, in terms of the cluster entropies:

$$S = N \sum_{n=1}^{m} z_n a_n S_n = -N k_B \sum_{n=1}^{m} z_n a_n \sum_{\vec{\sigma}_n} X_n(\vec{\sigma}_n) ln X_n(\vec{\sigma}_n), \tag{21}$$

where, as defined previously, n labels inequivalent clusters and m labels the maximum cluster. The coefficients a_n, obtained by inverting Eq. (19), are given by:[16]

$$\sum_{\beta \in \alpha}' a_\beta = 1, \tag{22}$$

where the equation is valid for each subcluster α of the maximum cluster, and where the sum runs over all subclusters β of the maximum clusters that contain or equal α. For example, in the tetrahedron approximation of the *fcc* lattice, the coefficients a_n are equal to 5, -1, 0 and 1 for the point, pair, triangle and tetrahedron clusters, respectively.

The total free energy functional (per atom) of the disordered alloy, including local volume relaxations and vibrational modes, is given by:

$$F_{tot} = \sum_{n=0}^{m} z_n V_n(\vec{r}, T) \xi_n + k_B T \sum_{n=1}^{m} z_n a_n \sum_{\vec{\sigma}_n} X_n(\vec{\sigma}_n) ln X_n(\vec{\sigma}_n). \tag{23}$$

Here the effective interactions $V_n(\vec{r}, T)$ are given by Eq. (15). Using the cluster expansion described in Section II.1, the cluster probability distributions can be written in terms of the multisite correlation functions:[10]

$$X_n(\vec{\sigma}_n) = \frac{1}{2^n} \left[1 + \sum_{n=1}^{4} \Theta_n(\vec{\sigma}_n) \xi_n \right], \tag{24}$$

Table I

Morse parameters for the Ni-Al system from total energies calculated in the ASW approximation. The energies of formation are referred to Ni(*fcc*) and Al(*fcc*) at their T=0 K equilibrium volume

System	r_0	λ (au^{-1})	C (Ryd)	A (Ryd)	B (Kbar)	Θ_0 (K)	γ
Ni (*fcc*)	2.5853	1.4047	0.3645	0.3645	2171	407	1.8158
Ni$_3$Al (L1$_2$)	2.6170	1.3217	0.3828	0.3481	1994	422	1.7294
NiAl (L1$_0$)	2.6885	1.1984	0.3973	0.3568	1656	424	1.6109
NiAl$_3$ (L1$_2$)	2.8167	1.1308	0.3436	0.3277	1217	413	1.5926
Al (*fcc*)	2.9564	1.1184	0.2665	0.2665	880	409	1.6532
Ni (*bcc*)	2.5835	1.2006	0.4577	0.4631	1993	390	1.5509
Ni$_3$Al (DO$_3$)	2.6237	1.2906	0.3959	0.3624	1961	419	1.6931
NiAl (B2)	2.6775	1.1309	0.4549	0.4003	1696	429	1.5140
NiAl (B32)	2.7043	1.1308	0.4270	0.3967	1576	415	1.5291
NiAl (B32)	2.8231	1.1892	0.2912	0.2804	1139	399	1.6786
Al (*bcc*)	2.9780	1.0983	0.2483	0.2524	785	387	1.6354

where the $\Theta_n(\vec{\sigma}_n)$ are the characteristic functions defined by as sums of products of the configurational variables σ_i for lattice sites i belonging to cluster n (see Eqs. (2) and (7)). Thus the free energy functional given by Eq. (23) is function only of the set of Wigner-Seitz radii $\vec{r}$ and of the correlation functions ξ_n.

At a given temperature and concentration, the latter being given by the point correlation ξ_1, the equilibrium free energy is obtained by minimizing the free energy functional with respect to the remaining correlation functions, ξ_n (with n equal 2, 3, and 4), and the set of Wigner-Seitz radii $\vec{r}$. As mentioned in Section II.3, the minimization with respect to $\vec{r}$ may be carried out using three different schemes: i) constraining r_k to be all equal, which results in global volume relaxation without allowing relaxation of local volumes; ii) varying the r_k independently which results in total relaxation of local volumes; and iii) subjection the r_k to the external constraints, such as that of Eq. (16), which gives partial relaxation of the local volumes.

III. The Ni-Al and Ni-Al-Ti Systems

In order to illustrate the capabilities and limitations of the first-principles statistical theory of alloy phase equilibrium we present results for the equilibrium phase diagrams for two alloy systems of significant technological interest: the binary Ni-Al and the ternary Al-Ni-Ti systems. The first example represent a relatively complex alloy system involving *bcc*- and *fcc*-phases while the second example stands out as the first application of the method to ternary alloys.

The total energy electronic structure calculations were carried out as function of volume for six *bcc*-based and five *fcc*-based structures. The structures of the different

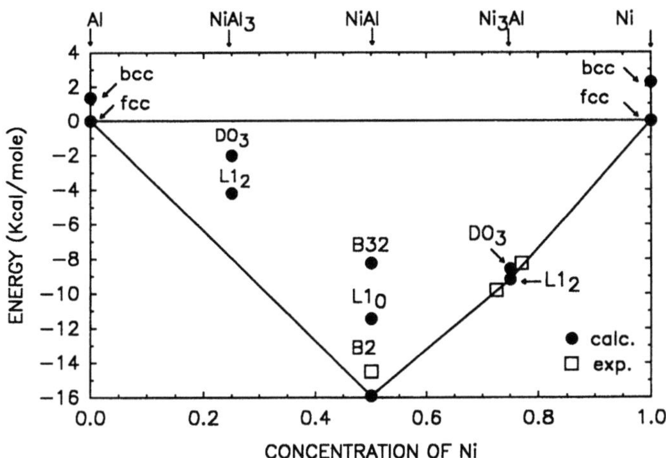

Figure 1. Calculated energies of formation at the equilibrium volume of *fcc-* and *bcc*-based compounds in the Ni-Al system.

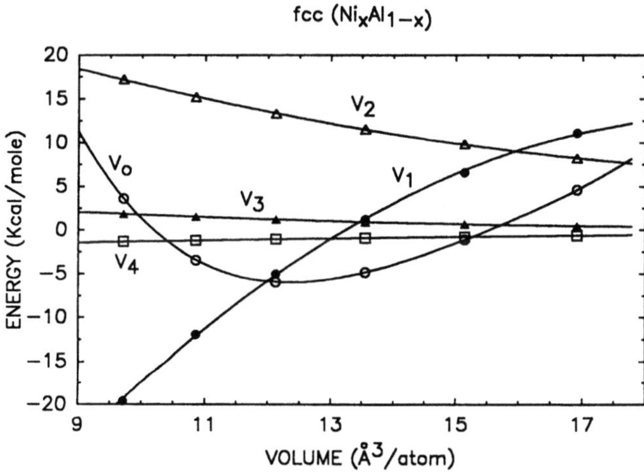

Figure 2. Calculated effective interactions as a function of average atomic volume for the empty (V_0), point (V_1), nearest-neighbor pair (V_2), triangle (V_3) and tetrahedron (V_4) clusters of the *fcc* lattice with global volume relaxations.

compounds together with the Morse parameters, Bulk moduli, Debye temperatures and Grüneisen constants are given in Table I.

The energy of formation at the equilibrium volume for each compound, calculated using the ASW and LDA approximations, is shown in Fig. 1. For comparison, available experimental data is also shown in the figure. The ground state diagram of Fig. 1 correctly reproduces, both qualitatively and quantitatively, the observed *fcc-* and *bcc*-based phases experimentally observed in Ni-Al.

Following the global relaxation procedure described previously, effective chemical interactions were obtained for both structures, *fcc* and *bcc*, and they are shown in

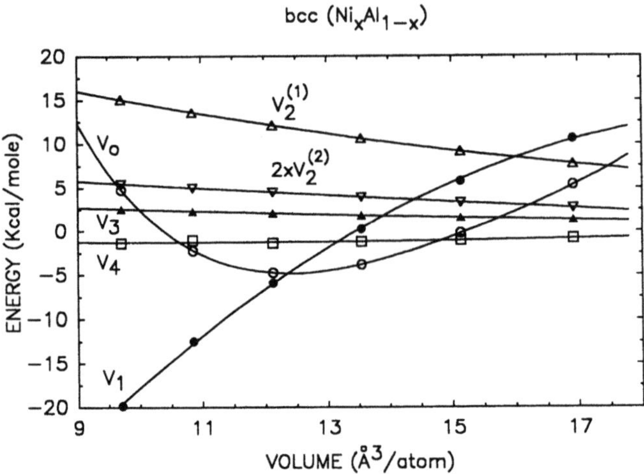

Figure 3. Calculated effective interactions as a function of average atomic volume for the empty (V_0), point (V_1), nearest-neighbor pair (V_2^1), next-nearest-neighbors (V_2^2), triangle (V_3) and tetrahedron (V_4) clusters of the *bcc* lattice with global volume relaxations.

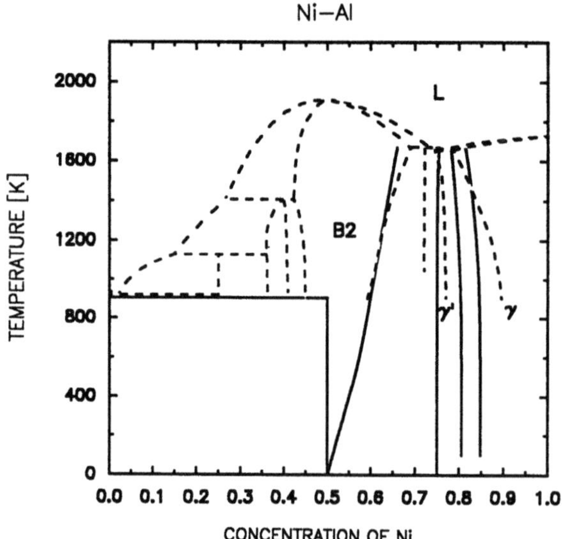

Figure 4. Calculated phase boundaries (solid lines) for the Ni-Al system compared to the experimental phase diagram (broken lines).

Figs. 2 and 3 as a function of average atomic volume. The interactions are for pair, triangle and tetrahedron clusters in both lattices. In the *fcc* lattice the range of interactions extend only to nearest-neighbors, whereas the *bcc* lattice also includes second-neighbors.

The calculated solid state portion of the phase diagram is compared in Fig. 4 with the experimentally determined diagram. The phase diagram calculations were carried out using the CVM in the tetrahedron-octahedron approximation for *fcc*-based and the tetrahedron approximation for *bcc*-based structures. Furthermore, vibrational modes

Table II

Morse parameters for *fcc*-based phases of the Ni-Al-Ti system from total energies calculated in the LMTO approximation. The energies of formation are referred to Ni(*fcc*), Al(*fcc*) and Ti(*fcc*) at their T=0 K calculated equilibrium volume. Binary phases are *fcc* (pure elements), L1$_2$ (A$_3$B and AB$_3$), L1$_0$ (AB) and tetragonal (A$_2$BC).

System	r_0	λ (au^{-1})	C (Ryd)	A (Ryd)	B (Kbar)	Θ_0 (K)	γ
Al	2.9470	1.2389	0.2142	0.2142	871	406	1.8255
Ni	2.5829	1.4800	0.3923	0.3923	2596	445	1.9113
Ti	3.0668	0.9583	0.5705	0.5705	1333	385	1.4695
Ni$_3$Al	2.6191	1.3722	0.4130	0.3745	2317	455	1.7969
NiAl	2.6829	1.3410	0.3747	0.3298	1960	461	1.7989
NiAl$_3$	2.7995	1.2748	0.2988	0.2765	1354	434	1.7844
Al$_3$Ti	2.9272	1.1525	0.3510	0.3199	1243	442	1.6868
AlTi	2.9709	1.0630	0.4613	0.4283	1369	434	1.5790
Al$_3$Ti	3.0097	1.0213	0.5117	0.4879	1384	411	1.5369
Ni$_3$Ti	2.6703	1.4454	0.3953	0.3540	2414	447	1.9298
NiTi	2.7967	1.2999	0.4171	0.3862	1967	423	1.8177
Ni$_3$Ti	2.9307	1.1048	0.4961	0.4813	1612	402	1.6189
Al$_2$NiTi	2.8218	1.1904	0.4282	0.3687	1678	452	1.6796
AlNi$_2$Ti	2.7734	1.2584	0.4202	0.3960	1873	433	1.7450
AlNiTi$_2$	2.8934	1.1245	0.4609	0.4391	1572	417	1.6268

are included. The solubility limits for the NiAl phase (B2) are reproduce quite well by the calculations. The calculated phase boundaries, although qualitatively correct, are less accurate for the Ni$_3$Al (L1$_2$) and *fcc* disordered phases. A likely reason for the disagreement is the fact that only nearest-neighbor interactions are included in the *fcc*-based phases. Nevertheless, the results are encouraging, particularly since the method is parameter free and uses only atomic numbers as input.

As an example of the application of the method to ternary systems, we consider here the *fcc*(γ)-L1$_2$(γ') phase equilibrium in the Ni-Al-Ti system. Allowing only nearest-neighbor interactions (pair and many-many body) interactions in the *fcc* lattice, it is necessary to calculate the total energy as a function of volume for 15 structures. The corresponding Morse potentials obtained from LMTO total energy calculations are shown in Table II. It should be noted that, for the NiAl system, the results obtained with the ASW and LMTO calculations are in general agreement (within 10%). The largest discrepancy is seen for the energy of formation of the metastable NiAl$_3$ compound (30%). The calculated $\gamma - \gamma'$ two phase boundaries at 1023 K are shown in Fig. 5. Although the two-phase region is considerably narrower than observed experimentally, the general trends and tie-line directions are well reproduced by the calculations.

IV. Conclusions

The calculation of phase diagrams from the knowledge of the electronic structure of compounds has, over the last few years, emerged as a potentially useful tool in alloy

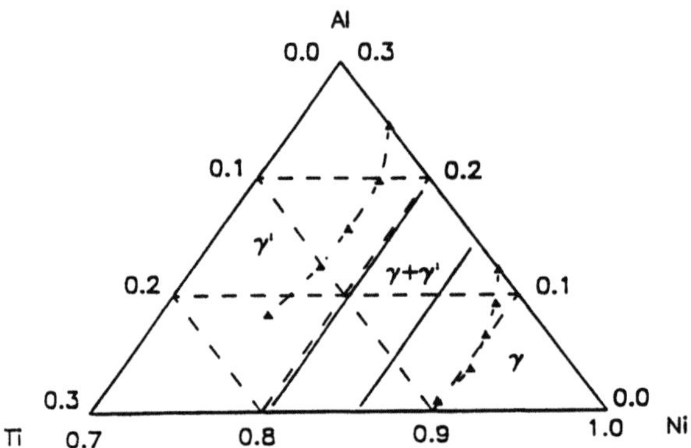

Figure 5. Calculated isothermal section at 1023 K showing the γ-γ' phases of the Ni-Al-Ti, compared to experimental results. The phase boundaries were calculated in the tetrahedron approximation of the CVM using a totally relaxed local volumes and vibrational modes are not included.

design. We have attempted to demonstrate that the basic theoretical principles, as well as the limitations of the method, are presently well understood. We have seen with two examples, Ni-Al and Ni-Al-Ti, that the theory is capable of giving a semiquantitative description using only atomic numbers as input. Thus, although much work remains to be done, it appears that the we are on the way to a truly predictive first-principles theory of alloy phase stability. Among the problems that are likely to be the focus of future work are the effect of elastic relaxations and applications to multicomponent systems.

Acknowledgments

This work was supported in part by NSF Grant No. DMR-91-14646 and by a Grant for International Joint Research Project from the NEDO, Japan.

References

1. J.M. Sanchez and D. de Fontaine, *Phys. Rev. B* **17**, 2926 (1978).
2. R. Kikuchi, J.M. Sanchez, D. de Fontaine, and H. Yamauchi, *Acta Metall.* **28**, 651 (1980).
3. C. Sigli and J.M. Sanchez, *Acta Metall.* **33**, 1097 (1985).
4. J.M. Sanchez, J.R. Barefoot, R.N. Jarret, and J.K. Tien, *Acta Metall.* **32**, 1519 (1984).
5. C.E. Dahmani, M.C. Cadeville, J.M. Sanchez, and J.L. Morán-López, *Phys. Rev. Lett.* **55**, 1208 (1985).
6. R. Kikuchi, *Phys. Rev.* **81**, 988 (1951); *J. Chem. Phys.* **60**, 1071 (1974).
7. W. Kohn and L.J. Sham, *Phys. Rev.* **140**, A1133 (1965); P. Hohenberg and W. Kohn, ibid. **136**, B864 (1964).
8. V.L. Moruzzi, J.F. Janak and A.R. Williams, *Calculated Electronic Properties of Metals* (Pergamon, New York, 1978).

9. O.K. Andersen, O. Jepsen, and D. Glötzel, *Highlights of Condensed Mater Theory, Proceedings of the International School of Physics Enrico Fermi*, (North-Holland, Amsterdam, 1985).

10. M.T. Yin and M.L. Cohen, *Phys. Rev. Lett* **45**, 1004 (1980).

11. A.R. Wiliams, C.D. Gelatt, and V.L. Moruzzi, *Phys. Rev. Lett.* **44**, 429 (1980).

12. C.D. Gelatt, A.R. Wiliams, and V.L. Moruzzi, *Phys. Rev. B* **27**, 2005 (1985).

13. J.S. Faulkner, *Prog. Mater. Sci.* **27**, 1, 1982 (Pergamon Press).

14. H. Winter and G.M. Stocks, *Phys. Rev. B* **27**, 882 (1982).

15. H. Winter, P.J. Durham, and G.M. Stocks, *J. Phys. F* **14**, 1047 (1984).

16. J.M. Sanchez, F. Ducastelle, and D. Gratias, *Physica* **128A**, 334 (1984).

17. J.W.D. Connolly and A.R. Williams, *Phys. Rev. B* **27**, 5169 (1983).

18. K. Terakura, T. Oguchi, T. Mohri, and K. Watanabe, *Phys. Rev. B* **35**, 2169 (1987).

19. A.A. Mbaye, L.G. Ferreira, and A. Zunger, *Phys. Rev. Lett.* **58**, 49 (1987).

20. A.E. Carlsson and J.M. Sanchez, *Solid State Comm.* **65**, 527 (1988).

21. T. Mohri, K. Terakura, T. Oguchi, and K. Watanabe, *Acta Metall.* **36**, 547 (1988).

22. A. Zunger, S.-H. Wei, A.A. Mbaye, and G.L. Ferreira, *Acta Metall.* **36**, 2239 (1988).

23. S. Takizawa, K. Terakura, and T. Mohri, *Phys. Rev. B* **39**, 5792 (1989).

24. L.G. Ferreira, S.-H. Wei, and A. Zunger, *Phys. Rev. B* **40**, 3197 (1989); ibid. **41**, 8240 (1990).

25. M. Sluiter, D. de Fontaine X.Q. Guo R. Podloucky, and A.J. Freeman, *Physical Rev. B* **42**, 10460 (1990).

26. J.D. Becker, J.M. Sanchez, and J.K. Tien, *Mat. Res. Soc. Symp. Proc.*, Vol. 213, p. 113–118, 1991.

27. J.M. Sanchez, J.P. Stark, and V.L. Moruzzi, *Phys. Rev. B* **44**, 5411 (1991).

28. J.M. Sanchez, J.D. Becker, and A.E. Carlsson, in: *Computer Aided Innovation of New Materials*, edited by M. Doyama, T.Suzuki, J. Kihara, and R. Yamamoto (Elsevier Science Publishers, 1991), p. 791.

29. V.L. Moruzzi, J.F. Janak, and K. Schwarz, *Phys. Rev B* **37**, 790 (1988).

30. J.M. Sanchez and D. de Fontaine, in: *Structure and Bonding in Crystals*, edited by M. O'Keeffe and A. Navrotsky (Academic Press, 1981), p. 117.

Nanostructured Materials

R.W. Siegel

Materials Science Division
Argonne National Laboratory
Argonne, Illinois 60439
U.S.A.

Abstract

The recently developed ability to synthesize materials from atomic precursors under controlled conditions on a nanometer size scale (below 100 nm) has the potential for revolutionizing materials science and engineering. Such nanostructured materials can now be synthesized with modulation dimensionalities from zero (clusters) to three (nanophase materials), each with their own particular advantages. These advantages stem from such diverse effects as, for example, quantum confinement, elastic strain accommodation, and grain size limitations. The general principles of nanostructured materials are considered and the particular opportunities for producing bulk nanophase materials with engineered properties, via the synthesis of metal and ceramic atom clusters followed by their *in situ* assembly under controlled conditions, are presented as an example.

I. Introduction

Increasing interest has focused on a variety of synthetic nanostructured materials, with average grain or other structural domain sizes below 100 nm, during the past several years with the anticipation that their properties will be different from, and often superior to, those of conventional materials that have phase or grain structures on a coarser size scale.[1] This interest has been stimulated not only by the recent efforts and successes in synthesizing a variety of beautifully symmetric and captivating atom clusters, zero-dimensionality quantum-well structures, and one-dimensionally modulated multilayered materials with nanometer scale modulations, but also by the exciting potential for synthesizing three-dimensionally modulated, bulk nanophase materials via the assembly of clusters of atoms.[2]

Nanophase materials are one of the broad class of nanostructured materials artificially synthesized with microstructures modulated in zero to three dimensions on

Advanced Topics in Materials Science and Engineering, Edited by
J.L. Morán-López and J.M. Sanchez, Plenum Press, New York, 1993

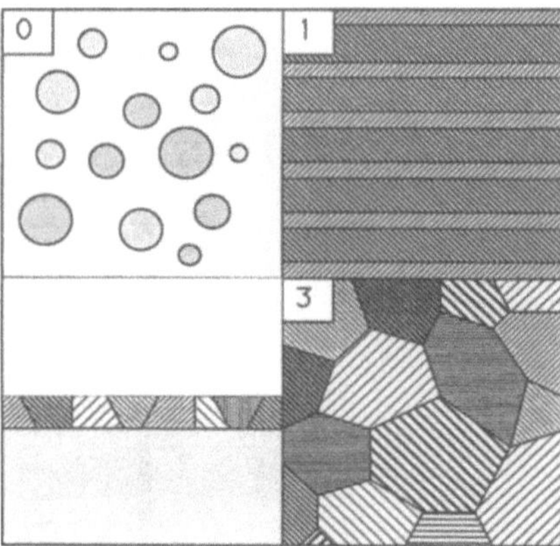

Figure 1. Schematic of the four types of nanostructured materials, classified according to integral modulation dimensionality: zero - clusters of any aspect ratio from 1 to ∞; one - multilayers; two - ultrafine-grained overlayers; three - nanophase materials. Intermediate dimensionalities can exist.

length scales less than 100 nm that it has become possible to create over the past few years. The various types of nanostructured materials share three features: atomic domains spatially confined to less than 100 nm, significant atom fractions associated with interfacial environments, and interactions between their constituent domains. Nanostructured materials thus include zero-dimensionality atom clusters and cluster assemblies, one-dimensionally modulated multilayers, and their three-dimensional analogues, nanophase materials, as indicated schematically in Fig. 1.

Atom clusters in the nanometer size regime, containing hundreds to tens of thousands of atoms, can now be produced in sufficient numbers by means of either physical or chemical processes that they can be assembled into materials that can be studied by a variety of conventional experimental methods. These materials can take advantage of and incorporate a number of size-related effects in condensed matter ranging from electronic effects (so-called "quantum size effects") caused by spatial confinement of delocalized valence electrons and altered cooperative ("many body") atom phenomena, such as lattice vibrations or melting, to the suppression of such lattice-defect mechanisms as dislocation generation and migration in confined grain sizes. The possibilities to assemble size-selected atom clusters into new materials with unique or improved properties may thus impact our ability to engineer a wide variety of controlled optical, electronic, mechanical, and chemical properties with attendant useful technological applications.

The present paper focuses on the general principles that underlie nanostructured materials and also the special opportunities presented by our ability to create nanophase materials assembled from atom clusters of metals and ceramics synthesized by means of the gas-condensation method.[3-7] This method appears to be the most generally applicable of the presently available avenues for producing size-selected atom clusters in the less-than-100-nm (nanostructure) regime, and thus appears to have very broad technological potential in the area of advanced materials.

II. General Principles

Generally, in synthesizing nanostructured materials from atomic or molecular precursors, one wants to be able to control a variety of microscopic aspects of the condensed ensemble. First, and probably foremost, is the size and size distribution of the constituent phases or structures. The desirable sizes are generally below 100 nm, since it is in this size range that various properties begin to change significantly owing to a variety of confinement effects. A property will be altered when the entity or mechanism (or combination thereof) responsible for that property is confined within a space smaller than some critical length associated with that entity or mechanism. So, for example, a metal which is coventionally ductile owing to the usual ease in creating and moving dislocations through its crystal lattice will become significantly harder when grain sizes are reduced to the point where dislocation sources are no longer able to operate at low levels of applied stress. Since the stress to operate a Frank-Read dislocation source is inversely proportional to the spacing between its pinning points, a critical length in this case is that for which the stress to operate this source becomes larger than the conventional yield stress for the given metal. Such confinement only appears to be different from that usually encountered in the technical literature, where for example the optical absorption properties of a so-called quantum-well semiconductor device are blue shifted (to shorter wavelengths) owing to the size of the well becoming comparable to and smaller than the effective size of the excitonic state responsible for this absorption. The specifics are indeed quite different, but the underlying general principle of confinement is not.

Second, the composition of the constituent phases in a nanostructured material is of crucial importance, as it invariably is to the performance of conventional materials. This can simply mean maintaining phase purity during synthesis in a single phase nanostructured material, such as an oxide or a metal, or it can mean controlling the impurity doping levels, the stoichiometries, the solute gradients, the phase mixtures, or combinations of these in more complex nanostructured materials. In this case, however, the length scales over which such composition control must be maintained can push the limits of our technical capabilities.

The third aspect of nanostructured materials that one would like to be able to control in their synthesis is the nature of the interfaces created between constituent phases and, hence, the nature of the interactions across the interfaces. These interfaces can, of course, be grain boundaries between the same phase with differing orientations, heterophase interfaces, or free surfaces. Since the number of interfaces present in nanostructured materials is large compared with conventional materials, this control can take on a much greater importance here. However, it is frequently rather difficult to create a nanostructured (or any other) material with prescribed interfaces. The greatest success to date has been achieved with one-dimensionally modulated nanoscale multilayers.

It is the interplay among these three features (size, composition, and interfaces) that determines the properties of nanostructured materials. In some cases, one or more of these features may dominate, as we will see in some of the examples given in this paper. Thus, one wants to be able to build nanostructured materials under controlled conditions, but with an eye to the particular property or properties of interest. The degree of control available, of course, depends upon the particular synthesis method being used to create the given nanostructured material. Discussion of the very wide variety of synthesis and processing methods for the creation of nanostructured materials is well beyond the scope of the present paper. An attempt to cover this subject in a somewhat more comprehensive form will appear elsewhere.[8] Previous reviews[1,6] can also be usefully consulted for such information. In the present paper, three-dimensionally modulated nanophase materials assembled from gas-condensed atom clusters will be used as a representative example of the broad range of nanostructured materials, since the author has been actively involved with these materials for the past several years and since they can be used to illustrate essentially all of the general principles just mentioned.

There are a number of advantages associated with the synthesis of materials from atom clusters. Some of these stem from the nanometer scale of the structures assembled and others arise from the inherent flexibility of dealing with clusters as the "building blocks" of these materials. Some of these advantages are as follows:[7]

(1) The ultrafine sizes of the atom clusters and their surface cleanliness allow the conventional restrictions of phase equilibria and kinetics to be essentialy overcome during material synthesis and processing by the combination of short diffusion distances, high driving forces, and uncontaminated surfaces and interfaces.

(2) The large fraction of atoms residing in the grain boundaries and interfaces of these materials allow interface atomic arrangements to constitute significant volume fractions of material, and thus novel materials properties may result.

(3) The reduced size scale and large surface-to-volume ratios of the individual nanophase grains can be predetermined and can alter and enhance a variety of physical and chemical properties.

(4) A wide range of materials and structures can be produced in this manner, including metals and alloys, intermetallic compounds, ceramics, and semiconductors.

(5) The possibilities for reacting, coating, and mixing *in situ* various types, sizes, and morphologies of clusters create a significant potential for the synthesis of a variety of new multicomponent and multifunctional composites with nanometer-sized microstructures and engineered properties. Most of the research carried out to date, nevertheless, has concentrated on single-phase metals and ceramics.

III. Synthesis and Structure of Nanophase Materials

The predominant feature of cluster-assembled nanophase materials, as in the case of all nanostructured materials, is their ultrafine grain size and, hence, the large fraction of their atoms that reside in grain boundaries or interfaces. For example, as indicated in Fig. 2, a nanophase material with a 5 nm average grain size will have from about 27 to 49% of its atoms associated with grain boundaries, assuming a simple grain boundary picture and an average grain boundary thickness of about 0.5 to 1.0 nm (*ca.* 2–4 nearest-neighbor distances). This percentage falls to about 14–27% for a 10 nm grain size, but is as low as 1–3% for a 100 nm grain size. The interface volume fraction is, of course, essentially negligible for conventional grain sizes of 1 μm and above. The simple structural model upon which Fig. 2 is based assumes only that the grain volumes scale as a length cubed and that the grain boundary volume includes the junctions between and among their boundaries. The properties of nanostructured materials are thus expected to be strongly influenced by their small grain sizes and the nature (atomic and electronic structure) of their internal boundaries, simply because of the very large number density of these interfaces.

Considerable effort has gone into the elucidation of the structure of nanophase grain boundaries.[6,10,11] Investigations of nanophase TiO_2 by Raman spectroscopy[12–14] and of nanophase Pd by high resolution, transmission electron microscopy[15,16] indicate

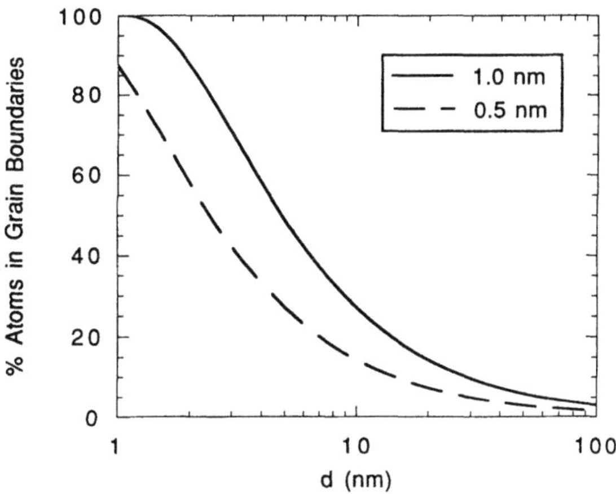

Figure 2. Percentage of atoms in grain boundaries of a nanophase material as a function of grain diameter, assuming that the average grain boundary thickness ranges from 0.5 to 1.0 nm (*ca.* 2 to 4 atomic planes wide).[9]

that the grain boundary structures in these materials are essentially low energy configurations that are rather similar to those in coarser grained, conventional materials.[17] The nanophase grain boundaries contain short-range ordered structural units representative of the bulk material and distortions that are localized to about ± 0.2 nm on either side of the grain boundary plane. These conclusions are also consistent with the results from complementary small-angle neutron scattering measurements,[18−20] and with the expectations for conventional high angle grain boundaries from condensed matter theory.[21,22] Whether or not there are indeed some local structural deviations in nanophase boundaries compared with conventional scale grain boundaries is an open question. However, no firm experimental or theoretical evidence presently exists that indicates that nanophase grain boundaries are actually different than their coarse-grained counterparts.

The synthesis of nanophase materials via the *in situ* consolidation of gas-condensed clusters has been described in detail elsewhere.[7] For the purposes of this paper it is sufficient to give only a brief outline of the process. A precursor material, either metal or compound, is evaporated in a gaseous atmosphere maintained at a few hundred Pa pressure in a back-filled, ultrahigh-vacuum chamber. The evaporated atoms or molecules lose energy via collisions with the gas atoms or molecules and undergo a homogeneous condensation to form atom clusters in the highly supersaturated vicinity of the precursor source, as indicated schematically in Fig. 3. In order to maintain small cluster sizes, by preventing further atom or molecule accretion, the clusters once nucleated must be removed rapidly from the region of high supersaturation. Since the clusters are already gas entrained, this is easily accomplished by moving the condensing gas. Such gas motion has generally resulted from natural convection under the action of gravity and the temperature difference between the precursor source and a thermophoretic cluster collector cooled by liquid nitrogen. However, a forced gas flow can also be used, and with significant advantage in terms of cluster size control and process efficiency. The clusters are then removed from the collector under vacuum and consolidated at room temperature at pressures up to about 1–2 GPa by means of a

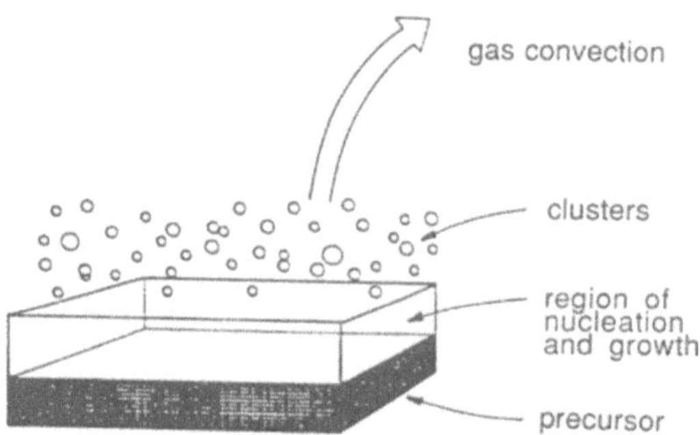

Figure 3. Conceptual model for the formation of atom clusters via gas condensation.[23]

piston-and-anvil device. The resulting samples are usually about 1 cm in diameter and about 0.1–0.5 mm thick, depending upon the amount of precursor material evaporated and the experiment planned. A typical evaporation in the laboratory now takes of order of about 1 h to make such a sample; fortunately, the process can be readily scaled up by several orders of magnitude.

Consideration of the model of Fig. 3 shows that the smallest available cluster sizes for a given metal, for example, will be obtained for a low precursor evaporation rate and condensation in a low pressure of a light inert gas, such as He.[23] These conditions lead to a lower supersaturation of precursor atoms in the gas, slower removal of energy from the evaporated atoms, via the lighter gas atoms at lower pressure, and more rapid convective gas flow owing also to the lower gas pressure. There are just three fundamental rates which essentially control the formation of the atom clusters in the gas-condensation process.[7] They are (1) the evaporation rate or rate of supply of atoms to the region of supersaturation where condensation occurs, (2) the rate of energy removal from the hot atoms via the condensing gas, and (3) the rate of removal of the clusters once nucleated from the supersaturation region. There are other factors that can also affect the clusters finally collected, particularly those that result in significant cluster-cluster coalescence, but these three rates represent the fundamental core of the gas-condensation process. It is therefore the control of these rates, relative to one another, that yield control over the sizes of the constituent grains in these materials and, hence, over their grain-size dependent properties.

The clusters that are collected on the surface of the collector, and which subsequently become the grains of the consolidated polycrystalline nanophase aggregate (see Fig. 4), have a rather narrow size distribution, usually with a full-width at half-maximum of about ±25% of the peak diameter, and the log-normal shape typical of clusters formed via gas-condensation.[23] This shape is rather typical for the grain size distribution in any of the nanophase materials thus far produced by this method. Transmission electron microscopy[15,16,24,25] has shown that the grains in nanophase compacts are rather equiaxed, similar to the atom clusters from which they were formed. On the other hand, the observations that the densities of nanophase materials consolidated from initially equiaxed clusters extend well beyond the theoretical limit (78%) for close packing of identical spheres indicate that an extrusion-like deformation of the clusters must result during the consolidation process, filling in (at least in part) the pores among the grains. However, essentially all of the nanophase materials consolidated from clusters at room temperature to date have invariably posessed a degree of

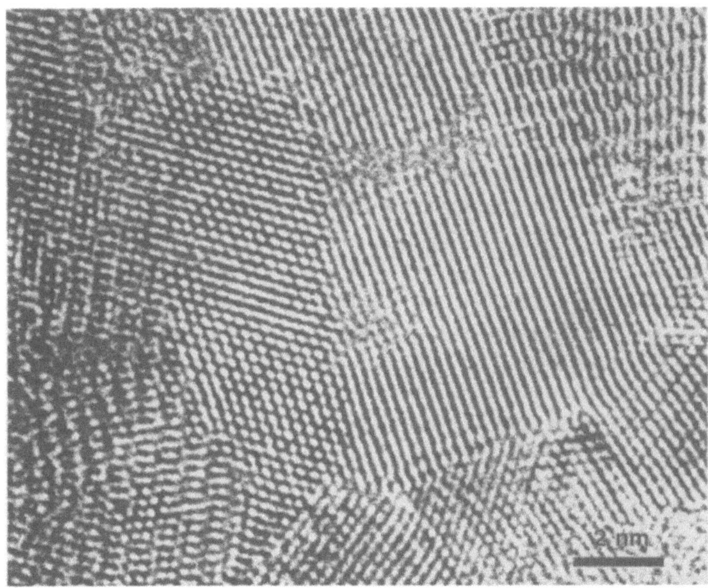

Figure 4. High resolution transmission electron micrograph of a typical area in nanophase Pd.[15]

porosity ranging from about 25% to less than 5%, as measured by Archimedes densitometry, with the larger values for ceramics and the smaller ones for metals. Evidence for this porosity was first obtained by positron annihilation spectroscopy[24,26,27] and more recently by precise densitometry[28] and porosimetry[29,30] measurements. These measurements have together shown that the porosity in as-consolidated nanophase metals and ceramics is primarily in the less than 100 nm size regime (although some larger porous flaws have been observed), and frequently of comparable sizes to the grains, but that the porosity is to a great extent interconnected and intersects with the specimen surfaces. Fortunately, it appears now that this porosity can be controlled in the synthesis steps and, if desired, can be removed completely during consolidation at elevated temperatures without sacrificing the ultrafine grain sizes in these materials.

The synthesis of elemental nanophase materials is rather straightforward using the gas-condensation method described above. Indeed, the purity of the material can not only be maintained in the process, but it can be frequently improved owing to the fact that the evaporation-condensation method is itself a form of distillation, with more volatile impurities being removed in the process. However, in the case of synthesizing more complex materials by this method, the necessary composition control can become more difficult. For example, in order to produce nanophase TiO_2 with a rutile structure,[24] Ti metal clusters condensed in He are first collected on a cold finger and subsequently oxidized by the rapid introduction of oxygen into the chamber. In this case, the material produced is found[14] to be initially deficient in oxygen, resulting in the substoichiometric compound $TiO_{1.89}$. However, oxygen stoichiometry can be subsequently controlled in such material. Raman spectroscopy has been a used[13,14] to calibrate the deviation from stoichiometry in as-produced nanophase titanium dioxide, which could then be easily oxidized to fully stoichiometric TiO_2, if desired, without sacrificing its small grain size (12 nm). Also, if intermediate deviations from stoichiometry were sought, in order to select particular properties of this material, they could be readily accessed as well. For other compounds, where maintaining composition control through thermal evaporation can be a problem, sputtering sources can be utilized in much the same way to produce atom clusters, with relative elemental concentrations of constituents intact, for consolidation into nanophase materials.[31,32]

An important aspect of nanophase materials assembled from atom clusters is their apparently inherent stability against grain growth.[33,34] Their grain sizes generally remain rather deeply metastable to elevated temperatures until about 0.4 to 0.5 of the absolute melting temperature is reached in single-phase nanophase materials. This stability (actually, deep metastability) is rather typical for the nanophase oxides investigated and for nanophase metals also. It appears that the narrow grain size distributions normally observed in these cluster-assembled materials coupled with their relatively flat and faceted grain boundaries (and also enhanced by their multiplicity of grain boundary junctions) place these nanophase structures in a local minimum in energy from which they are not easily extricated. They are thus analogous to a variety of closed-cell foam structures, which are stable (really, deeply metastable) despite their large stored surface energy. Under such conditions, only at temperatures above which bulk diffusion distances are comparable to or greater than the mean grain size will this metastability give way to global energy minimization via rapid grain growth. One could, of course, intentionally stabilize against grain growth by appropriate doping or composite formation in the grain boundaries to prevent or to retard their migration. It might also be possible to further stabilize even pure single-phase nanophase materials against grain growth, if it were possible to control the precise nature of the grain boundaries formed. Unfortunately, at this time such control over the assembly of clusters via consolidation is not feasible, and essentially random ensembles of grains and high-angle grain boundaries result.

IV. Properties

Cluster-assembled nanophase materials, as other nanostructured materials, have a variety of properties that are considerably different and improved in comparison with those of conventional coarser-grained structures. These result from the impact of various combinations of the three important aspects (size, composition, and interfaces) of their nature cited in Sec. II. Several examples will be given from this particular class of nanostructured materials to try and give a sense of how the interplay among these aspects can significantly alter some useful technological properties of these materials.

First of all, the processing of nanophase materials is dramatically improved by the combination of ultrafine cluster (grain) sizes, short diffusion distances, and cleanliness of cluster surfaces prior to consolidation. For example, nanophase TiO_2 (rutile) exhibits significant improvements in both sinterability and resulting mechanical properties relative to conventionally synthesized coarser-grained rutile.[24,29,35,36] Nanophase TiO_2 with a 12 nm initial mean grain diameter has been shown[24] to sinter under ambient pressures at up to 600°C lower temperatures than conventional coarser-grained rutile, and without the need for any compacting or sintering aid, such as polyvinyl alcohol, which is usually required. Furthermore, it has more recently been demonstrated[29] that sintering the same nanophase material under pressure (1 GPa), or with appropriate dopants such as Y, can further reduce the sintering temperatures,

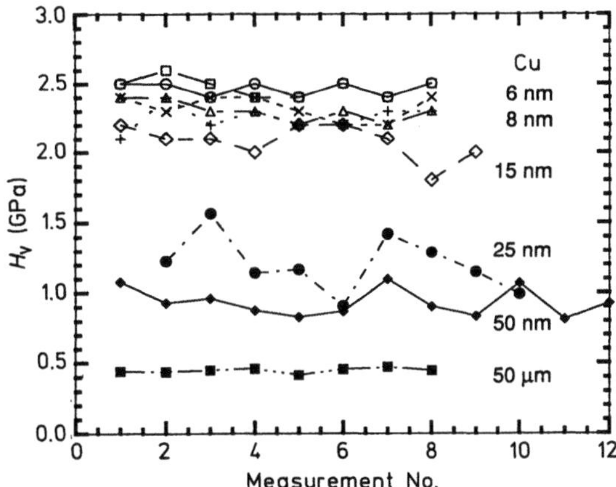

Figure 5. Vickers microhardness measurements at a number of positions across several nanophase Cu samples ranging in grain size from 6 to 50 nm, compared with similar measurements from an annealed conventional 50 μm grain size Cu sample.[28]

while suppressing grain growth as well, thus allowing for the unique possibility to sinter nanophase ceramics to full density while retaining their ultrafine grain size. The resulting fracture characteristics[35,37,38] developed for sintered nanophase TiO_2 are as good, or in some aspects improved, relative to those for conventional coarser-grained rutile.

Another property that is altered and can significantly affect the processing of nanostructured materials is diffusion. Atomic diffusion in nanophase materials, which can also have a significant bearing on their mechanical properties, such as creep and superplasticity, and electrical properties as well, has been found to be very rapid compared with conventional materials. Measurements of self- and solute-diffusion[35,39−42] in as-consolidated nanophase metals and ceramics indicate that atomic transport is orders of magnitude faster in these materials than in coarser-grained polycrystalline samples. This very rapid diffusion in as-consolidated nanophase materials appears to be intrinsically coupled with the porous nature of the interfaces in these materials, since the diffusion can be suppressed back to conventional values by sintering samples to full density.[35] Nonetheless, there exist considerable possibilities for efficiently doping nanophase materials at relatively low temperatures via the rapid diffusion available along their ubiquitous grain-boundary networks, with only short diffusion paths remaining into their grain interiors, to synthesize materials with tailored optical, electrical, or mechanical properties.

The predominant mechanical property change resulting from reducing the grain sizes of nanophase metals is the significant increase in their strength.[28,43,44] Fig. 5 shows recent microhardness results for several samples of nanophase Cu compared with results for a coarser-grained sample. The smallest grain size (6 nm) sample is seen to exhibit about a 500% increase in hardness over the coarser-grained (*ca.* 50 μm) sample. Nanophase Pd samples with 5–10 nm grain sizes have also been observe to exhibit up to about a 500% increase in hardness over coarser-grained (*ca.* 100 μm) samples,[43] with concomitant increases in their yield stress. The common strengthening behavior found in nanophase Cu and Pd indicates that this response is generic to nanophase metals, at least those with a *fcc* structure. The likelihood that such mechanical behavior is more broadly generic to nanophase metals in general is enhanced by the

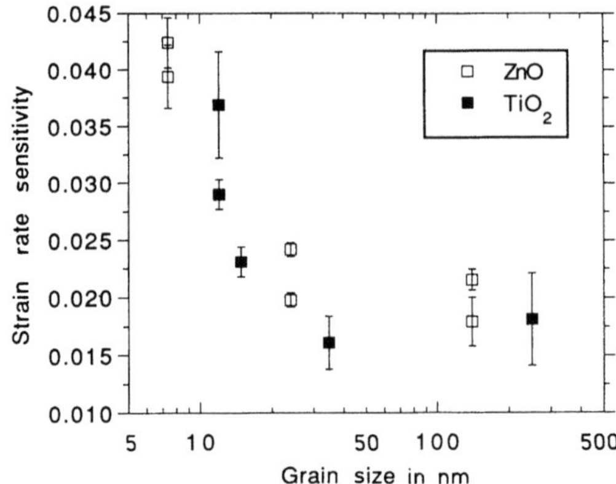

Figure 6. Strain rate sensitivity of nanophase TiO$_2$ and ZnO as a function of grain size.[36,52] The strain rate sensitivity was measured by a nanoindentation method and the grain size was determined by dark-field transmission electron microscopy.

observations that nanophase metals and alloys produced via mechanical attrition also exhibit significantly enhanced strength. For example, Koch and coworkers[45,46] have found hardness increases of factors of 4 to 5 in nanophase Fe and a factor of about 1.2 in nanophase Nb$_3$Sn when the grain size drops from 100 nm to 6 nm.

The increased strength observed in ultrafine-grained nanophase metals, although apparently analogous to conventional Hall-Petch[47,48] strengthening observed with decreasing grain size in coarser grained metals, must result from fundamentally different mechanisms. As discused in Sec. II, it can be expected that as nanophase grain sizes decrease and fall below the critical length scale for a given mechanism to operate, the associated property will be significantly changed. The grain sizes in the nanophase metals considered here are smaller than the necessary critical bowing lengths for Frank-Read dislocation sources to operate at the stresses involved and smaller also than the normal spacings between dislocations in a pile-up. It is therefore clear that an adequate description of the mechanisms responsible for the increased strength observed in nanophase metals will need to accommodate to the ultrafine grain-size scale in these materials. As this scale is reduced, and conventional dislocation generation and migration become increasingly difficult, it is apparent that the energetic hierarchy of microscopic deformation mechanisms will become successively accessed. Thus, easier mechanisms (such as dislocation generation from Frank-Read sources) will become frozen out at sufficiently small grain sizes and more energetically costly mechanisms will become necessary to effect deformation. Hence, grain confinement appears to be the dominant cause for the increased strength of nanophase metals. An additional contribution to the observed strengthening may also be the elastic strain accommodation[28] resulting from cluster consolidation.

Also in the area of mechanical behavior, it has been observed that nanophase ceramics are easily formed.[24,49−51] Furthermore, nanoindenter measurements on nanophase TiO$_2$ and ZnO have recently demonstrated[36,52] that a dramatic increase of strain rate sensitivity occurs with decreasing grain size, as shown in Fig. 6. Since this strong grain-size dependence is found for sets of samples in which the porosity is changing very little, it appears that this increased tendency toward ductility is

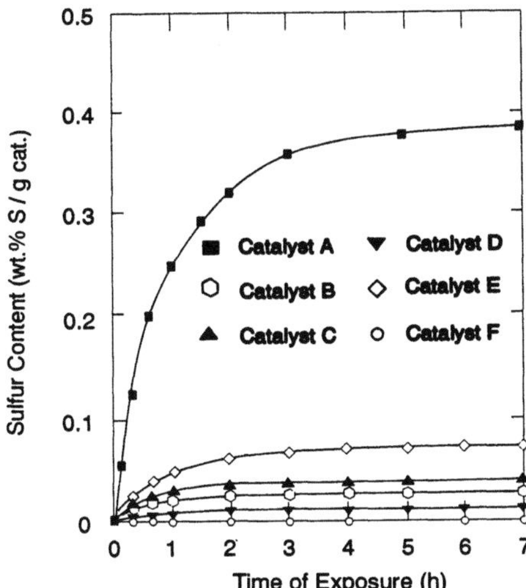

Figure 7. Activity of nanophase TiO_2 for H_2S decomposition as a function of exposure time at 500°C compared with that from several commercially available TiO_2 materials and a reference (A: 76 m^2/g nanophase rutile; B: 61 m^2/g anatase; C: 2.4 m^2/g rutile; D: 30 m^2/g anatase; E: 20 m^2/g rutile; F: reference alumina).[54]

an intrinsic property of these ultrafine-grained ceramics. The strain rate sensitivity values at the smallest grain sizes yet investigated (12 nm in nanophase TiO_2 and 7 nm in ZnO) thus indicate ductile behavior of these nanophase ceramics, as well as a significant potential for increased ductility at even smaller grain sizes and elevated temperatures. The maximum strain rate sensitivities measured in these studies, about 0.04, are already approximately one-quarter that for lead at room temperature, for example. However, no superplasticity has yet been observed in nanophase materials at room temperature, which would yield strain rate sensitivity values about an order of magnitude higher than the maximum observed. Nevertheless, it seems clear that at smaller grain sizes and/or at elevated temperatures, superplasticity of these materials will indeed be observed. Indeed, some limited evidence for such behavior has recently appeared.[51]

The enhanced strain rate sensitivity at room temperature found in the nanophase ceramics TiO_2[36] and ZnO[52] shown in Fig. 6. appears to result from increased grain boundary sliding in this material, aided by the presence of porosity, ultrafine grain size, and probably rapid short-range diffusion as well. This behavior is therefore dominated by the presence of the numerous interfaces in these materials and the very short diffusion distances involved in effecting the necessary atomic healing of incipient cracks for grain boundary sliding to progress at the strain rates utilized without fracturing the sample. Extrapolating from this apparently generic behavior, one can expect that grain boundary sliding mechanisms, accompanied by short-range diffusion assisted healing events, would be expected to increasingly dominate the deformation of a wide range of nanophase materials. Enhanced forming and even superplasticity in a wide range of nanophase materials, including intermetallic compounds, ceramics, and semiconductors might become a reality. Consequently, increased opportunities for high deformation or superplastic near-net-shape forming of a very wide range of even conventionally rather brittle and difficult to form materials could result.

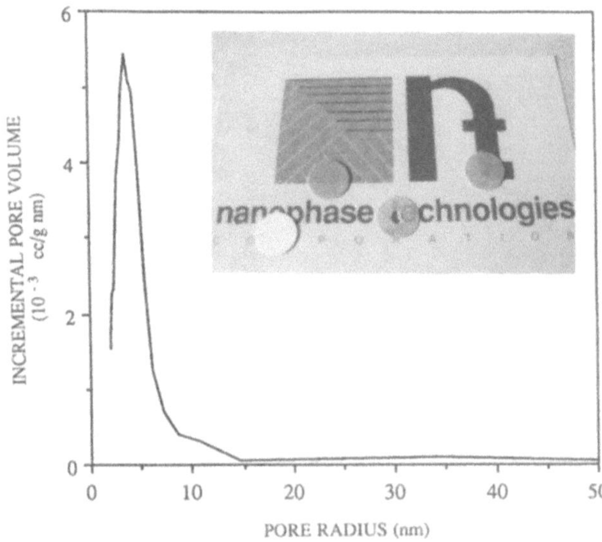

Figure 8. Pore size distribution in nanophase Y_2O_3 consolidated at room temperature under 500 Pa pressure to a density of between 65–75% of the theoretical value. A similar set of samples is shown in the inset along with an opaque sample with larger pores.[55]

The possibilities for plastic forming nanophase ceramics to near-net shape appear to be well on their way to realization. Karch and Birringer[50] have recently demonstrated that nanophase TiO_2 could be readily formed to a desired shape with excellent detail below 900°C, and the fracture toughness was found to increase by a factor of two as well. The ability to extensively deform nanophase TiO_2 at elevated temperatures (*ca.* 800°C) without cracking or fracture has been clearly demonstrated by Hahn and coworkers.[51,53] While these latter demonstrations have been accompanied by significant grain growth in the samples at the elevated temperatures employed, it can be expected that lower temperature studies (below 0.4-0.5 of the absolute melting temperature) in the future will also allow for near-net-shape forming of nanophase ceramics with both their ultrafine grain sizes and their attendant properties retained.

The chemical reactivity of nanophase materials, with their high surface areas compared to conventional materials, can also be rather striking. Since the gas-condensed, high-surface-area clusters are assembled via consolidation in the method described in this paper, there can be an excellent degree of control over the total available surface area in these self-supported ensembles. Also, as pointed out earlier, composition control can be readily achieved. Recently completed measurements[54] of the decomposition of H_2S over lightly consolidated nanophase TiO_2 at 500°C demonstrate the enhanced chemical reactivity of nanophase materials rather well. Fig. 7 shows the catalytic activity for S removal from H_2S via dissiciative adsorption for nanophase TiO_2 (rutile) compared with that for a number of other commercially available forms of TiO_2 having either the rutile or anatase structure. It can be seen that the nanophase sample was far more reactive than any of the other samples tested, both initially and also after extended exposure to the H_2S. This greatly enhanced activity was shown to result from a combination of unique and controllable features of the nanophase material, its high surface area combined with its rutile structure and its oxygen deficient composition.

Rather limited research has been carried out so far on the physical properties of cluster-assembled nanophase materials. However there appear to be interesting prospects, based upon what little has been accomplished and on the expectations for confined systems of atoms in which the sizes of constituent domains fall below the critical length scales pertinent to a given property. Fig. 8 shows an example of how the ability to control the porosity in the synthesis of nanophase ceramics will apparently lead to some rather interesting optical applications in the near future. It has been recently demonstrated[55] that the oxidation step in the synthesis of cluster-consolidated nanophase Y_2O_3 can be so controlled that the resulting "green-state" porosity of 25–35% has a size distribution sufficiently small and narrow, as shown in Fig. 8, that the as-consolidated material is effectively transparent. Samples of this material are shown in the inset of Fig. 8 along with a similar opaque sample containing a larger population of pores. The ability to thus reduce the size of the porosity in such material to well below the wavelengths of visible or other radiation, coupled with the possibilities for efficient doping of these materials cited previously, should lead to a variety of interesting optical properties and related applications.

V. Conclusions

There appear to be tremendous opportunities for synthesizing nanostructured materials with new architectures at nanometer length scales from atomic or molecular precursors via the assembly of atom clusters and by a myriad of other techniques now becoming available. The keys to the future of nanostructured materials, however, will be both in our ability to significantly change for the better the properties of materials by artificially structuring them on these nanometer length scales and in developing the methods for producing these materials in commercially viable quantities. The future appears to hold great promise for nanostructured materials, based upon the limited knowledge that has already been accumulated, but much work remains to be done.

The examples presented in Sec. IV have demonstrated that, even in our crude initial attempts to change and control the properties of materials via nanostructuring, it is already possible to dramatically alter a variety of technologically important properties of materials in this manner. It was shown that the strength of normally very ductile pure metals can be increased by orders of magnitude simply by reducing their grain sizes into the range where dislocation generation and migration becomes confined and difficult. On the other hand, normally very brittle oxide ceramics can be rendered more ductile and formable by reducing their grain sizes into the range where grain boundary sliding becomes facilitated by the high number density of internal interfaces and the rapid atomic diffusion therein. Since cluster "building blocks" can already be assembled with some modicum of control over the remaining porosity in the nanophase compact, it was also demonstrated that significant positive effects on both the chemical reactivity of a self-supported nanophase "catalyst" and the optical transparency of a nanophase ceramic "window" can result. The possibilities for engineering both the catalytic and optical properties of nanostructured materials will clearly be further enhanced by the ability to control the compositions and defect structures in these materials. Such examples suggest that even more exciting opportunities may be available as we learn to assemble more sophisticated multifunctional composite nanostructured materials in the future.

Combining the various aspects of nanometer scale and composition and interface control in nanostructured materials to create a variety of new technological materials with atomically engineered properties will be a great challange. Fortunately, we are beginning to learn something about the relationships between nanostructure and material properties and how these might be changed to control material performance. It is a challenge, therefore, that appears to be well worth the effort.

Acknowledgements

This work was supported by the U.S. Department of Energy, BES-Materials Sciences, under Contract W-31-109-Eng-38. This paper is dedicated to the memory of Francisco Mejía-Lira and John K. Tien who, with their colleagues, were instrumental in organizing the First México-USA Symposium: The Frontiers in Materials Science held in Ixtapa in September 1991, at which this paper was presented.

References

1. B.H. Kear, L.E. Cross, J.E. Keem, R.W. Siegel, F. Spaepen, K.C. Taylor, E.L. Thomas, and K.-N. Tu, *Research Opportunities for Materials with Ultrafine Microstructures*, (National Academy, Washington, DC, 1989), Vol. NMAB-454.
2. R.P. Andres, R.S. Averback, W.L. Brown, L.E. Brus, W.A. Goddard, III, A. Kaldor, S.G. Louie, M. Moskovits, P.S. Peercy, S.J. Riley, R.W. Siegel, F. Spaepen, and Y. Wang, *J. Mater. Res.* **4**, 704 (1989).
3. H. Gleiter, in: *Deformation of Polycrystals: Mechanisms and Microstructures*, edited by N. Hansen *et al.*, (RisøNational Laboratory, Roskilde, 1981), p. 15.
4. R. Birringer, U. Herr, and H. Gleiter, *Suppl. Trans. Jpn. Inst. Met.* **27**, 43 (1986).
5. R.W. Siegel and H. Hahn, in: *Current Trends in the Physics of Materials*, edited by M. Yussouff, (World Scientific Publ. Co., Singapore, 1987), p. 403.
6. H. Gleiter, *Progress in Materials Science* **33**, 223 (1989).
7. R.W. Siegel, in: *Materials Science and Technology*, Vol. 15, edited by R.W. Cahn, (VCH, Weinheim, 1991), p. 583.
8. R.W. Siegel, in: *Mechanical Properties and Deformation Behavior of Materials Having Ultra-Fine Microstructures*, edited by M.A. Nastasi *et al.*, (Kluwer, Dordrecht, 1993), in press.
9. R.W. Siegel, *Ann. Rev. Mater. Sci.* **21**, 559 (1991).
10. R. Birringer and H. Gleiter, in: *Encyclopedia of Materials Science and Engineering*, Suppl. Vol. 1, edited by R.W. Cahn, (Pergamon Press, Oxford, 1988), p. 339.
11. R.W. Siegel, in: *Materials Interfaces: Atomic-Level Structure and Properties*, edited by D. Wolf and S. Yip, (Chapman and Hall, London, 1992), p. 431.
12. C.A. Melendres, A. Narayanasamy, V.A. Maroni, and R.W. Siegel, *J. Mater. Res.* **4**, 1246 (1989).
13. J.C. Parker and R.W. Siegel, *J. Mater. Res.* **5**, 1246 (1990).
14. J.C. Parker and R.W. Siegel, *Appl. Phys. Lett.* **57**, 943 (1990).
15. G.J. Thomas, R.W. Siegel, and J.A. Eastman, *Mater. Res. Soc. Symp. Proc.* **153**, 13 (1989).
16. G.J. Thomas, R.W. Siegel, and J.A. Eastman, *Scripta Metall. et Mater.* **24**, 201 (1990).
17. R.W. Siegel and G.J. Thomas, *Ultramicroscopy* **40**, 376 (1992).

18. J.E. Epperson, R.W. Siegel, J.W. White, T.E. Klippert, A. Narayanasamy, J.A. Eastman, and F. Trouw, *Mater. Res. Soc. Symp. Proc.* **132**, 15 (1989).

19. E. Jorra, H. Franz, J. Peisl, G. Wallner, W. Petry, R. Birringer, H. Gleiter, and T. Haubold, *Phil. Mag. B* **60**, 159 (1989).

20. J.E. Epperson, R.W. Siegel, J.W. White, J.A. Eastman, Y.X. Liao, and A. Narayanasamy, *Mater. Res. Soc. Symp. Proc.* **166**, 87 (1990).

21. S.R. Phillpot, D. Wolf, and S. Yip, *MRS Bulletin* **XV**(10), 38 (1990).

22. D. Wolf and J.F. Lutsko, *Phys. Rev. Lett.* **60**, 1170 (1988).

23. C.G. Granqvist and R.A. Buhrman, *J. Appl. Phys.* **47**, 2200 (1976).

24. R.W. Siegel, S. Ramasamy, H. Hahn, Z. Li, T. Lu, and R. Gronsky, *J. Mater. Res.* **3**, 1367 (1988).

25. W. Wunderlich, Y. Ishida, and R. Maurer, *Scripta Metall. et Mater.* **24**, 403 (1990).

26. H.E. Schaefer, R. Würschum, M. Scheytt, R. Birringer, and H. Gleiter, *Mater. Sci. Forum* **15-18**, 955 (1987).

27. H.E. Schaefer, R. Würschum, R. Birringer, and H. Gleiter, *Phys. Rev. B* **38**, 9545 (1988).

28. G.W. Nieman, J.R. Weertman, and R.W. Siegel, *J. Mater. Res.* **6**, 1012 (1991).

29. H. Hahn, J. Logas, and R.S. Averback, *J. Mater. Res.* **5**, 609 (1990).

30. W. Wagner, R.S. Averback, H. Hahn, W. Petry, and A. Wiedenmann, *J. Mater. Res.* **6**, 2193 (1991).

31. H. Oya, T. Ichihashi, and N. Wada, *Jpn. J. Appl. Phys.* **21**, 554 (1982).

32. H. Hahn and R.S. Averback, *J. Appl. Phys.* **67**, 1113 (1990).

33. J.A. Eastman, Y.X. Liao, A. Narayanasamy, and R.W. Siegel, *Mater. Res. Soc. Symp. Proc.* **155**, 255 (1989).

34. R.W. Siegel, *Mater. Res. Soc. Symp. Proc.* **196**, 59 (1990).

35. R.S. Averback, H. Hahn, H.J. Höfler, J.L. Logas, and T.C. Chen, *Mater. Res. Soc. Symp. Proc.* **153**, 3 (1989).

36. M.J. Mayo, R.W. Siegel, A. Narayanasamy, and W.D. Nix, *J. Mater. Res.* **5**, 1073 (1990).

37. H.J. Höfler and R.S. Averback, *Scripta Metall. et Mater.* **24**, 2401 (1990).

38. Z. Li, S. Ramasamy, H. Hahn, and R.W. Siegel, *Mater. Lett.* **6**, 195 (1988).

39. J. Horváth, R. Birringer, and H. Gleiter, *Solid State Commun.* **62**, 319 (1987).

40. J. Horváth, *Defect and Diffusion Forum* **66-69**, 207 (1989).

41. H. Hahn, H. Höfler, and R.S. Averback, *Defect and Diffusion Forum* **66-69**, 549 (1989).

42. S. Schumacher, R. Birringer, R. Straub, and H. Gleiter, *Acta Metall.* **37**, 2485 (1989).

43. G.W. Nieman, J.R. Weertman, and R.W. Siegel, *Scripta Metall.* **23**, 2013 (1989).

44. G.W. Nieman, J.R. Weertman, and R.W. Siegel, *Scripta Metall. et Mater.* **24**, 145 (1990).

45. J.S.C. Jang and C.C. Koch, *Scripta Metall. et Mater.* **24**, 1599 (1990).

46. C.C. Koch and Y.S. Cho, *Nanostructured Mater.* **1**, 207 (1992).

47. E.O. Hall, *Proc. Phys. Soc. London* **B64**, 747 (1951).

48. N.J. Petch, *J. Iron Steel Inst.* **174**, 25 (1953).

49. J. Karch, R. Birringer, and H. Gleiter, *Nature* **330**, 556 (1987).

50. J. Karch and R. Birringer, *Ceramics International* **16**, 291 (1990).

51. H. Hahn, J. Logas, H.J. Höfler, P. Kurath, and R.S. Averback, *Mater. Res. Soc. Symp. Proc.* **196**, 71 (1990).

52. M.J. Mayo, R.W. Siegel, Y.X. Liao, and W.D. Nix, *J. Mater. Res.* **7**, 973 (1992).

53. M. Guermazi, H.J. Höfler, H. Hahn, and R.S. Averback, *J. Amer. Cer. Soc.* **74**, 2672 (1991).
54. D.D. Beck and R.W. Siegel, *J. Mater. Res.* **7**, 2840 (1992).
55. G. Skandan, H. Hahn, and J.C. Parker, *Scripta Metall. et Mater.* **25**, 2389 (1991).

Theoretical Studies of Physico-Chemical Properties of Nanostructures

J.L. Morán-López, J. Dorantes-Dávila, F. Aguilera-Granja
and J.M. Montejano-Carrizales

Instituto de Física
"Manuel Sandoval Vallarta"
Universidad Autónoma de San Luis Potosí
78000 San Luis Potosí, S.L.P.
MEXICO

Abstract

Atomic agregates with dimensions in the range of the nanometer scale possess unexpected physico-chemical properties. Those very special properties stem mainly from two facts: i) a large fraction of the total number of atoms lie at the surface and ii) the unfilled electronic bonding of these atoms makes them extremely reactive. In general it is expected that nanostructures should give the link between atomic, surface and bulk matter physico-chemistry. We discuss briefly only some properties that are subjects of research in our group; *i.e.*, 1) chemical ordering in bimetallic nanostructures with compact shapes, 2) magnetic properties of transition metal clusters, and 3) electronic and structural properties of semiconductor nanostructures.

I. Introduction

In the past several years, clusters with dimensions in the nanometer scale comprising from two to several hundred atoms, have become the subject of intensive experimental and theoretical investigations. On general grounds, it is expected that clusters of a few atoms or molecules exhibit molecular properties whereas large clusters show properties resembling those of bulk materials. It has been observed, however, that middle size clusters possess properties that far from linear interpolations between the two extremes.

The very particular cluster properties are determined mainly by the very high ratio (called dispersion D) of the number of surface atoms to the total number of atoms

Advanced Topics in Materials Science and Engineering, Edited by
J.L. Morán-López and J.M. Sanchez, Plenum Press, New York, 1993

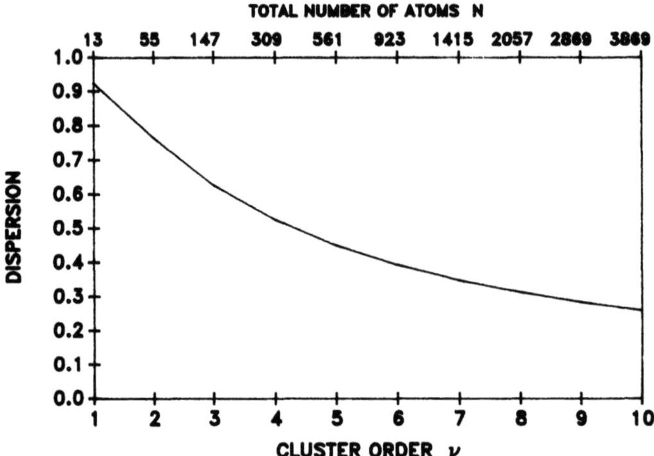

Figure 1. The dispersion as a function of the total number of atoms in icosahedral or cubo-octahedral clusters.

in the cluster. This quantity is shown in Fig. 1 as a function of the total number of atoms in clusters with compact shapes (icosahedron and cubo-octahedron).[1,2] One can notice that for clusters with one thousand atoms, about four hundred of them are located at the surface. Important as well is the fact that the surface atoms are low-coordinated and therefore some of the bonds are unsaturated. This influences the cluster reactivity and gives the very particular catalytic properties.[3-5]

The theoretical understanding of the properties of small clusters as a function of size, composition, structure, etc. is an intellectually stimulating challenge. However, what is perhaps more important is that these systems have high potential applications in material science and engineering, *i.e.*, in catalysis, electronic and magnetic materials, optical systems, etc. It is expected that in the near future one would be able to tailor specific properties of new materials through a judicious selection of the structure and composition of the nanostructures.[6,7]

The forces responsible for the stability of clusters varies widely, depending on the nature on the materials involved. Although there is not a clear cut one could recognize that Van der Waals binding is characteristic of inert elements (Ar, Ne, etc.), while, semiconductor (Si, C) and transition metals (Cu, Ni, Fe, etc.) are tight through covalent and metallic bonds, respectively.

Metal clusters with a small number of atoms, usually adopt compact shapes. The icosahedron (I) and the cubo-octahedron (C) are the most common.[8,9] Clusters with less that 150-200 atoms crystallize in the form of icosahedra. In larger clusters, that structure becomes unstable and they transform to the cubo-octahedra structure. That change in structure is expected, since in contrast to the icosahedron, the cubo-octahedron is formed by a subset of points that belong to a crystalline lattice (the face-centered-cubic).

Compact geometries have been also proposed for semiconductor clusters with small number of atoms (< 20). For example, a capped octagon[10] and a distorted tetracapped triangular prism[11] have been proposed for the Si_{10} nanostructure. These structures minimize surface area and dangling bonds. Larger clusters with a smaller dispersion may adopt more open structures.

The first successfull attempt to produce small clusters was carried out in the fifties. In those early experiments an oven was used to vaporize a metal, which was then precipitated as clusters on a substrate.[12] The technique used now vaporizes the solid metal with a laser.[13] A pulsed laser light is focused on the metal sample and evaporates its atoms into an extremely hot plasma. A gust of helium, released previously cools the vapor so that it condenses and forms clusters of various sizes. It sweeps the clusters along into an evacueted chamber, where the pressure differential causes the spay to expand supersonically. Collisions that take place during this expansion cool the clusters to a temperature near absolute zero, stabilizing them for further study.

The center of the spray pases through an aperture on the opposite side of the chamber. The resulting beam is then irradiated by an ultraviolet laser. Thereby the clusters are ionized and can be accelerated in an electric field. A time-of-flight mass spectrometer sorts the clusters into different packets. The sizes that appear particulary often are thought to be associated with particulary stable structures.

Among the very rich variety of cluster properties, we report in this contribution results obtained in three particular cluster problems that have been studied in our group: 1) the chemical ordering in bimetallic nanostructures with compact structures, 2) the magnetic properties of transition metal clusters and 3) the electronic and structural properties of semiconductor clusters.

II. Chemical Ordering in Bimetallic Nanostructures

Bimetallic nanostructures play an important role in heterogeneous catalysis. Despite to the fact that their properties are not yet completly understood, these systems are widely used now in the petroleum industry.[3,4] For example, copper-ruthenium or copper-osmium nanostructures are used because of their high selectivity in the conversion of cyclohexane to benzene. Also in these catalysts, the presence of copper inhibits the hydrogenolysis to alkanes but affects only very slightly their dehydrogenation properties.[5] Furthermore, transition metal clusters also catalyze the oxidation of hydrocarbons to form carbon dioxide and water.

The excellent catalytic properties of those systems, stem from the fact that a large number of atoms (those at the surface) are exposed to chemical reactions. In addition, since the surface atoms have a low coordination some bonds are not saturated. This makes them very reactive. Ideally, one would like to have a large number of nanoclusters to get the largest possible reacting area. However, practically one observes that the exposure of the clusters to high temperatures, tends to sinter them, reducing the reacting area. This problem is usually solved by embedding the catalysts in an inactive support, typically of silica or alumina. This inhibits their getting together and fusing with a consequent loss in surface area.

For obvious reasons, one would like to know how catalytic activities and selectivities depend on the support and the characteristics of the nanostructures (size, shape, atomic distribution, etc.). Under these circumstances we feel that a necessary step towards the understanding of the catalytic behavior of small bimetallic clusters, is the knowledge of the positions that the two chemical species occupy in the cluster as a function of the cluster characteristics and the temperature.

In recent publications,[2,14,15] we presented a systematic study of the general aspects and ground state of bimetallic nanoclusters with icosahedral and cubo-octahedral structures,[2] the spatial atomic distribution as function of the temperature for clean surfaces with applications to 55-atom icosahedral and 147-atom cubo-octahedral clus-

Figure 2. Cubo-octahedron (a) and icosahedron (b) with 147 atoms. In each cluster, all equivalent sites on the surface have the same shading. The numbers represent the shells on which they lie.

ters of CuPd, NiPt, and CuNi,[14] and how chemisorption affects the results obtained for clean surfaces.[15] In this paper, we give a short account of the results obtained in the case of CuPd, NiPt, and CuNi clusters with a clean surface and when oxygen is chemisorbed.

The equilibrium spatial distribution is obtained by minimizing the free energy calculated within the regular solution model.[16] To take into account the chemisorption effects we assume that the surface is covered by oxygen chemisorbed at sites that have been observed on surfaces of macroscopic systems.

II.1 The Model

We studied the icosahedral and cubo-octahedral nanostructures. The geometric characteristics of the cubo-octahedral and icosahedral clusters up to $N = 147$ atoms, are summarized in Tables 1 and 2 of Ref. 2, respectively. The icosahedron and cubo-octahedron with 147 atoms are illustrated in Fig. 2. In each cluster equivalent sites on the surface have the same shading and the numbers represent the shells on which they lie. A shell is formed by all the atoms that are equidistant from the central atom. It is worth noticing that the surface of the cluster is formed by various shells.

The equilibrium configuration in a bimetallic cluster with N_A atoms of type A and $N_B = N - N_A$ atoms of type B (there are no vacancies) is obtained by minimizing the free energy with respect to the positions that occupy both kinds of atoms. We take into account only interactions between nearest neighbor atoms, denoted by $U_{i,j}$, $i, j = A, B$. Due to the symmetry of the cluster all sites in a given shell are equivalent and the concentration of the component A in the i-th shell is defined by:

$$x_i = N_{A,i}/N_i, \tag{1}$$

where $N_{A,i}$ and N_i denote the number of A atoms and the total number of sites in the i-th shell, respectively. The internal energy of the system can be written in terms of the various concentrations as follows:

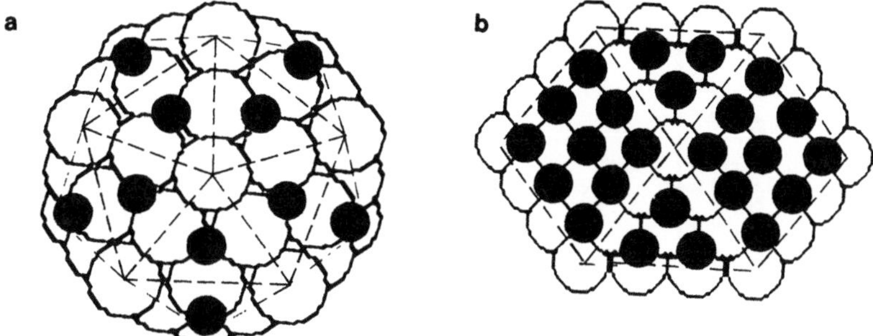

Figure 3. Chemisorbed oxygen (black circles) on triangular and four coordinated sites on (a) 55-atom icosahedral, and (b) 147-atom cubo-octahedral cluster.

$$U = \frac{W}{2} \sum_{ij} N_i Z_{ij} x_i x_j + \frac{1}{2}(\Delta - W) \sum_{ij} N_{A,i} Z_{ij} + \frac{U_{BB}}{2} \sum_{ij} N_i Z_{ij}$$

$$+ \sum_{ss} N_i Y_{ai} x_i (\xi_{aA} - \xi_{aB}) + \sum_{ss} N_i Y_{ai} \xi_{aB}, \tag{2}$$

where, the values of Z_{ij} are given in Tables 1 and 2 of Ref. 2, and Y_{ai} is the coordination of the chemisorbed atom to the i-th shell located at the surface of the cluster. The quantity $W = U_{AA} + U_{BB} - 2U_{AB}$ is the heat of mixing, and $\Delta = U_{AA} - U_{BB}$ is identified as the difference in the cohesive energy of the pure components. Notice that $\sum_{ij}$ is the sum over all the shells around the central atom, and $\sum_{ss}$ is the sum only over the shells that form the cluster surface.

The last two terms of Eq. (2) correspond to the energy associated to the chemisorbed atoms. The interactions between the atoms located at the surface and the chemisorbed atoms a, are denoted by ξ_{ak}, $k = A, B$.

The configurational entropy of the binary system is given by the expression:

$$S = \sum_i ln \frac{N_i!}{(N_i - N_{A,i})! N_{A,i}!}, \tag{3}$$

where the sum is carried out over all the shells around the central atom.

The equilibrium values of the concentration at the various shells x_i, are obtained by minimizing the free energy

$$F = U - TS, \tag{4}$$

with the constraint that the total number of A- and B-atoms remain constant and that the chemisorbed species do not diffuse into the cluster.

The chemisorption phenomenon is a complicated issue. It is well established that chemisorption depends on the chemisorbed atoms, coverage, temperature, surface atoms and their geometry, etc. However, to mantain the model treatable we assume that the chemisorbed atoms cover all the surface of the cluster and consider in the case of oxygen,for simplicity, only triangular and four-coordinated chemisorption sites. In Fig. 3a we illustrate the O (black circles) chemisorption on a 55-atom icosahedral

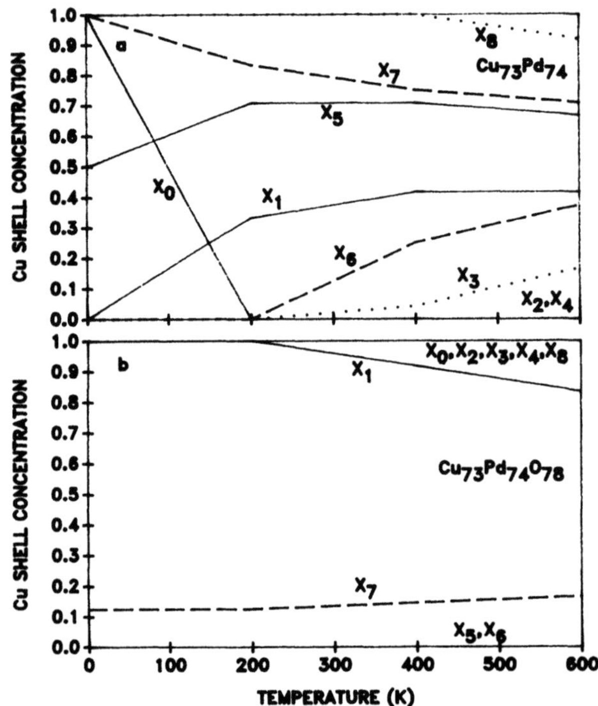

Figure 4. Temperature dependence of the Cu concentration at the various shells, x_i, i =0–8 of the 147-atom cubo-octahedral particles $Cu_{73}Pd_{74}$, in the cases of: a) clean surface, b) O-chemisorption. The total number of O atoms chemisorbed at the cluster surface is 78.

cluster, and in Fig. 3b on a 147 cubo-octahedral clusters.This assumption is supported by experimental observations of O chemisorption on (111) and (100) Ni surfaces.[17,18]

We present specific applications to the case of O-chemisorbed at the surface of CuPd, NiPt, and CuNi clusters. Based on the data of the adsorption energy,[19] and the estimated values of W and Δ, for the mentionated systems,[20,21] we obtained the results present below. Furthermore, we assumed in the calculation that the ratio between long and short bonds in the icosahedral nanocluster is $\alpha = 0.9$. For this value of α the equilibrium shape of the clusters with 147-atoms is cubo-octahedral. In all the calculations we assumed that O is chemisorbed in triangular and on four-coordinated central sites depending on the geometry of the various faces of the cluster surface. The desorption temperature of O in transition metals is between 300 and 500 K. We present results up to 600 K.

II.2 The CuPd System.

To have a clear idea of this system let us recall some of its main features. The CuPd alloy is an ordering system with *fcc* crystal structure.[20] The surface segregation, in clean macroscopic surfaces, has been studied under various conditions. Equilibrium segregation studies[22] reveal that Pd segregates to the surface. On the other hand induced by surface bombardment with energetic particle beams,[23] shows the opposite.

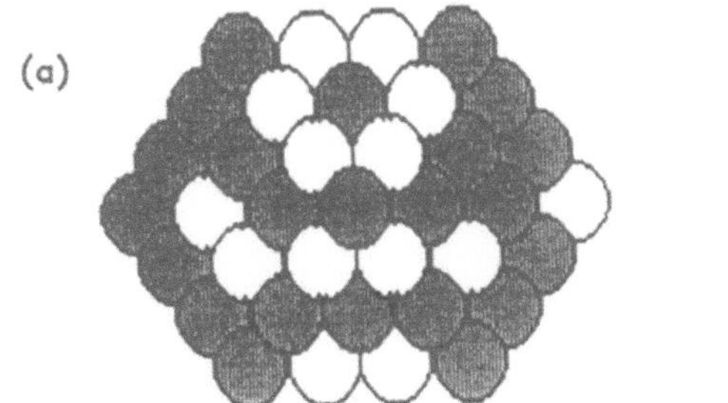
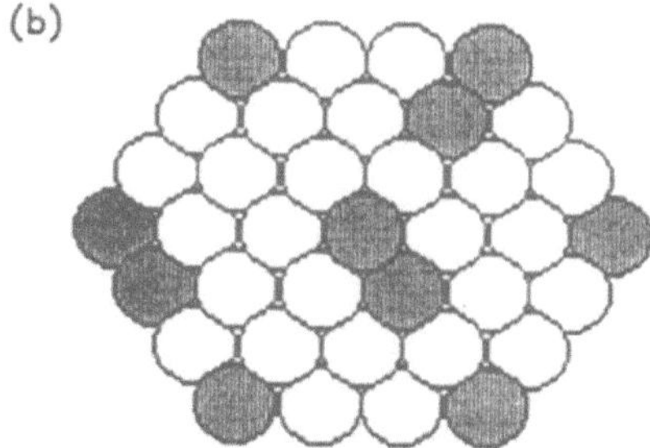

Figure 5. Equilibrium spatial distribution at $T = 600$ K, of the $Cu_{73}Pd_{74}$ cubo-octahedral cluster a) without chemisorbed species, b) with O-chemisorption. The chemisorbed atoms are not shown for clearness. Filled (empty) circles represent Cu (Pd) atoms.

That behaviour may be due to the fact that the difference in the cohesive energy of the pure systems is small (0.436 eV/atom).[24] On theoretical grounds,[25,26] since the cohesive energy of copper (3.5 eV/atom) is smaller that the one of palladium (3.936 eV/atom), one would expect Cu segregation.

The adsorption energy of O on Pd is larger than the one to Cu.[19] In Fig. 4 we show the temperature dependence of the Cu concentration x_i in the various shells of the 147-atom cubo-octahedral cluster $Cu_{73}Pd_{74}$. Figs. 4a and 4b contain the results in the absence and presence of chemisorbed oxygen. One observes that for a clean surface, at low temperatures, the Cu atoms segregate to the surface shells; with exception of the central site, the core is formed of Pd atoms. At higher temperatures the Cu and Pd atoms are distributed over all the cluster but Cu occupies preferently the most external shells.

In the case of O-chemisorption the opposite is observed. The core and the 6th shells are occupied by Cu atoms. The occupancy of the shell 7 is approximatly the same in the whole range of temperatures. One can see that the effect of O-chemisorption is to invert the spatial ordering, *i.e.* for a clean surface we found Cu segregation, and for O-chemisorption Pd segregates to the surface.

In Fig. 5 we show the equilibrium spatial distribution at $T = 600$ K, of the cubo-octahedral cluster $Cu_{73}Pd_{74}$ without (Fig. 5a) and with O-chemisorbed at the surface (Fig. 5b). The chemisorbed atoms are not shown for clearness. Filled (empty) circles represent Cu (Pd) atoms.

II.3 The NiPt System.

The main characteristics of this system are: it is an ordered alloy with *fcc* crystal structure.[14] According to bond-breaking,[22] surface enthalpies and heats of solution,[24,27] theories of surface segregation, the Ni atoms should segregate to the surface in macroscopic systems. A more elaborate theory, based on the electronic structure of the components,[19] predicts that in the diluted NiPt alloy, Pt segregates. The opposite is predicted to occur in the diluted PtNi alloys. Experimentally, it has been found that the surface of clean platinum-nickel alloys is always enriched with platinum.[22] That behaviour can be reversed to a Ni enrichment when the sample is subject to oxidation. The discrepancy between theory and experiment might be due mainly to relaxation effects. We found[8] that in the absence of chemisorbed species Ni segregates at all temperatures.

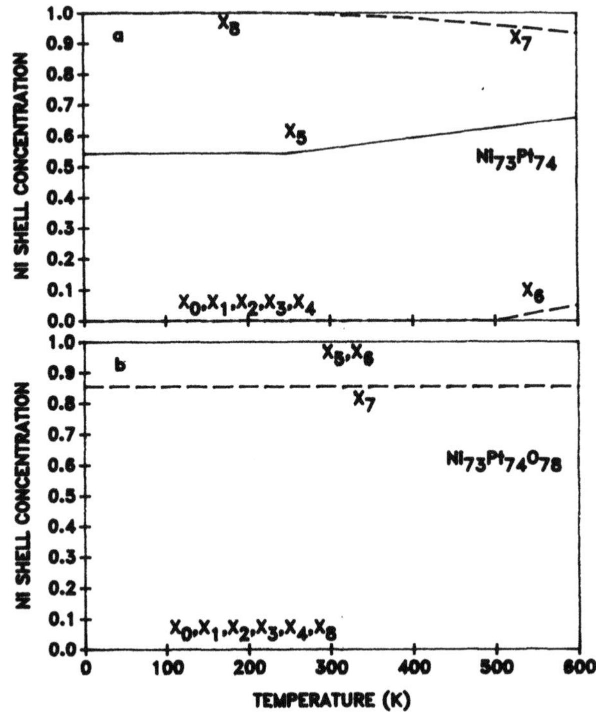

Figure 6. Temperature dependence of the Ni concentration at the various shells, x_i, i =0–8 of the 147-atom cubo-octahedral particles $Ni_{73}Pt_{74}$, in the cases of: a) clean surface and b) O-chemisorption. The total number of O atoms chemisorbed at the cluster surface is 78.

In this system the adsorption energy of O to Ni is larger than the one to Pt.[23] We present in Fig. 6a the temperature dependence of the Ni concentration x_i in the various shells for the cubo-octahedral 147-atom $Ni_{73}Pt_{74}$ cluster, in the absence of chemisorbed species. The results of O-chemisorption are presented in Fig. 6b. The effect of O-chemisorption is to enhance the surface segregation of Ni.

II.4 The CuNi System.

The CuNi alloys are some of the most studied systems. The CuNi alloys phase separate at low temperatures.[23] This system crystalizes with a *fcc* structure and presents a miscibility gap that goes up to 400°C. For surface segregation, experiments and theories, CuNi alloys represent a prototype system. Equilibrium segregation experiments[30,31] show the segregation of copper to the surface. On the other hand, it has been observed that chemisorbed atoms like H, S and O enhance the Ni concentration at the surface.[32,33] In the case of clean surfaces, all the theoretical models predict the segregation of copper to the surface. In nanostructures, we found[8] strong Cu segregation for cubo-octahedral 147-atom nanoclusters.

In this system the difference in adsorption energies for O-chemisorption on Cu and Ni in 2.822 eV. The temperature dependence of the Cu concentrations x_i in the

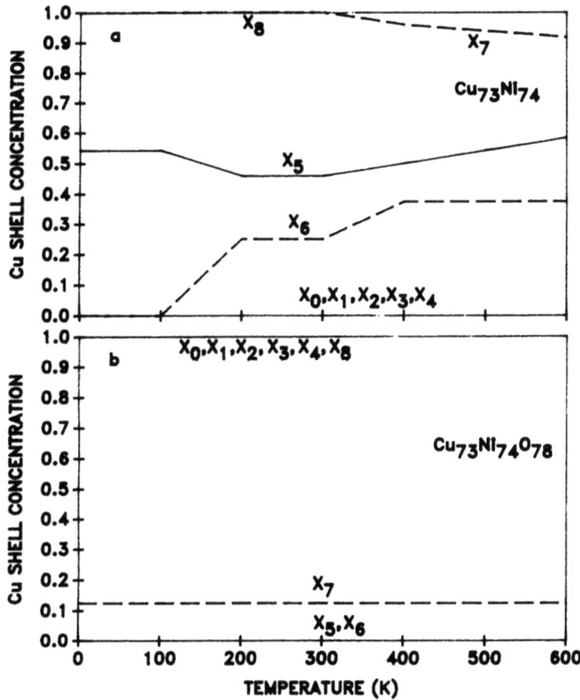

Figure 7. Temperature dependence of the Cu concentration at the various shells, of the 147-atom cubo-octahedral particle $Cu_{73}Ni_{74}$, in the cases of: a) clean surface and b) O-chemisorption. The total number of O atoms chemisorbed at the cluster surface is 78.

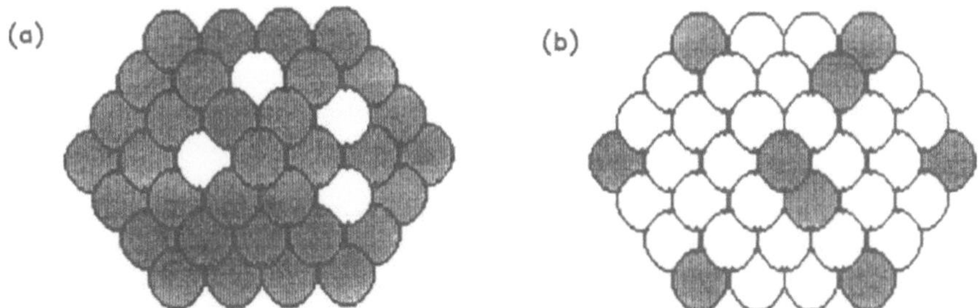

Figure 8. Equilibrium spatial distribution at $T = 300$ K, of the $Cu_{73}Ni_{74}$ cubo-octahedral cluster a) without chemisorbed species and b) with O-chemisorption. The chemisorbed atoms are not shown for clearness. Filled (empty) circles represent Cu (Ni) atoms.

various shells for the 147-atom cluster $Cu_{73}Ni_{74}$ are presented in Fig. 7. The clean surface case is presented in Fig. 7a and the O-chemisorption results are shown in Fig. 7b. In the absence of chemisorbed species one finds again a strong surface segregation of Cu atoms. On the other hand, we see that for O-chemisorption, the core consists of Cu atoms, and the rest of them occupy partially shell 7, and shell 8. As temperature increases, no changes are observed in the case of O-chemisorption.

In Fig. 8 we show the equilibrium spatial distribution at $T = 300$ K, of the 147-atom cubo-octahedral clusters without (Fig. 8a) and with O-chemisorbed at the surface (Fig. 8b). The chemisorbed atoms are not shown for clearness . Filled (empty) circles represent Cu (Ni) atoms.

II.5 Summary.

Based on a simple model, we have presented results of the spatial atomic ordering in bimetallic nanostructures in the presence and absence of chemisorbed species. We found that chemisorption may invert the surface segregation observed in clean clusters. We applied the theory to the chemisorption of O on CuPd, NiPt, and CuNi nanoclusters. We obtained in the first case that chemisorption inverts the surface segregation of Cu. In the second case chemisorption enhances the surface segregation of Ni. Lastly in the case CuNi clusters, chemisorption inverts the segregation of Cu characteristic of clean cluster surfaces.

The general trends in bimetallic nanostructures are similar to those in macroscopic surfaces. However we may notice the following differences: i) in a small cluster the number of non-equivalent sites (all the sites of shells that form the surface cluster) is larger than in a macroscopic surface; therefore, the local atomic environment offered to chemisorbed species is richer in small agregates. ii) Due to the the high dispersion ratio in nanostructures the segregation effects are more important in these systems than in macroscopic surfaces.

Due to the various approximations made in this model it is not appropriate to compare to experimental data. Our results should be taken as an attempt to understand the main trends of the chemisorpion effects on the spatial atomic distribution in bimetallic clusters.

III. Magnetic Properties of Transition Metal Clusters

The magnetic properties of low dimensional systems have been studied intensively in the recent years.[34,35] This is driven by the great potential that these systems have for important technological applications as well as for deeper understanding of basic magnetic interactions.

The impresive developments in molecular beam technology offer now unique opportunities to study clusters that contain from a few to thousands of atoms. Recently, small transition metal[36-40] and rare-earth[41] clusters produced by supersonic jet expansion, were deflected in an inhomogeneous magnetic field, and thereby the magnetic properties investigated. Cox *et al.*[36] carried out for the first time magnetic measurements in free Fe clusters. They observed that the average magnetic moment per atom $\bar{\mu}_n$ of Fe$_n$ for clusters with a small number of atoms ($n \leq 17$) is close to the bulk moment ($\mu \approx 2\mu_B$). More recently, Fe$_n$ clusters in a size range $15 \leq n \leq 650$,[37] and Co$_n$ clusters with a size between $20 \leq n \leq 200$ atoms [38], were investigated. In contrast to previous results, the effective magnetic moment per atom deduced from the cluster deflection is much lower as compared with the bulk moment for all the investigated cluster sizes.

Furthermore, the results obtained from the Stern-Gerlach deflection profile, as a function of field strength and cluster size, are similar for both materials. The measured magnetic moment increases linearly as a function of the magnetic field and of cluster size.

In this section we report some results of the magnetic properties of clusters as obtained by means of the tight-binding Hubbard Hamiltonian[42] in the unrestricted Hartree-Fock approximation. This method, in contrast to first-principle-methods is not limited to small clusters with high symmetry.[43,44] We were able to calculate the size dependence of the cluster average magnetic moment, local magnetic moments at the various inequivalent sites, the cohesive energy, and the average bond lenght.[45-47]

III.1 Theory

Here we give a short outline of the theory. A detailed account can be found in Ref. 45. The Hubbard Hamiltonian is given by

$$H = \sum_{\substack{\alpha,\beta,\sigma \\ i \neq j}} t_{ij}^{\alpha\beta} c_{i\alpha\sigma}^{+} c_{j\beta\sigma} + H_I, \tag{5}$$

where $c_{i\alpha\sigma}^{+}(c_{i\alpha\sigma})$ refers to the creation (annhilation) operator of an electron with spin σ at atomic site i in the orbital α ($\alpha = xy,\ yz,\ zx,\ x^2 - y^2,\ 3z^2 - r^2$) and $t_{ij}^{\alpha\beta}$ to the hopping integral between sites i and j. The interaction Hamiltonian H_I in the unrestricted Hartree-Fock approximation is given by:

$$H_I = \sum_{i\sigma} \epsilon_{i\sigma} \hat{n}_{i\sigma} - E_{dc}, \tag{6}$$

with,

$$\epsilon_{i\sigma} = \epsilon_d^0 + U\Delta n(i) - \frac{1}{2}\sigma J\mu(i). \tag{7}$$

In equation (6),

$$E_{dc} = \frac{1}{2}\sum_{i\sigma} < \hat{n}_{i\sigma} > (\epsilon_{i\sigma} - \epsilon_d^0), \tag{8}$$

refers to the correction due to double counting, and $\hat{n}_{i\sigma}$ to the electron number operator. The exchange and effective direct intra-atomic Coulomb integrals, are taken to be independent of the cluster size and are given by $J = (U_{\uparrow\downarrow} - U_{\uparrow\uparrow})$, $U = (U_{\uparrow\downarrow} + U_{\uparrow\uparrow})/2$, where $U_{\uparrow\downarrow}$ and $U_{\uparrow\uparrow}$ are the Coulomb interactions between electrons with opposite and parallel spins incluiding exchange, respectively.

The number of d-electrons $n(i)$, and the local magnetic moment $\mu(i)$ at at particular site site i, given by

$$\begin{aligned} n(i) &=< \hat{n}_{i\uparrow} > + < \hat{n}_{i\downarrow} > \\ \mu(i) &=< \hat{n}_{i\uparrow} > - < \hat{n}_{i\downarrow} >, \end{aligned} \tag{9}$$

are determined in a self-consistent manner.

In general, in small systems where not all the geometrical sites are equivalent, the number of electrons is location dependent. As a consequence charge transfer between different atomic sites i will result and the atomic levels are fixed by imposing global charge neutrality:

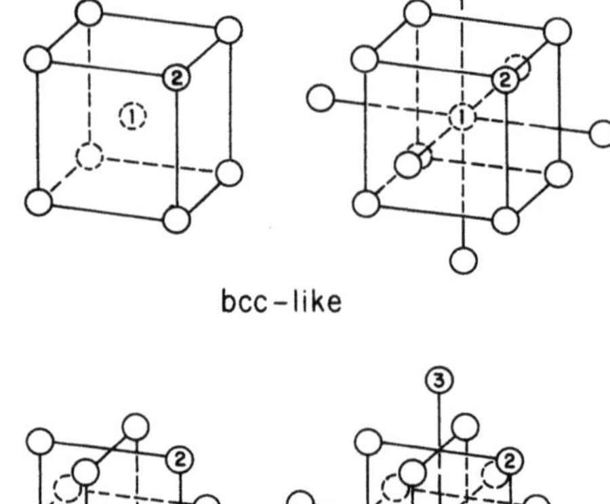

bcc–like

fcc–like

Figure 9. Illustration of the *bcc* and *fcc*-like cluster geometry. The numbers denote the various shells around the central atom.

$$q = \frac{1}{n} \sum_i q_i = \frac{1}{n} \sum_i n(i). \tag{10}$$

The spin polarized local density of states $N_{i\sigma}$ is calculated by using the recursion method.[48] We assume for the hopping integrals the canonical values varying as the inverse fifth power of the interatomic distance.[49] For *bcc* and icosahedral like structures, nearest- and next-nearest-neighbor hopping integrals are included. The average magnetic moment $\bar{\mu}_n$, given by

$$\bar{\mu}_n = \frac{1}{n} \sum_{i=1}^{n} \mu(i), \tag{11}$$

represents the average magnetization of the cluster at T=0.

The cohesive energy per cluster atom is given by

$$E_{coh}(n) = E_{band}(1) - E_{band}(n) - E_R, \tag{12}$$

where $E_{band}(n)$ refers to the electronic d-band contribution per atom,

$$E_{band}(n) = \frac{1}{n} \sum_{\substack{i=1 \\ \sigma}}^{n} \int_{-\infty}^{E_F} \epsilon N_{i\sigma}(\epsilon) d\epsilon - E_{dc}. \tag{13}$$

Table I

Results for the size and structural dependence of the cohesive energy $E_{coh}(n)$ (in Ev), average bond length d_n (d_b = bulk bond length), average magnetic moment $\bar{\mu}_n$, and local magnetic momemnt $\mu(i)$ (in units of μ_B), of small Fe_n clusters ($n \leq 8$). Different symmetry atoms i are indicated in the assumed geometries. Results for the unrelaxed geometries are given in parentheses.

	Structure	$E_{coh}(n)$	d_n/d_b	$\bar{\mu}_n$	$\mu(1)$	$\mu(2)$	$\mu(3)$
Fe_2		0.92 (0.81)	0.90	3.00 (3.00)	3.00 (3.00)		
Fe_3		1.21 (1.12)	0.91	2.33 (3.00)	2.33 (3.00)		
Fe_3		0.80 (0.74)	0.94	3.00 (3.00)	2.98 (2.98)	3.01 (3.01)	
Fe_4		1.31 (1.26)	0.97	3.00 (3.00)	3.00 (3.00)		
Fe_4	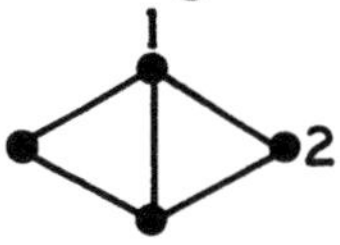	1.32 (1.26)	0.96	3.00 (3.00)	2.98 (2.99)	3.02 (3.01)	
Fe_5	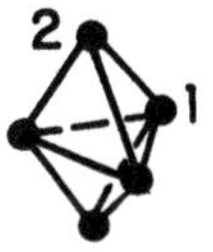	1.39 (1.37)	0.97	3.00 (3.00)	2.98 (2.99)	3.02 (3.02)	
Fe_5	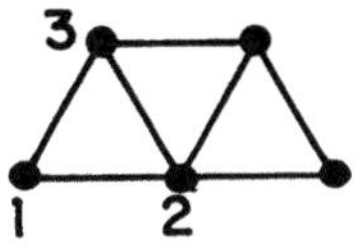	1.29 (1.25)	0.96	3.00 (3.00)	3.03 (3.02)	2.98 (2.99)	2.98 (2.99)
Fe_6	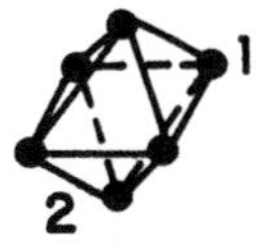	1.26 (1.23)	0.99	3.00 (3.00)	3.01 (3.01)	2.97 (2.97)	
Fe_7	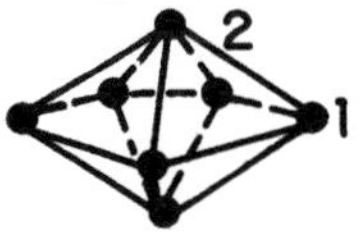	1.50 (1.49)	0.99	3.00 (3.00)	3.00 (3.00)	2.99 (3.00)	
Fe_8	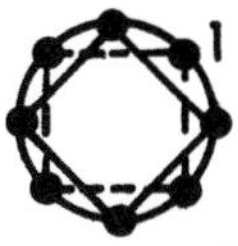	1.42 (1.41)	0.99	3.00 (3.00)	3.00 (3.00)		
Fe_8	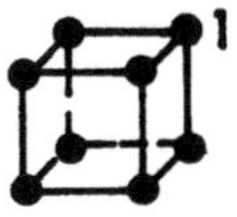	1.16 (1.14)	0.98	3.00 (3.00)	3.00 (3.00)		

Table II

Results as in Table I for larger Fe$_n$ clusters (n>9).
Different shells i are indicated as in Fig. 9.

Structure		E_{coh}(n)	d_n/d_b	$\bar{\mu}_n$	$\mu(1)$	$\mu(2)$	$\mu(3)$	$\mu(4)$	$\mu(5)$
Fe$_9$	*bcc*	1.37	0.91	2.33	0.40	2.57			
		(1.08)		(3.00)	(2.96)	(3.01)			
Fe$_{11}$	*bcc*	1.41	0.97	3.00	2.96	3.02	2.96		
		(1.36)		(3.00)	(2.96)	(3.02)	(2.97)		
Fe$_{13}$	*bcc*	1.62	0.97	2.54	0.42	2.65	2.85		
		(1.55)		(3.00)	(2.95)	(3.01)	2.98		
Fe$_{13}$	*bcc*	1.42	0.98	1.92	−1.71	2.23			
		(1.41)		(2.08)	(−1.79)	(2.40)			
Fe$_{13}$	*bcc*	1.41	0.98	2.08	−1.98	2.41			
		(1.39)		(2.23)	−2.20	2.60			
Fe$_{15}$	*bcc*	1.63	0.97	2.60	0.48	2.67	2.86		
		(1.60)		(2.73)	(1.28)	(2.88)	(2.76)		
Fe$_{19}$	*fcc*	1.42	0.99	1.84	−1.09	1.76	2.49		
		(1.41)		(1.95)	(−1.28)	(1.91)	(2.56)		
Fe$_{27}$	*bcc*	(1.41)		(2.85)	(2.88)	(2.67)	(2.85)	(2.97)	
Fe$_{43}$	*fcc*	(1.70)		(1.23)	(−1.37)	(−0.90)	(0.89)	(2.49)	
Fe$_{51}$	*bcc*	(1.73)		(2.45)	(1.28)	(1.87)	(1.57)	(2.62)	(2.83)
bulk	*bcc*	2.00	1.00	2.21	2.21				

The Born-Meyer repulsive energy E_R is calculated from

$$E_R = \frac{1}{2n} \sum_{i=1}^{n} z_i A\exp(-p(\frac{d}{d_b} - 1)), \tag{14}$$

where z_i, refers to the coordination number at a site i and d_b to the bulk interatomic distance. The parameters A and p are fitted to the bulk equilibrium condition and compressibility modulus. By allowing uniform relaxation, the energy is minimized with respect to d. Thus one obtains the average equilibrium bond lenght d_n, the cohesive energy $E_{coh}(n)$, and the magnetic moment for each assumed structure.

III.2 Iron Clusters

Although we have calculated the magnetic properties for iron, chromium and nickel clusters,[45−47] here we report only the results for iron nanostructures. The parameters used in the calculation are: n_d=7, $J = 0.73$ eV is fitted to the bulk magnetic moment, the bulk band-width (6.0 eV) is taken from band structure calculations,[50] and $U = 5.4$ eV, is estimated from atomic spectroscopic data.[51]

For small clusters ($n < 9$) we assume the structures shown in Table I, and for larger clusters, *bcc*-, *fcc*-, or icosahedral-like structures obtained by adding to a central atom the successive shells of the first, second, etc. neighbors, as shown in Fig. 9.

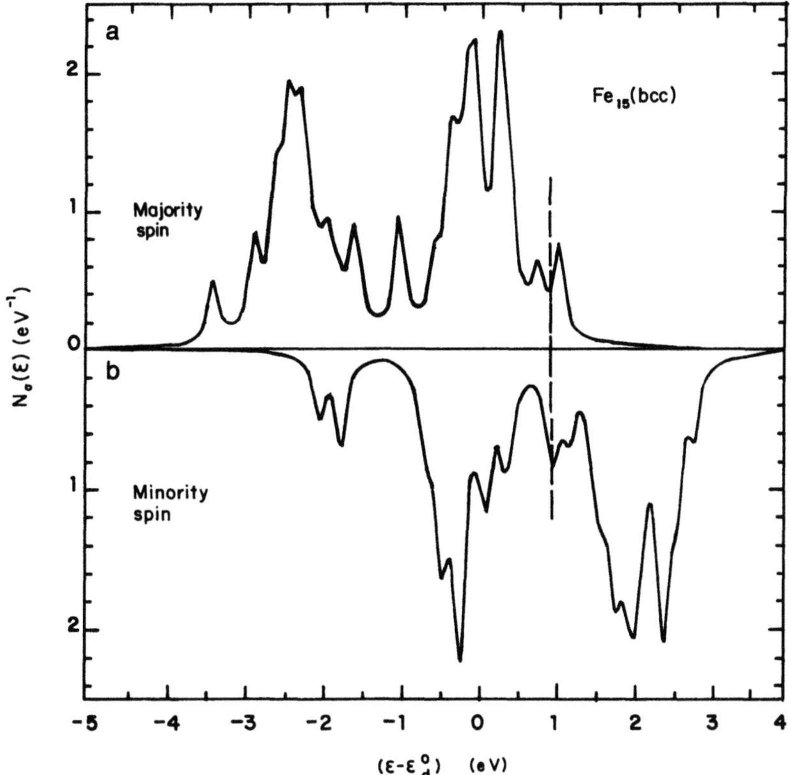

Figure 10. a) Local electronic densities of states $N_{i\sigma}(\epsilon)$ of bcc-Fe$_{15}$. The numbers refer to the various shells ordered by increasing distance to the cluster center as in Fig. 9. b) Average density of states of bcc-Fe$_{15}$. A Lorenzian was used to broaden the cluster energy levels.

The results for clusters with less than 9 atoms are presented in Table I. Using the bulk bond length (no relaxation), we obtain saturated magnetic moments for all these clusters, $\mu(i) = [10 - n_d(i)]\mu_B \simeq 3.0\mu_B$, i. e., fully polarized d-band. On the other hand when uniform relaxation is allowed, the cluster bond length contracts in order to minimize the total energy. The value of the contraction varies from about 10% for the dimer and trimer, to 1–2% for larger clusters with closed packed structures. No significant change of $\overline{\mu}_n$ is obtained upon contraction. An exception is provided by the triangular trimer, where a 9% contraction causes the total magnetic moment to change from $9\mu_B$ to $7\mu_B$. A similar effect is observed for $\overline{\mu}_9$.

We obtain for the difference in the cohesive energy between the various structures assumed, $\Delta E_{coh} \simeq 0.01 - 0.09$ eV for $n = 3 - 5$. These values are to small to conclude safely about the most stable geometrical structure at T=0, but rather indicate that a strong coupling between electronic and translational degrees of freedom can be expected at finite temperatures.

The results for larger clusters ($9 \leq 51$) with bcc structure are given in Table II. These clusters show in all cases ferromagneticlike order. By using the bulk bond length (results in parentheses) we obtain for $n \leq 13$ saturated local moments. For larger clusters, the local magnetic moments $\mu(i)$ show interesting environmental dependence.

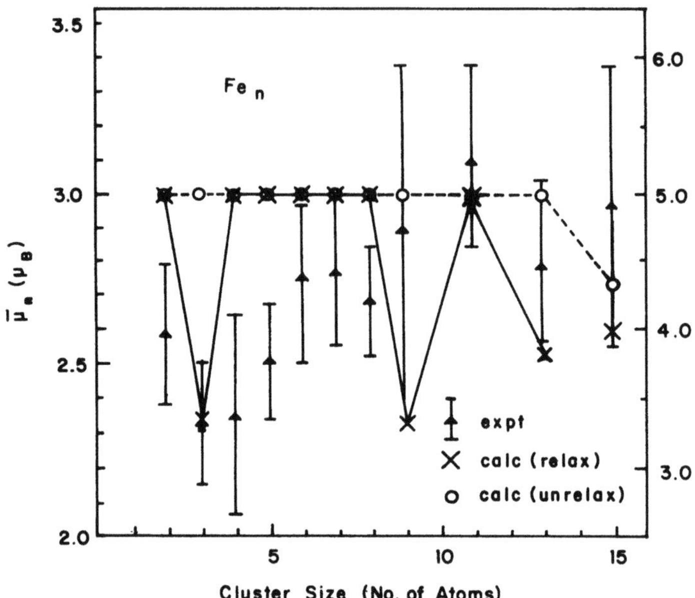

Figure 11. Average magnetic moment $\overline{\mu}_n$ of Fe$_n$ clusters as a function of n. Crosses (circles) refer to calculations for relaxed (unrelaxed) clusters. Experimental results for the depletion factor (approximately proportional to $\overline{\mu}_n$) are indicated by vertical bars.[36]

For example for unrelaxed Fe$_{15}$ we obtain (in units of μ_B) $\mu(1) = 1.28$, $\mu(2) = 2.88$, and $\mu(3) = 2.76$.

A crucial test on the accuracy of the tight binding calculation is provided by the LDO of Fe$_{15}$ shown in Fig. 10. The shape of the total DOS indicates that a resemblance between the general features of the distribution of the energy levels for the clusters and the bulk starts to develope. For example, the typical bonding and antibonding broad peaks of *bcc* bulk separated by a valley near the center of the band are already present.

In Table II we present results for *fcc*- and icosahedral-like Fe$_n$ clusters. The clusters show antiferromagnetic with the magnetic moments in the core pointing in the opposite directions to that of the outermost shells (Fe$_{13}$: ↑↓↑, Fe$_{19}$: ↑↑↓↑↑, Fe$_{43}$: ↑↑↓↓↓↑↑). Similar behaviour seems to be observed in γ-Fe particles[52] and is probably related to the antiferromagnetic ordering observed in bulk γ-Fe.[53]

Concerning the structural stability, we obtain that the *bcc*-like Fe$_n$ clusters are more stable that the *fcc*-like clusters for $n=$ 13–19, in agreement with previous results[54] derived from the size dependence of the ionization energy. In Fig. 11 we show a plot for $\overline{\mu}_n$ as a function of cluster size. Since for the unrelaxed clusters with $n \leq 13$ one obtains saturated magnetic moments, $\overline{\mu}_n$ exhibits a very weak dependence on the geometrical structure. However for relaxed clusters we find that $\overline{\mu}_n$ changes strongly for $n=$ 3, 9 and $n=$ 13–15. This shows that the magnetic moment may depend sensitively on the bond length and cluster geometry. For well defined structures our results are in good agreement with experiment.[36]

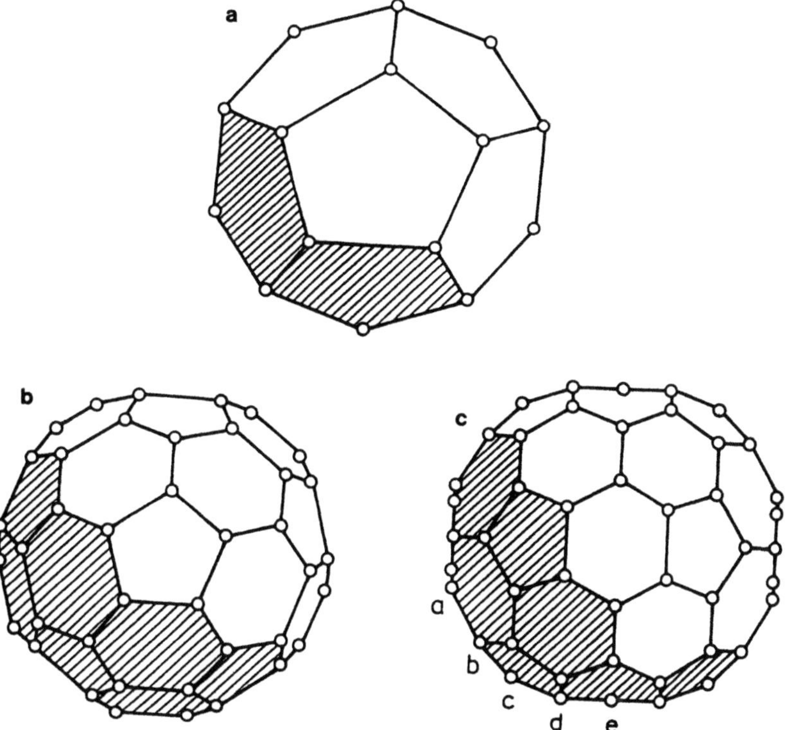

Figure 12. The dodecahedron (a), the truncated icosahedron (b), and the 70-atom Fullerene structures. The five inequivalent sites in the 70-atom cluster are denoted by *a-e*.

III.3 Summary

The size and structural dependence of the magnetic properties of Fe$_n$ clusters were determined by means of a tight-binding Hubbard Hamiltonian in the unrestricted Hartree-Fock approximation. The average magnetic moment $\bar{\mu}_n$, the local magnetic moments $\mu(i)$, magnetic ordering, cohesive energy, and average bond lenght were calculated a $T = 0$. For all studied clusters we obtain larger magnetic moments than for bulk material. Interesting dependence of magnetic order within the cluster on the structure was obained: *bcc*-Fe is ferromagnetic, whereas *fcc*-Fe is antiferromagnetic.

IV. Electronic Structure of Semiconductor Cluster

In the last years, a large number of experimental[55-64] and theoretical[10,11,65,66] studies have been carried out in semiconductor neutral and charged nanostructures. The main interest is to know when and how the properties of the nanostructures approach those in the bulk material as the cluster size is increased. Due to the covalent nature of their bonding with sp^3 hibridization directional bonds, one would expect that they would condense in clusters with open structures, similar to those of bulk materials. However, it has been found that the unsatisfied bonds at the surface force closed

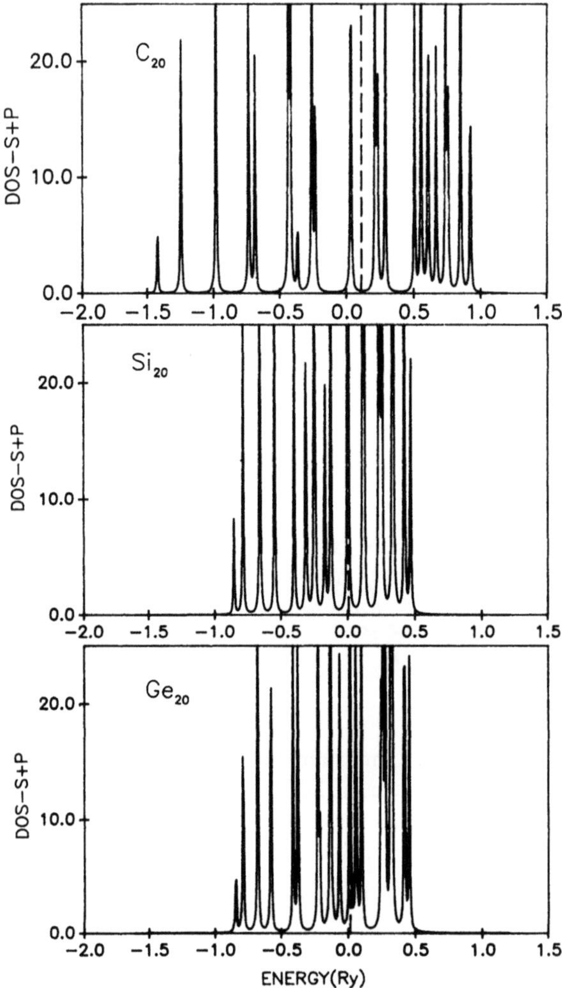

Figure 13. The total electronic density of states per atom of the C_{20}, Si_{20}, and Ge_{20} with dodecahedral structure. The Fermi energy is narked by the dashed line.

packed structures for systems with a small number of atoms. Several models for the mechanism of cluster formation and cluster structures have been proposed.[55-59]

The geometrical structure and stability of small silicon clusters Si_n ($n < 12$), has been the subject of various theoretical studies.[10,11,65,66] Originally an open structure for the Si_{10} called adamantane was proposed.[65] However, it was shown later than closed structures, the capped octahedron[10] and the distorted tetracapped triangular prism[11] are more stable. More recently, based on the observation that there is a dramatic variation on the reactivity of silicon clusters with ammonia and methanol as a function of clusters size with an aparent periodicity in units of six atoms,[58] lead to the proposition that clusters in the range of $20 \leq n \leq 60$ are arranged in stacked, six membered rings.[66,67]

Initially, the high reactivity of semiconductor surfaces, motivated to study mainly silicon and germanium clusters. However, the syntesis of macroscopic quantities of

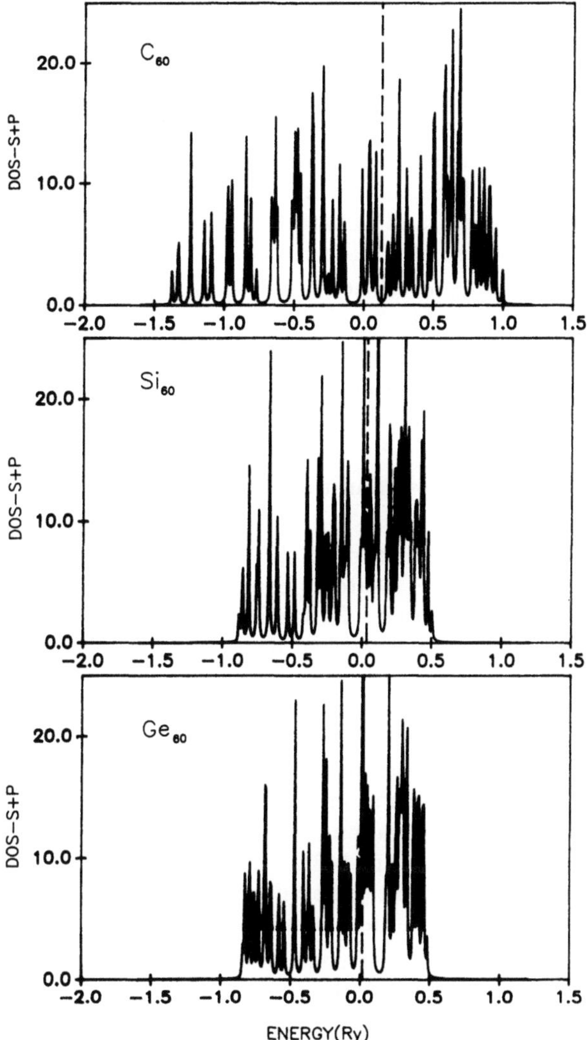

Figure 14. The total electronic density of states per atom of the C_{60}, Si_{60}, and Ge_{60} with truncated icosahedral structure. The Fermi energy is marked by the dashed line.

carbon bucky balls (cage-shaped carbon molecules),[60] lead to an intense study of their physical and chemical properties in the last years. Their potential application as new materials is the major driving force.

Studies of the optical properties of interstellar mater motivated the laboratory production of small carbon molecules.[61] It was observed that only even-atom clusters could be produced and simple physicochemical considerations led to the conclusion that hollow cages formed by pentagons and hexagons were the best candidate structures. These hollow cages, called now bucky balls or fullerenes, can be formed by 20, 24, 32, 36, 50, 60, and 70 atoms.[62] The most stable of all seems to be C_{60} and has been obtained both, in molecular form and macroscopy crystals with face-centered-cubic structure. We show in Fig. 12 the 20, 60 and 70-atom Fullerenes.

Experiments on the C_{60} solid reveal that is a non-conductor and band structure calculations performed within the local-density approximation in the density functional theory, indicate that the minimum bandgap is about 1.5 eV.[63] This system when dopped with alkali atoms turn the solid into a superconductor with transition temperatures as high as 33 K in $CsRb_2C_{60}$.[64]

Here we are reporting calculations of the electronic structure of the 20, 60 and 70 semiconductor bucky balls by means of the real space solid state method, described in sec. III and elucidate the s and l-character of the electronic states. We compare the clusters results with those of infinite structures: *i.e.* hexagonal and diamond.

To study the semiconductor clusters we consider the Hamiltonian (Eq. 5) but now the states are s and p. The local density of states $\rho_{i\sigma}^{\alpha}$ is calculated by using the recursion method.[48] For the parametrized Slater-Koster hopping integrals $t_{ij}^{\alpha\beta}$, up to third neighbors, we take the values chosen to fit the bulk band structure of the semiconductor materials as reported by Papaconstantopoulos.[68]

IV.1 Carbon, Silicon, and Germanium Fullerenes

We calculated the electronic structure of three bucky balls: the dodecahedron, the truncated icosahedron, and the 70-atoms complex. These structures are shown in Fig. 12. The dodecahedron contains 20 vertices, 30 edges and 12 pantagonal faces. On the other hand, the truncated icosahedron contains 60 vertices, 90 edges, 12 pentagonal faces and 20 hexagonal faces. Assuming that the semiconductor atoms A, are located in the vertices, these structures correspond to A_{20} and A_{60}. All the atomic sites are equivalent and are coordinated to three nearest neigbors. In the dodecahedron each atom belongs to three pentagonal faces and the dihedral angle is of 116°34'. In the truncated icosahedron each atom belongs to one pentagonal and two hexagonal faces. In this case, the dihedral angles between pentagonal and hexagonal faces and between two hexagonal faces are 142°37' and 138°11', respectively. The 70-atom Fullerene is also a network of pentagons and hexagons. The number of pentagonal faces is also 12 but the number of hexagonal faces is 25.

The total density of states ($s + p$-contributions) for the dodecahedral cluster are shown in Fig. 13. The results for carbon, silicon, and germanium are presented in Figs. 13a, 13b, and 13c, respectively. The Fermi energy is marked with a dashed line. In this case all the sites are equivalent. The energy range between the lowest and highest energy levels (band-width) W_A ($A = $ C, Si, Ge) is wider for carbon that for the other elements. One can observe that the Fermi energy in the case of carbon falls in a gap. On the other hand the density of states close to E_F in Si and Ge is more dense. In all the cases the low energy part is mainly of s-character and that the p-electronic states ocuppy the high energy region.[69]

The results for the C_{60}, Si_{60}, and Ge_{60} bucky balls are shown in Fig. 14. Only in the case of the carbon molecule the Fermi energy falls in a well defined gap. This feature is what makes the C_{60} Fullerene very stable. From the partial s- and p-local density of states[69] one can see in a neater way that the electronic states near the Fermi energy are of p-type. The bottom of the band is populated mainly with s-electrons. These characteristics coincide with photoemission experiments performed in graphite and diamond,[70,71] in which the authors conclude that the electronic states that falls within $\sim$ 10 eV of the Fermi energy are of p-type and that deeper states have s-character. We obtain that in the case of C_{60} the electronic states occupy an energy range of approximately 2.5 Ry.

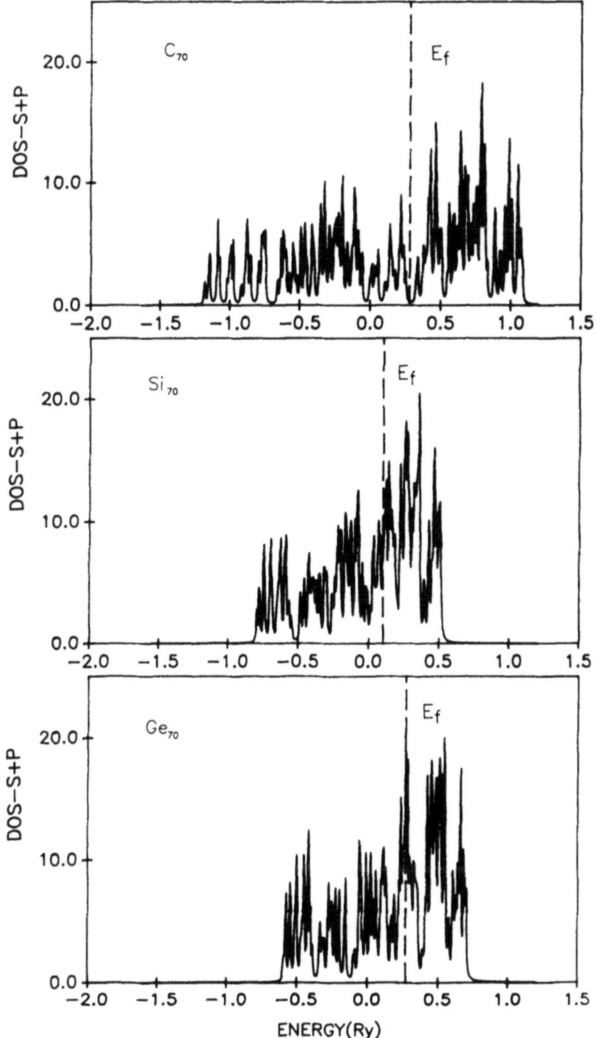

Figure 15. The total electronic density of states per atom of the C_{70}, Si_{70}, and Ge_{70} with Fullerene structure. The average of the five inequivalent sites is presented. The Fermi energy is marked by the dashed line.

We show in Fig. 15 the total density of states per atom for the 70-atom bucky ball for the three elements. In this case there are five inequivalent sites and the self-consistent results are obtained under the constraint of global charge neutrality. The partial s- and p-local density of states will be published somewhere else. One can notice that in the C molecule the Fermi energy falls also in a minimum. That is not the case for Si and Ge, where the density of states at E_F is in a very populated region.

The results obtained for C_{60} and C_{70} are in good agreement with other more elaboarted calculations[72,73] and with photoemission and inverse photoemission experiments.[74,75] In the case of C it seems that the lowest energy structures are the bucky balls. That might not be the case for Si and Ge, in which the diamond struc-

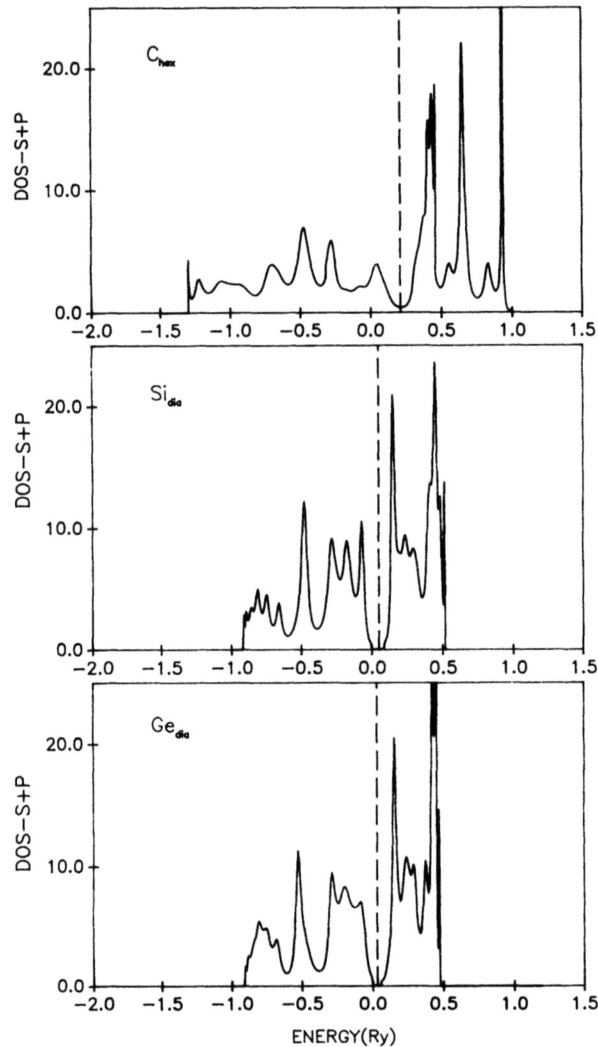

Figure 16. The total electronic density of states per atom of the infinite carbon haxagonal lattice. b) The total electronic density of states for the infinite silicon diamond lattice. c) The total electronic density of states per atom for the infinite germanium diamond lattice. The Fermi energy is marked by the dashed line.

ture is the most stable in bulk samples. A comparison of the Fullerene structures with other geometries based on stacked planar rings is in progress.

Finally we show in Fig. 16 the electronic density of states for the infinite hexagonal lattice (carbon) and the infinite diamond lattice (silicon and germanium). In the hexagonal lattice one obtains a small density of states for the p-type electrons in contrast to the bucky balls. The results for bulk silicon and germanium are very similar to those obtained by pseudopotential methods.[76] The similarity of the results by the two methods give us confidence in the method used here.

Acknowledgements

This work was partially suported by Dirección General de Investigación Científica y Superación Académica de la Secretaría de Educación Pública through Grants C910724-001-268 and C910724-001-963.

References

1. A.L. Mackay, *Acta Cryst.* **15**, 916 (1962).
2. J.M. Montejano-Carrizales and J.L. Morán-López, *Surf. Sci.* **239**, 169 (1990).
3. J.H. Sinfelt, *Rev. Mod. Phys.* **51**, 569 (1979).
4. V. Poneç, *Adv. Catal.* **32**, 149 (1983).
5. J.H. Sinfelt, *J. Catal.* **29**, 308 (1978).
6. P. Jena, B.K. Rao, and S.N. Khanna, Eds., *Physics and Chemistry of Small Clusters*, NATO ASI Series, Vol. 158 (Plenum Press, New York, 1987).
7. P. Jena, S.N. Khanna, and B.K. Rao, Eds., *Physics and Chemistry of Finite Systems: From Clusters to Crystals*, NATO ASI Series, Vol. 374 (Kluwer Academic Publications, Dordrecht, 1992).
8. M. Yacamán, in: *Catalytic Materials, Relationship Between Structure and Reactivity*, edited by T.E. Whyte, R.A. Betta, E.G. Deroudane, and R.T.K. Baker (American Chemical Society, New York, 1983), p. 341.
9. A. Renou and M. Guillet, *Surf. Sci.* **106**, 27 (1981).
10. D. Tománek and M.A. Schlüter, *Phys. Rev. Lett.* **56**, 1055 (1986).
11. J.R. Chelikowsky and J.C. Phillips, *Phys. Rev. Lett.* **63**, 1653 (1989).
12. M.A. Duncan and D.H. Rouvray, *Scientific American*, December, 1989, p. 60.
13. T.G. Dietz, M.A. Duncan, D.E. Powers, and R.E. Snalley, *J. Chem. Phys.* **74**, 6511 (1981).
14. J.M. Montejano-Carrizales and J.L. Morán-López, *Surf. Sci.* **239**, 178 (1990).
15. J.M. Montejano-Carrizales and J.L. Morán-López, *Surf. Sci.* **265**, 209 (1992).
16. R.A. Swalin, *Thermodynamics of Solids*, (Wiley, New York, 1972).
17. J.E. Demuth and T.N. Rhodin, *Surf. Sci.* **45**, 249 (1975).
18. H. Conrad, G. Ertl, J. Kuppers, and E.E. Latta, *Surf. Sci.* **57**, 475 (1976).
19. I. Toyoshima and G.A. Somorjai, *Catal. Rev. Sci. Eng.* **19**, 105 (1979).
20. M. Hansen, *Constitution of Binary Alloys*, (Mc. Graw-Hill, 1958).
21. C. Kittel, *Introduction to Solid State Physics*, (Wiley, New York, 1971).
22. A.D. van Langeveld, H.A.C.M. Hendricks, and B.E. Nieuwenhuys, *Thin Solid Films*, **109**, 179 (1983).
23. G. Betz, J. Marton, and P. Braun, *Nucl. Inst. Methods*, **168**, 541 (1980).
24. J.R. Chelikowsky, *Surf. Sci.*, **139**, L197 (1984).
25. S. Mukherjee and J.L. Morán-López, *Surf. Sci.* **188**, L742 (1987).
26. F.L. Williams and D. Nason, *Surface Sci.*, **45**, 377 (1974).
27. A.R. Miedema, *Z. Metallk.* **69**, 455 (1978).
28. R. Baudoing, Y. Gauthier, M. Lundberg, and J. Rundgren, *J. Phys.* **C19**, 2825 (1986).
29. J. Vrijen and S. Radelaar, *Phys. Rev.* **B17**, 409 (1978).
30. H. Shimizu, M. Ono, and K. Nakayama, *Surf. Sci.*, **36**, 817 (1973).
31. L.E. Rehn, H.A. Hoff, and N.Q. Lam, *Phys. Rev. Lett.*, **57**, 780 (1986).
32. S. Modak and B.C. Khanra, *Chem. Phys. Lett.* **134**, 39 (1987).
33. J.H. Sinfelt, J.L. Carter, and D.J.C. Yates, *J. Catal.*, **24**, 283 (1972).

34. *Magnetic Properties of Low Dimensional Systems*, edited by L.M. Falicov and J.L. Morán-López, (Springer-Verlag, Heidelberg, 1986).

35. *Magnetic Properties of Low Dimensional Systems II: New Developments*, edited by L.M. Falicov, F. Mejía-Lira, and J.L. Morán-López (Springer-Verlag, Heidelberg, 1990).

36. D.M. Cox, D.J. Trevor, R.L. Whetten, E.A. Rohlfing, and A. Kaldor, *Phys. Rev. B* **32**, 7290 (1985).

37. W.A. de Herr, P. Milani, and A. Chatelain, *Phys. Rev. Lett.* **65**, 488 (1990).

38. J.P. Bucher, D.C. Douglas, P. Xia, B. Haynes, and L.A. Bloomfield, *Phys. Rev. Lett.* **66**, 3052 (1991).

39. P. Milani and W.A. de Heer, *Phys. Rev. B***44**, 8346 (1991).

40. D.C. Douglas, J.P. Bucher, D.B. Haynes, and L.A. Bloomfield, in: *Physics and Chemistry of Finite Systems: From Clusters to Crystals*, edited by P. Jena, S.N. Khanna, and B.K. Rao (Klewer Academic Publishers, Dodrecht, 1992), p. 799.

41. J.P. Bucher, in: *Physics and Chemistry of Finite Systems: From Clusters to Crystals*, edited by P. Jena, S.N. Khanna, and B.K. Rao (Klewer Academic Publishers, Dodrecht, 1992), p. 799.

42. G.M. Pastor, J. Dorantes-Dávila, and K.H. Bennemann, *Physica B+C* **149B**, 22 (1988).

43. K. Lee, J. Callaway, and S. Dhar, *Phys. Rev. B***30**, 1724 (1984).

44. D.R. Salahub and R.P. Messmer, *Surf. Sci.***106**, 415 (1981).

45. G.M. Pastor, J. Dorantes-Dávila, and K.H. Bennemann, *Phys. Rev. B***40**, 7642 (1989).

46. J. Dorantes-Dávila, G.M. Pastor, and K.H. Bennemann, in: *Magnetic Properties of Low Dimensional Systems II: New Developments*, edited by L.M. Falicov, F. Mejía-Lira, and J.L. Morán-López (Springer-Verlag, Heidelberg, 1990) p. 222.

47. J. Dorantes-Dávila, H. Dreysee, and G.M. Pastor, in: *Physics and Chemistry of Finite Systems: From Clusters to Crystals*, edited by P. Jena, S.N. Khanna, and B.K. Rao (Klewer Academic Publishers, Dodrecht, 1992), p. 767.

48. R. Haydock, in: *Solid State Physics* (Academic, London 1980), Vol. **35**, p. 215.

49. V. Heine, *Phys. Rev.* **153**, 673 (1967).

50. V.L. Moruzzi, J.F. Janak, A.R. Williams, in: *Calculated Electronic Properties of Metals*, (Pergamon, New York, 1978).

51. L.W. Bos and D.W. Lynch, *Phys. Rev. B***2**, 4567 (1970).

52. W. Keune, R. Halbauer, U. Gonser, J. Lauer, and D.L. Williamson, *J. Mag. Mag. Mat.* **6**, 192 (1977).

53. J. Kübler, *Phys. Lett.* **81A**, 81 (1981).

54. G.M. Pastor, J. Dorantes-Dávila, and K.H. Bennemann, *Chem. Phys. Lett.* **148**, 459 (1988).

55. J.R. Heath, Y. Liu, S.C. O'Brian, R.F. Curl, F.K. Tittel, and R.E. Smalley, *J. Chem. Phys.*, **84**, 4074 (1986).

56. L.A. Bloomfield, R.R. Freeman, and W.R. Brown, *Phys. Rev Lett.*, **54**, 2246 (1985).

57. S.C. O'Brian, Y. Liu, Q. Zhang, J.R. Heath, F.K. Tittel, R.F. Kurl, and R.E. Smalley, *J. Chem. Phys.*, **84**, 4074 (1986).

58. J.L. Elkind, J.M. Alford, F.D. Weiss, R.T. Laaksonen, and R.E. Smalley, *J. Chem. Phys.*, **87**, 2397 (1987).

59. Q. Zhang, Y. Liu, R.F. Kurl, F.K. Tittel, and R.E. Smalley,*J. Chem. Phys.*, **88**, 1670 (1988).

60. W. Krätschmer, L.D. L.amb, K. Fostiropoulos, and D.R. Huffmann, *Nature* (London) **347**, 354 (1990).

61. H.W. Kroto, J.R. Heath, S.C. O'brien, R.F. Kurl, and R.E. Smalley, *Nature* (London) **318**, 162 (1985).

62. H. Kroto, *Science,* **242**, 1139 (1988).

63. S. Saito and A. Oshiyama, *Phys. Rev. Lett.,* **66**, 2637 (1991).

64. K. Tanigaki, T.W. Ebbesen, S. Saito, J. Mizuki, J.S. Tsai, Y. Kubo, and S. Kuroshima, *Nature* (London), **352**, 222,(1991).

65. J.C. Phillips, *J. Chem. Phys.,* **83**, 3330 (1985).

66. J.C. Phillips, *J. Chem. Phys.,* **88**, 2090 (1988).

67. D.A. Jelski, Z.C. Wu, and T.F. George, *J. Cluster Sci.,* **1**, 143 (1990).

68. D.A. Papaconstantopoulos in: *Handbook of the Band Structure of Elemental Solids,* (new York and London: Plenum Press 1986).

69. J. Ortíz–Saavedra, F. Aguilera-Granja, J. Dorantes-Dávila, and J.L. Morán-López, *Solid State Commun.,* submitted.

70. R.F. McFeeley, S.P. Kowalczyk, L. Ley, R.G. Cavell, R.A. Pollak, and D.A. Shirley, *Phys. Rev. B* **9**, 5268 (1974).

71. A. Bianconi, S.B. Hagström, and R.Z. Bachrach, *Phys., Rev. B* **16**, 5543 (1977).

72. S. Saito and A. Oshiyama, *Phys. Rev. Lett.* **66**, 2637 (1991).

73. S. Saito and A. Oshiyama, *Phys. Rev. B* **44**, 11532 (1991).

74. J.H. Weaver, J.L. Martins, T. Komeda, Y. Chen, T.R. Ohno, G.H. Kroll, N. Troullier, R.E. Haufler, and R.E. Smalley, *Phys. Rev. Lett.* **66**, 1741 (1991).

75. M.B. Jost, N. Troullier, D.M. Poirier, J.L. Martins, J.H. Weaver, L.P F. Chibante, and R.E. Smalley, *Phys. Rev. B* **44**, 1966 (1991).

76. M.L. Cohen and J.R. Chelikowsky, in: *Handbook on Semiconductors,* edited by W. Paul (North-Holland, Amsterdam, 1982), p. 219.

Synthesis and Processing of Nanostructured W-Base Materials

B.H. Kear[1], L. Wu[1], N.C. Angastiniotis[1], and L.E. McCandlish[2]

[1] *Dept. of Mechanics and Materials Science*
Rutgers University
P.O. Box 909, Piscataway, NJ 08855-0909

[2] *Nanodyne Inc.*
11 Industrial Drive
New Brunswick, NJ 08901
U.S.A.

Abstract

Chemical processing is a more direct route for making materials than traditional metallurgical processing. It offers the potential for lower cost production of novel materials with homogeneous ultrafine structures (nanostructures) and improved properties.

High surface area nanocrystalline W and nanostructured WC-Co composite powders have been synthesized by thermochemical conversion of homogeneous precursor compounds. Reductive decomposition of WO_3 at T > 625°C yields nanocrystalline α-W (*bcc* structure), whereas the same treatment at $\sim$ 500°C gives nanocrystalline β-W (A15 structure). The elemental form of β-W has an A15 defect structure, from which all or nearly all of the oxygen atoms have been removed. Diffusional disordering (amorphization) of high surface area elemental β-W in ammonia occurs at temperatures 350°C. The amorphous state of W(N) decomposes at higher temperatures in the range 650 – 850°C to form nanocrystalline W_2N.

Reductive decomposition of an homogeneous spray dried powder of ammonium metatungstate and cobalt nitrate yields a high surface area reactive intermediate of W-Co, which can be gas phase carburized to nanostructured WC-Co. The carburization treatment may be performed in CO/CO_2 gas mixtures of controlled carbon activity in the range 0.35 – 0.95. Rapid carbon uptake gives rise initially to W_2C-Co, which gradually transforms to the thermodynamically stable WC-Co composition. However, the fastest carburization kinetics at the lowest permissible temperatures, with control of uncombined carbon formation, occurs with CO/H_2 gas mixtures.

Advanced Topics in Materials Science and Engineering, Edited by
J.L. Morán-López and J.M. Sanchez, Plenum Press, New York, 1993

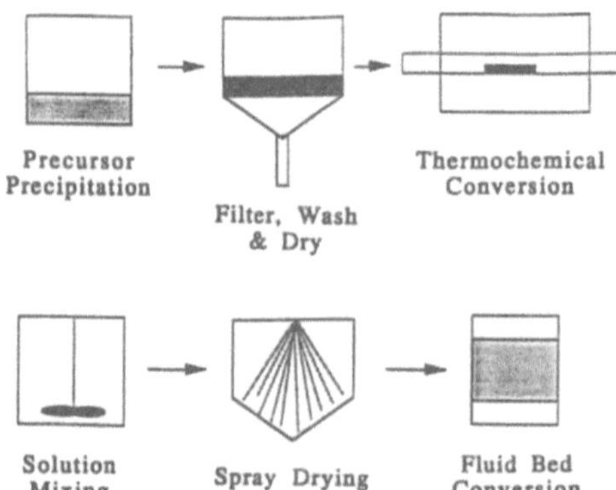

Figure 1. Schematic diagrams of (a) laboratory scale process, (b) industrial scale process for the thermochemical processing of nanostructured powders, starting from aqueous solution mixtures.

Nanostructured WC-Co powders have been consolidated by low pressure plasma spraying and by cold compaction and liquid phase sintering. The use of VC as a grain growth inhibitor is essential to mitigate the WC grain growth during liquid phase sintering. Theoretically dense VC-doped WC-Co materials display superior hardness, wear resistance and cracking resistance.

I. Introduction

The processing of materials from chemical precursors offers the potential for lower cost production of novel materials with homogeneous ultrafine microstructures (nanostructures) and improved properties. The technology gained its initial prominence as the preferred route for synthesizing ceramic materials. Today, there is growing interest in the applicability of chemical processing technology to the production of metallic materials.

At Rutgers University, we have been developing new capabilities for the synthesis and processing of nanostructured powders, starting from water soluble precursor compounds.[1] After an extensive evaluation of alternative synthesis routes, "Spray Conversion Processing" has emerged as the most versatile and reproducible. The new processing method consists of three coordinated steps:

    (1) preparation and mixing of aqueous solutions of the precursor compounds to fix the composition of the starting solution;

    (2) spray drying of the starting solution to form a chemically homogeneous precursor powder;

    (3) thermochemical conversion of the precursor powder to the desired nanostructured end-product powder.

The latter step may be performed in a fixed bed reactor when the amount of powder being processed is small. However, for the thermochemical processing of large quantities of powder, it is advisable to use a fluid bed reactor, so as to ensure a uniform conversion rate for all the particles in the bed, Fig. 1. All three steps in the process are readily scaleable. An integrated manufacturing technology for the production of nanostructured composite powders has been developed by Nanodyne Inc.

Spray drying is an essential step in the process when dealing with starting solutions that contain two or more precursor compounds. Rapid drying of the aerosol droplets, accompanied by rapid precipitation of the solute, produces chemically homogeneous precursor powders, even from complex starting solutions. In other words, spray drying tends to suppress phase separation, which would normally occur during conventional crystallization of the solution mixtures. Typically, the spray dried precursor powders are spherical shells about 10–50 microns in diameter, and have amorphous or microcrystalline structures.

Thermochemical conversion of the precursor powder in a fluid bed reactor is also an important step in the integrated process. This is because the local environment with respect to temperature and gas concentration in the fluid bed reactor is the same for all parts of the bed, which ensures uniform conversion of the precursor powder to the end-product powder. This is not the case in a fixed bed reactor, where uniformity of gas percolation and temperature is difficult to maintain throughout the powder aggregate.

A considerable amount of research on nanostructured materials has been done using the Spray Conversion Processing method. In this paper, we will describe some highlights of our research on (1) formation and alloying of nanocrystalline W, and (2) synthesis and processing of nanostructured WC-Co.

II. Formation and Alloying of Nanocrystalline W

The stable structure of tungsten is body centered cubic (α-W), but a second form (β-W, with the A15 structure) has long been recognized.[2] Our research has demonstrated the feasibility of synthesizing nanocrystalline α and β phases, starting from spray dried ammonium metatungstate (AMT).[3] Furthermore, it has been found that under special circumstances an amorphous structure can be formed.

High surface area WO_3 powder formed by pyrolysis of AMT $(NH_4)_6(H_2W_{12}O_{40})$ $\cdot 4H_2O$ was the starting material for all experiments. The pyrolysis was carried out in flowing argon at 500°C. Initial experiments on the formation of nanocrystalline W by hydrogen reduction of the pyrolyzed product were conducted in a TGA unit. Because of the unusually high surface area of the powders and their susceptibility to rapid oxidation upon exposure to air, some type of passivation treatment was necessary in order to preserve the samples for XRD analysis. Typically, the powders were passivated in 2% O_2/Ar at room temperature, or by using paraffin, which was vapor transported onto the powder prior to opening the reactor. More critical experiments were conducted in a controlled atmosphere, high temperature XRD unit, which avoided contamination altogether.

II.1 Synthesis of α-W

Hydrogen reduction of the high surface area WO_3 powder above 625°C invariably yielded WO_2, which subsequently reduced to α-W. Reduction followed by re-oxidation and re-reduction at a specific temperature yielded a characteristic grain size for both α-W and WO_3. Such cyclic oxidation and reduction experiments were performed at 800°C, 700°C, and 625°C. At each temperature, three cyclic treatments were required before a characteristic grain size was obtained for both WO_3 and α-W. The characteristic grain size decreased with temperature, as was clear from the increased width of the most prominent diffraction peaks of the two phases.

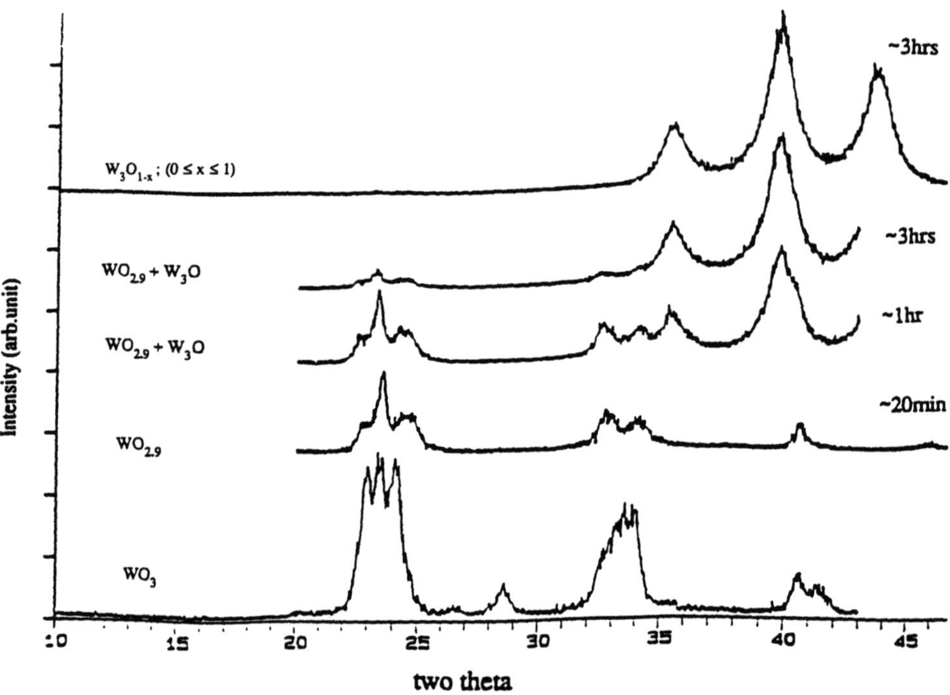

Figure 2. Formation of W_3O_{1-x} (β-W; A15 structure) by hydrogen reduction of high surface area WO_3 at T $\sim$ 500°C.

II.2 Synthesis of β-W

Hydrogen reduction of WO_3 at 500°C followed an entirely different phase sequence, where WO_3 quickly reduced to $WO_{2.9}$, followed by a gradual transformation to W_3O. Ultimately, W_3O reduced to the metallic suboxide W_3O_{1-x} ($x \sim .85$). The reductive oxide phase sequence is illustrated in Fig. 2.

The W_3O phase (A15 structure) has been surrounded by controversy since its discovery in 1931,[2] when it was suggested that it was an allotropic form of tungsten. Since then, many authors have argued over whether it is a suboxide or a metallic phase of tungsten. We have concluded on the basis of structure factor calculations that W_3O can exist as a stable oxide, and also as a metastable metallic tungsten W_3O_{1-x}, in which most of the oxygen atoms are removed. Further, that the actual structure of the β-W phase (W_3O_{1-x}) is an A15 defect structure, with vacancies randomly distributed on both corner and face center sites, Fig. 3.

The first series of *in situ* experiments carried out in an He atmosphere revealed that heating W_3O rapidly to 700°C in purified He (99.99%) yielded α-W and WO_2. The relative amount of WO_2 with respect to α-W correlated with the reduction time at 500°C. The longer the reduction at 500°C, the smaller the oxygen content, and therefore the amount of WO_2 formed. This finding is in agreement with previous results[4] that have stated that β-W can be prepared with less than 0.01 oxygen atoms

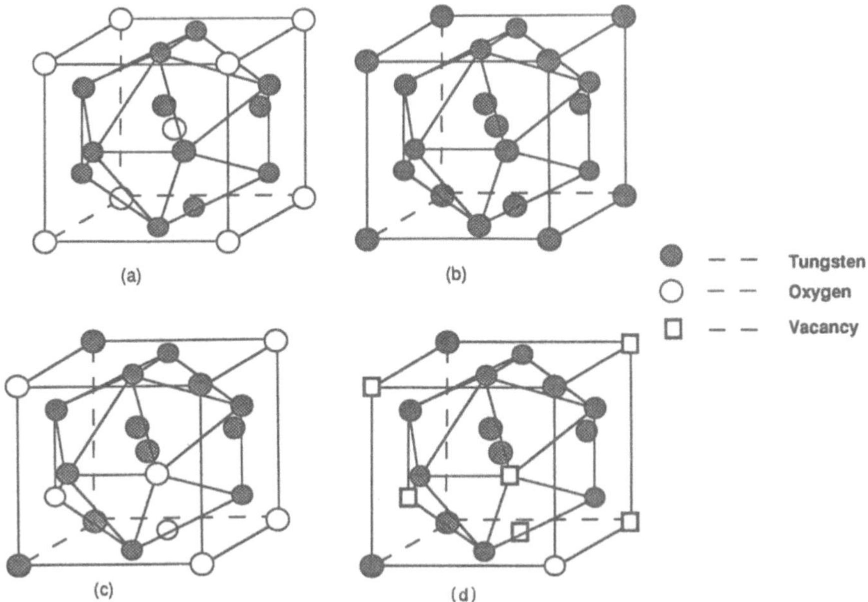

Figure 3. Unit cells of (a) ideal (ordered) A15 structure of W_3O, isostructural with Nb_3Sn; (b) hypothetical low temperature allotropic form of tungsten with A15 structure; (c) actual structure of W_3O with random distribution of tungsten and oxygen in the A15 structure; (d) actual structure of β-W.

per atom of tungsten. Our TGA experiments in H_2 gas also confirmed gradual oxygen loss at 500°C, as the W_3O transformed to $W_3O_{(1-x)}$.

Further studies on the transformation kinetics of $W_3O_{(1-x)}$ indicated stability up to a maximum temperature of 575°C, at which point a sluggish transformation to α-W took place. At this point, it was realized that we could take advantage of the slow transformation of β-W to α-W to remove any residual oxygen by rapid thermal cycling to 625°C. Indeed, several thermal spikes from 500°C to 625°C in flowing H_2 produced the fully reduced state of the β-W structure, which contained little or no residual oxygen. We consider this oxygen- depleted structure to be elemental β-W, Fig. 3. It should be emphasized that rapid heating of the β-W in a He atmosphere to 700°C yielded X-ray pure α-W, *i.e.*, no evidence for WO_2. These findings indicate that elemental β-W has a disordered A15 defect structure, with vacant lattice sites where the original oxygen atoms resided.

II.3 Diffusional Amorphization of β-W

The reactivity of the high surface area nanocrystalline β-W with air, as manifested by its pyrophoricity, suggested to us that W-base compounds might be formed by reactions with other gases. Also the disordered defect structure of β-W raised the possibility of doing this at low temperatures.

A good illustration of the low temperature alloying behavior of β-W is provided by the reaction sequence observed with gaseous ammonia as the reactant gas. Fig. 4

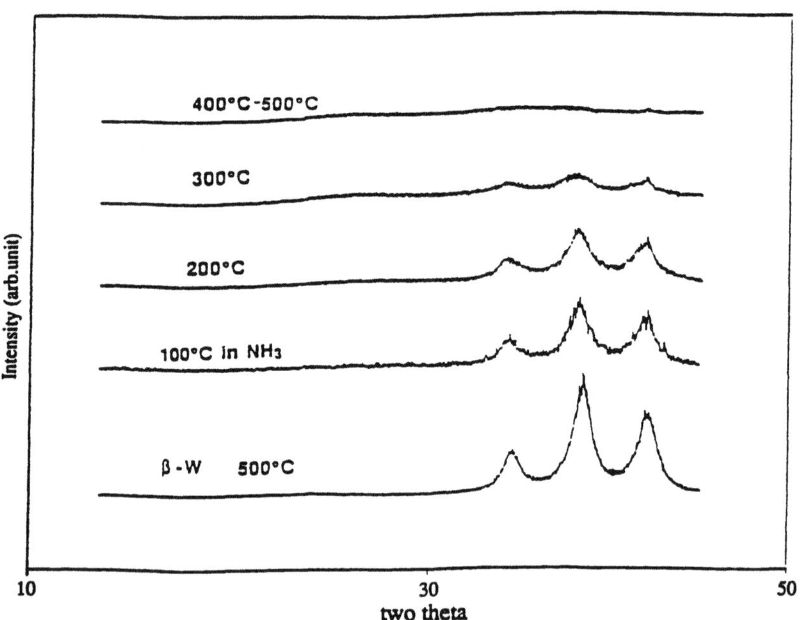

Figure 4. Diffusional disordering of elemental β-W (A15 defect structure) by exposure to flowing ammonia at 100° – 400°C.

shows the effect of exposure to ammonia, starting at 100°C, and after incrementing the temperature at intervals of one hour up to 200°C, 300°C, 400°C and 500°C. The remarkable consequence of this treatment is the complete amorphization of the original β-W, and the appearance of a very broad diffuse peak centered at 25–30° 2θ. We believe that at these low temperatures the ammonia catalytically decomposes on the high surface area tungsten forming atomic nitrogen, hydrogen and free radical species. The hydrogen removes any residual oxygen in the W, and the reaction of the nitrogen with the tungsten causes amorphization. Such diffusional amorphization is to be expected when the temperature is too low for crystallization of the equilibrium phase, here assumed to be W_2N. An extensive discussion of diffusional amorphization in crystalline solids is presented in a recent review article by Johnson.[5]

The diffusionally disordered state of tungsten with nitrogen is surprisingly stable. As shown in Fig. 5, a broad peak in diffracted intensity at about 36° 2θ only begins to make its appearance at about 600°C. With increasing temperature the peak gradually sharpens and eventually two peaks centered at 36° and 43° 2θ become apparent. These are the characteristic peaks of nanocrystalline W_2N. At higher temperatures 900°C, α-W begins to make an appearance (Fig. 5). We interpret this to mean that W_2N is unstable in flowing ammonia (1:3/N:H) at 900°C. Further evidence for this is provided by the fact that when pure hydrogen is substituted for the ammonia immediate crystallization to α-W occurs.

In complete contrast, α-W (formed at 625°C) does not diffusionally disorder (amorphize) when exposed to ammonia at low temperatures. On the contrary, it forms a passivation layer of nitride phase, which precludes further reaction of the tungsten with the ammonia gas. This behavior clearly indicates that the nanocrys-talline β-W defect structure is more susceptible to alloying than the nanocrystalline

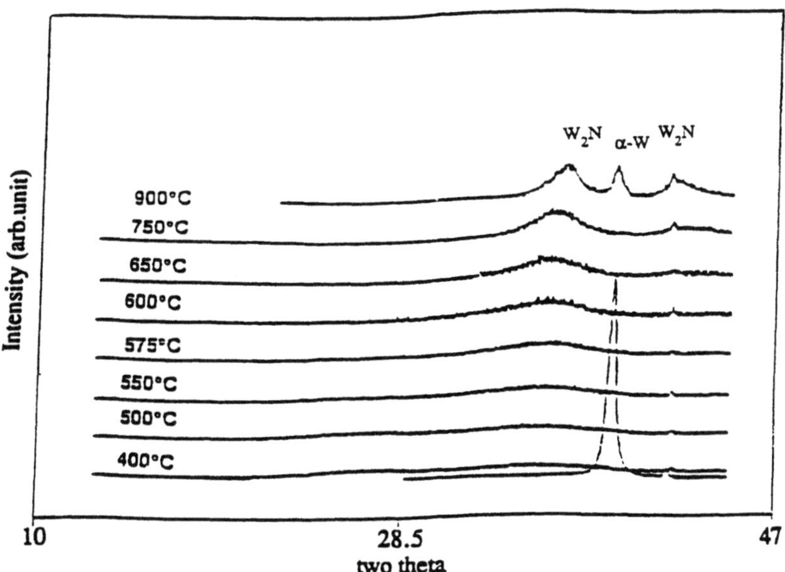

Figure 5. Decomposition of amorphous W(N) into nanocrystalline W_2N at $600°$ – $900°C$. The W_2N phase yields α-W in flowing H_2 at $900°C$ (superimposed sharp diffraction peak).

α-W. A similar behavior has been found for other reactive gas species, which will be discussed at a later date.

To summarize, reduction of WO_3 at temperatures higher than $625°C$ yields nanocrystalline α-W (*bcc* structure). Reduction of WO_3 at $500°C$ yields nanocrystalline W_3O_{1-x} ($0 \leq x \leq .85$; A15 structure), which can be oxygen depleted by rapid thermal transients to $625°C$, taking advantage of the slow transformation from A15 to *bcc*. The highly defective elemental form of β-W can be amorphized with suitable gases (gas-solid alloying). Successful amorphization has been achieved with ammonia and other reactive gases.

III Synthesis and Processing of Nanostructured WC-Co

III.1 Powder Synthesis

Spray conversion processing technology evolved quite naturally from basic research on chemical synthesis of WC-Co powders. An appreciation for the difficulties encountered and overcome in moving from a limited laboratory-scale process to a versatile industrial-scale process may be understood by considering the chronology of our research activities in this area:

(1) The original laboratory-scale process made use of a single chemical precursor compound, cobalt tris(ethylenediamine) tungstate, ($Co(en)_3WO_4$), which after thermochemical conversion gave WC-23 wt. % Co.[6] Processing was carried out in a carefully controlled gas atmosphere in a small tube furnace or TGA unit. Typical batch sizes were limited to a few grams. Even though the batch sizes were small, and the WC-23 wt. % Co composition contained more cobalt than most commercially useful compositions, this phase of the research nevertheless proved

the principle of thermochemical processing of homogeneous precursor powders to form nanostructured WC-Co powders.

In a typical procedure, $Co(en)_3WO_4$ powder was crystallized from solution and reduced in flowing argon/hydrogen to yield nanoporous/nanophase W-Co. This high-surface-area reactive intermediate was then converted directly to nanostructured WC-Co powder by gas phase carburization in flowing CO/CO_2 ($a_c = 0.9$). The resulting powder particles had the same morphology as the original $Co(en)_3WO_4$ particles, but the size of each particle was reduced by about 50%. In this process, the scale of the powder particle microstructure may be controlled from nanometer up to micrometer dimensions by adjusting the temperature of the carburization reaction, the time at temperature, and the carbon activity of the gas phase.

(2) In order to extend the range of compositions available to those of commercial interest (3–30 wt. % Co), we decided that sufficient flexibility could be obtained by preparing chemically homogeneous precursor powders by rapid spray drying of aqueous solutions of W and Co salts. In spray drying, the solvent phase is rapidly evaporated in a hot gas stream, which results in rapid precipitation of the solute mixture; at sufficiently high rates of precipitation phase separation can be avoided. Tests on spray drying of $Co(en)_3WO_4$-H_2WO_4 solutions showed that we could produce amorphous or microcrystalline precursor powders without phase separation. We also demonstrated that these homogeneous precursor powders could be thermochemically converted to various nanophase WC-Co compositions. Economic considerations led us to search for alternative lower cost starting materials. Today we routinely produce homogeneous precursor powders from AMT and $CoCl_2$, $Co(NO_3)_2$, or $Co(CH_3COO)_2$, all of which are readily available commercially.

By spray drying solution mixtures of AMT and cobalt salts, we have adjusted the Co/W ratio to 1.0, 0.63, 0.21 and 0.1. After thermochemical conversion, the resulting nanophase WC-Co powders had respectively 23, 15, 6 and 3 wt. % Co binder phase. The microstructures of these powders were substantially the same as that obtained from crystalline $Co(en)_3WO_4$ powder, Fig. 6.

(3) There is a limitation on the amount of powder that can be thermochemically processed in a fixed bed laboratory-scale reactor, because of inherent difficulties in maintaining a uniform temperature and gas percolation rate throughout the powder bed. Typically, only a few grams of material can be processed in a single run. To overcome this limitation, we have adopted fluid bed reactor technology for thermochemical processing of the powders. Fluidization involves levitation of the precursor powder particles in a flowing gas stream. The gross circulation of particles in the bed, the uniform gas flow about each particle, and the uniform bed temperature provide an uniform environment for the control of chemical reactions in the powder bed. Thus, a fluid bed reactor is ideal for thermochemical conversion of powders. Furthermore, fluid bed processing is a proven, scaleable technology.

(4) Until quite recently, all fluid bed carburization treatments of the precursor powders were conducted in CO/CO_2 gas mixtures with carbon activities in the range 0.50 to 0.95. In all cases, the rapid initial uptake of carbon gave rise to a metastable phase (W_2C), prior to its conversion to the thermodynamically stable WC-Co composition. This had two consequences. First, the metastability of the intermediate phase prolonged the overall conversion time and increased the cost. Second, the longer reaction time permitted particle coarsening and limited the ultimate structural scale attainable. For example, starting with a nanophase (10 nm) mix-

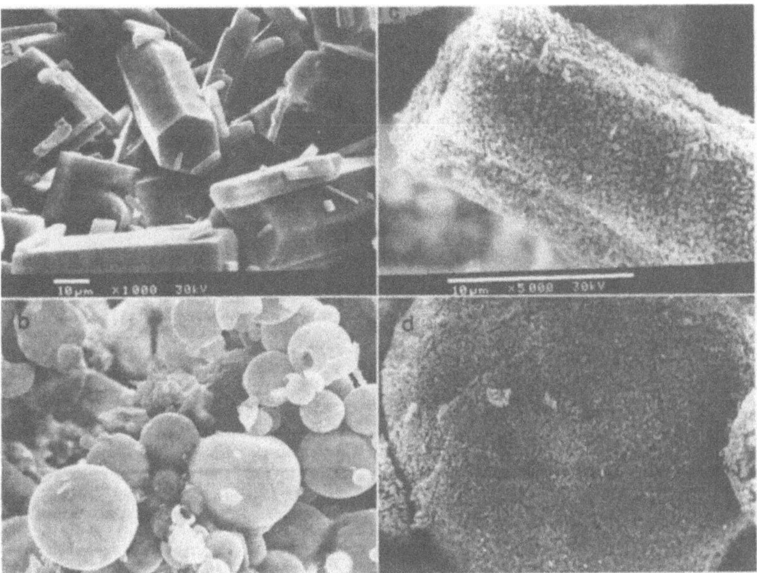

Figure 6. Comparison of powder structures, before and after thermochemical processing; (a) and (b) - precipitated powder; (c) and (d) - spray dried powder.

ture of W and Co, after reduction of the precursor powder, the carburization step in the process can result in a WC grain size $\sim$ 500 nm.

Rapid carbon uptake in the first stage of carburization is the key to reducing the nanostructural scale of the WC-Co powder. Such uptake of carbon occurs at unexpectedly low temperatures, apparently because of catalytic decomposition of CO by the nanodispersed cobalt and tungsten phases. Using pure CO, rather than CO/CO_2 mixtures, we have found that the carburization rate can be increased significantly, which has the desirable effect of reducing the WC grain size to < 100 nm. Unfortunately, the faster conversion kinetics of W to WC is accompanied by the production of uncombined or free carbon. After the carburization treatment, therefore, the excess free carbon must be removed by selective gasification in a CO/CO_2 gas mixture of 0.5 carbon activity.

(5) Carbothermic reaction processing can also be carried out in CO/Ar and CO/H_2 gas mixtures. An advantage of the CO/H_2 mixture is that it permits better control of the formation of uncombined, or free carbon, while maintaining a high carburization rate. For example, using a 30 $CO/70$ H_2 gas mixture the carburization rate is high, and the amount of uncombined carbon is negligible. An interesting feature of the CO/H_2 treatment is that it permits both precursor reduction and carburization simultaneously, which adds to the efficiency of the process.

Fig. 7 compares the carburization rates for pure CO, 30 $CO/70$ H_2, and CO/CO_2 ($a_c \sim 0.9$) at 750°C. The curves are normalized to the calculated weight gain of carbon for stoichiometric carbide formation. Thus, the first and second plateaus of two of the curves correspond to W_2C and WC formation, respectively. In contrast, the curve for pure CO shows only an inflection point at the W_2C level. In this case, the rapid carbon uptake is due mainly to the formation of uncombined carbon. As

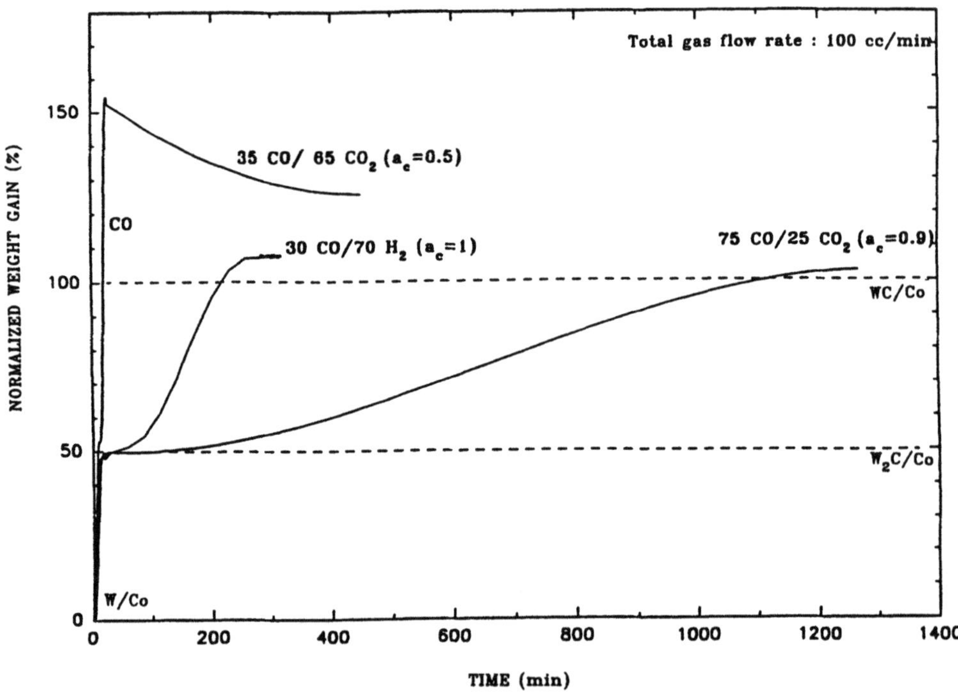

Figure 7. Comparison of carburization rates of reduced powders of W/Co using (1) pure CO,(2) 30 Co/70 H_2 ($a_c = 1$), and (3) 75 Co/25 CO_2 ($a_c = 0.9$). Note formation at W_2C/Co (first plateau), prior to formation of WC/Co (second plateau). All tests on WC/11 wt. % Co at 750°C.

indicated, the excess free carbon can be selectively gasified in CO/CO_2 ($a_c \sim 0.5$). The effectiveness of the CO/H_2 gas mixture in the carburization step of the process can be understood in terms of the following equilibria. The essential reaction is the formation of carbon by the low temperature catalytic decomposition of CO in the presence of nanodispersed metallic cobalt.

$$2CO = CO_2 + C \tag{1}$$

If H_2 is present, the water gas reaction becomes important.

$$CO + H_2 = H_2O + C \tag{2}$$

Subtraction of Eq. (2) from Eq. (1) gives the water shift reaction, which governs the equilibrium composition of the gas phase.

$$CO + H_2O = H_2 + CO_2 \tag{3}$$

Thus, the carbon that has been deposited on the powder surface can either diffuse into the particles and react with W to form WC, or it can be gasified by reaction with CO_2 or H_2O. The reaction limits the build-up of uncombined carbon during the carburization step, but does not eliminate it. The final uncombined carbon content of the WC-Co powder can be adjusted via the equilibrium reaction, (1) by fixing the CO/CO_2 ratio of the gas stream, and the total gas pressure. By adjusting the duration of the CO/CO_2 treatment, various levels of uncombined carbon can be obtained.

III.2 Powder Processing

A wide variety of powder processing methods are in use today for consolidating powders. Most of our research on nanostructured WC-Co powders has been directed towards the formation of (1) low porosity coatings by thermal spraying, and (2) theoretically dense forms by cold compaction and liquid phase sintering.

(1) Thermal spray methods for preparing coatings from conventional WC-Co powders include low velocity combustion (LVC), high velocity oxygen fuel (HVOF), conventional plasma (CP), and high energy plasma (HEP). The effectiveness of these coating methods can be rated on the basis of the amount of retained WC in the coating, which correlates with wear performance. The highest amount of retained WC, and therefore the lowest wear rate, occurs for the HVOF process. This is because the lower flame temperature and higher particle velocity, compared with the other processes, minimizes the decomposition rate of WC in the oxidizing environment of the flame or plasma. The oxidative decomposition involves decarburization of the WC phase to form W.

Similar oxidative decomposition reactions occur in thermal spraying of nanostructured WC-Co powders, except that the problem is exacerbated by the high surface area of the as- synthesized powders. The severity of this problem can be diminished to some degree by densification of the powder particles prior to thermal spraying. Even so, particle degradation in the flame or plasma persists, so that much of the benefit of having an ultrafine structure in the coating is not realized. Thus, both conventional and nanostructured WC-Co powders are susceptible to decarburization during thermal spray deposition in an ambient air environment. To date, the only practical solution to this generic problem has been to conduct the thermal spraying operation in vacuum, as in low pressure plasma (LPP) spraying. This technique has been used successfully for depositing WC-Co coatings on various substrates, using both conventional and nanostructured powders as feedstocks.

An important distinction between thermal sprayed conventional and nanostructured powders is their inherently different melting and solidification characteristics, Fig. 8. Conventional powder particles experience surface melting only, accompanied by slow and limited dissolution of the WC particles in the liquid Co, as the temperature is increased above the pseudo-binary eutectic ($\sim 1350°C$). The resulting spray deposited coating layer, therefore, tends to be somewhat porous, since the presence of the relatively large WC grains in the partially melted particles impedes fluid flow on the substrate surface. Nanostructured powder particles, due to the high surface area of contact between the Co and WC phases, undergo homogeneous or "bulk" melting, accompanied by rapid and extensive dissolution of the WC nanograins with increasing superheat above the eutectic. In this case, the resulting coating is much denser, owing to the facility with which the nanodispersed semi-solid or "mushy" particles can spread out over the substrate surface. Although a direct comparison of the porosity of coatings prepared from the two types of powders of the same composition has not yet been made, there have been

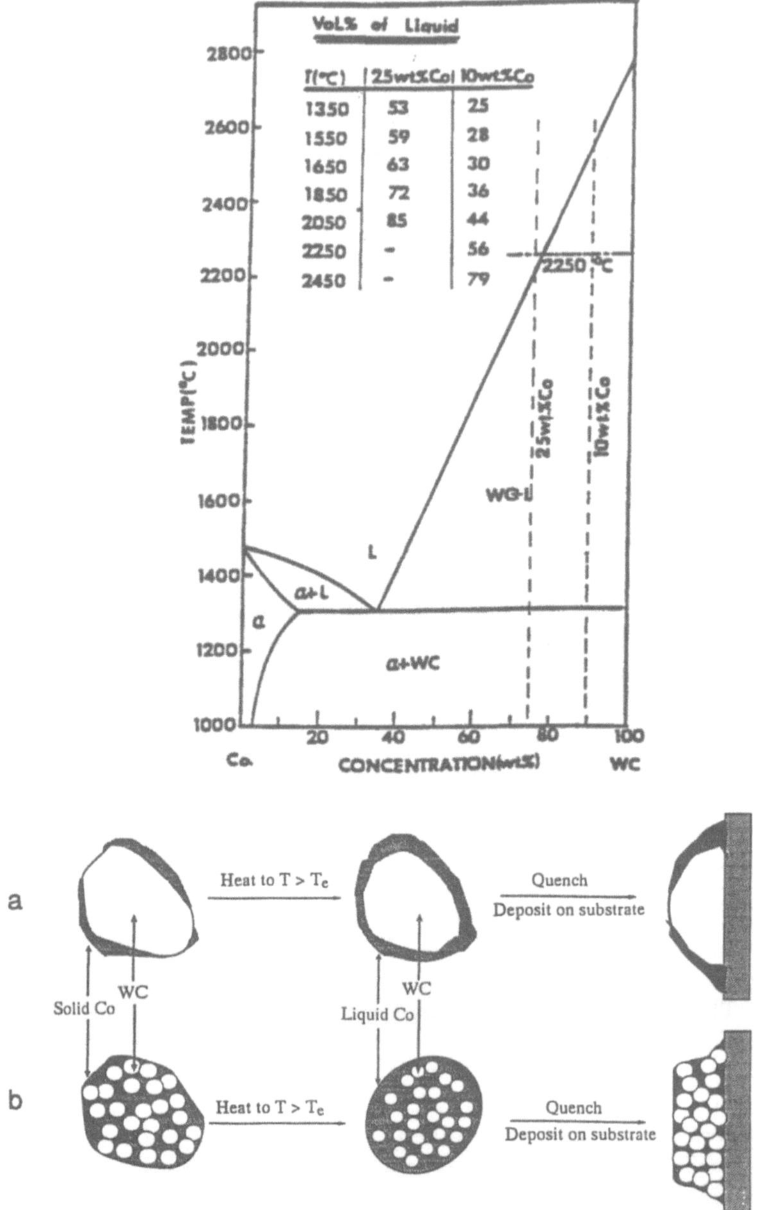

Figure 8. Comparison of melting and solidification characteristics of thermal sprayed (a) conventional and (b) nanostructured WC-Co powders.

indications that the enhanced fluidity of the mushy particles does promote the formation of dense coatings.

Another interesting consequence of the rapid solidification experienced by the mushy particles as they spread out over the substrate surface is the tendency for the supersaturated Co-rich liquid to transform to an amorphous solid. Such an effect has, in fact, been observed in an LPP-coating of nanostructured WC-23 wt. % Co alloy.[7] Both X-ray and TEM observations have shown that the spray deposited structure consists of a nanodispersion of WC grains in an amorphous Co-rich matrix phase. Such a structure exhibited both reduced friction coefficient and wear rate. The average coefficient of friction increased slightly with load from 0.13 to 0.17 as the load was increased from 300 g to 900 g; the friction coefficient for conventional WC-Co ranges from 0.2 to 0.4 for the same load range. Wear rate was about 10^{-4} mm^3/Nm, which compares with 10^{-2} mm^3/Nm for conventional material of the same composition. These encouraging results have prompted further research on the tribological properties of LPP coatings, extending over the range of compositions of commercial interest (5 – 30 wt. % Co).

Although loss of WC in the spray deposited coatings can be avoided by the LPP coating method, there is an urgent need to have the same capability for the much simpler, less expensive, more versatile and much more widely used flame and plasma spraying methods, which are conducted in an ambient air environment. To accomplish this desirable objective, we are investigating the efficacy of a sacrificial coating of carbon as a means to mitigate WC particle degradation by oxidative reaction during plasma spraying. This approach looks promising, but is far from being optimized.

(2) Liquid phase sintering is widely used for consolidating WC-Co powder. After cold compaction in a high pressure hydraulic press, the powder compact is heated in vacuum or hydrogen to a temperature above the pseudo-binary WC-Co eutectic where liquid phase sintering occurs. Theoretically dense structures are routinely produced by this method.

As a result of extensive research on this subject, many factors that influence the sinterability of the powder compacts have been identified. Important factors are (1) carbon content of the powders, which should be sufficient to maintain stoichiometric WC and avoid η-carbide formation, (2) purity of the starting materials, *e.g.* Ca, Mg, and Al should be avoided, since oxide particles of these elements are mainly responsible for residual porosity in the sintered material, (3) particle size, distribution and morphology, since these parameters strongly influence the packing density and uniformity of packing in the cold compact, (4) use of lubricants/binders to prevent the build-up of stress gradients in the powder compact, which otherwise can cause cracking, and (5) use of grain growth inhibitors, such as VC, TaC, and NbC, which effectively mitigate WC particle growth during liquid phase sintering. All these factors have proved to be just as important in the sintering of nanostructured WC-Co powders, but in addition there is a need to minimize the time spent at the peak sintering temperature in order to avoid significant particle coarsening. In this regard, the use of grain growth inhibitors has proved to be indispensible.

The mechanism of liquid phase sintering is reasonably well understood. Two main stages of sintering have been identified:

(a) Stage 1-matrix melting and particle rearrangement:
    In this first stage of liquid phase sintering, densification is driven by strong surface tension or capillary forces. Upon melting of the matrix phase, wetting of the solid particles by the liquid phase promotes rapid densification of the powder compact. The degree of consolidation depends on the volume fraction of the liquid phase. Assuming ideal close packing of identical spherical particles (*fcc* packing), the minimum volume fraction of liquid phase required to achieve a theoretically dense composite is about 26%. Larger volume frac-

tions of liquid result in uniformly spaced particles, or local regions where the particles are more densely packed, depending on the surface energies of the various interfaces. Smaller volume fractions of liquid leave residual porosity, which can only be eliminated by further sintering (stage 2).

The driving force, P, for particle rearrangement by capillary forces is inversely proportional to the particle radius, r:

$$P = 2\gamma r^{-1} cos\theta \tag{4}$$

where γ is the surface energy, and θ is the contact angle between the liquid and solid. With good wetting and sufficiently small particles, a theoretically dense structure can be realized almost instantaneously. Chemically synthesized nanostructured WC-Co is an ideal system because of the high interface area between the two phases, the ultrafine scale of the refractory phase, and the high purity of the powders. Furthermore, the small size of the WC grains facilitates particle rearrangement to form dense packing in the liquid matrix. The time spent at the incipient melting temperature will determine the degree of microstructural coarsening. To maintain a nanoscale structure it is essential to keep this time to a minimum.

(b) Stage 2-solution/reprecipitation and particle coalescence:
Rapid growth of the solid phase particles in the liquid phase can occur by the classical solution/reprecipitation mechanism (Ostwald ripening), where larger particles grow at the expense of smaller particles. This is a result of the difference in solubility of small and large particles in the liquid phase, which is governed by the Gibbs-Thomson equation:

$$c - c_0 = 2\gamma_{s,l} c_0 \Omega (RT)^{-1} r^{-1} \tag{5}$$

In this equation c is the solubility of a particle of the solid phase with radius r, c_0 is the solubility of a flat surface of the solid, Ω is the atomic volume of a solute atom, and $\gamma_{s,l}$ is the solid-liquid interfacial energy. This equation strictly applies to systems in which the distances between solid particles are equal to or larger than the particle diameters. More properly, for high solid volume fraction systems, the coarsening mechanism is particle coalescence. If the contact angle is positive, grain boundaries will be formed between aggregates of particles. High stresses at contact points will promote particle rearrangement and densification. In time, a skeletal structure will develop, in which the refractory phase is interconnected in three dimensions. Even though this structure makes complete densification more difficult, it is mechanically advantageous because any applied load is carried mainly by the refractory component of the composite. Thus, the skeleton can contribute substantially to the strength of the composite material. By varying the volume fraction of the matrix phase in the range where such bicontinuous structures are formed, it is possible to trade-off such critical properties as fracture toughness and strength.

In order to derive the full benefit from the nanostructure (10 – 30 nm) of the starting WC-Co powder, it is essential to minimize WC particle coarsening during liquid phase sintering. One approach is to limit the time spent at the sintering temperature, say to a few minutes. This has some merit for particle dispersed alloys (< 65 vol. % WC), since matrix melting and particle arrangement (Stage I sintering) controls the densification process. For

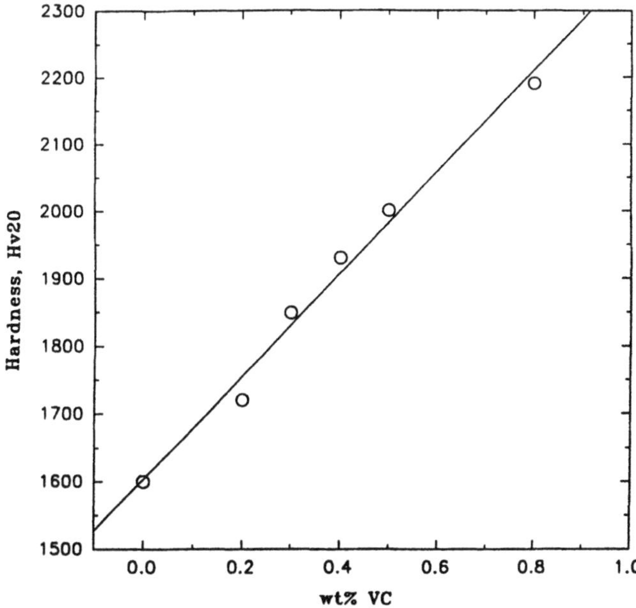

Figure 9. Increase in hardness of VC-doped WC-7 wt. % Co alloys, after sintering at 1400°C for 60 minutes. Increase in VC concentration correlates with reduced WC grain size (after Eason *et al.*[8])

network strengthened alloys (> 75 vol. % WC), however, the elimination of microporosity by a solution/reprecipitation and particle coalescence mechanism (Stage 2 sintering) is a prerequisite, so that much longer sintering times are needed. For this case, we have found that additions of known grain growth inhibitors, such as VC, TaC and NbC, are essential. Sintering tests performed on VC-doped WC-7 wt. % Co have shown a systematic increase in hardness with VC concentration, Fig. 9. These data correlated with a reduced mean free path for the cobalt binder phase (*i.e.*, reduced WC grain size), as determined by standard coercivity measurements. This is clear evidence for the potency of VC as a WC grain growth inhibitor in liquid phase sintering of WC-Co alloys. Measurements of crack resistance (Palmquist tests) of the nanostructured material, compared with conventional material, are shown in Fig. 10. The most striking feature is the linear dependence of crack resistance on hardness in the nanostructured material, and the wide scatter in the data for conventional material. Metallographic and SEM examinations have shown that this difference in behavior is a reflection of the much more uniform structure of the nanograined material. More recent data has also demonstrated that an increase in hardness of WC-Co does not necessarily cause a reduction in transverse rupture strength (TRS). On the contrary, as shown by Bock *et al.*, when the structure is uniform and essentially flaw free the transverse rupture strength is essentially independent of hardness in VC-doped alloys. Finally, we note that TEM observations on several samples have revealed a typical bicontinuous structure, in which the faceted WC grains are bonded together

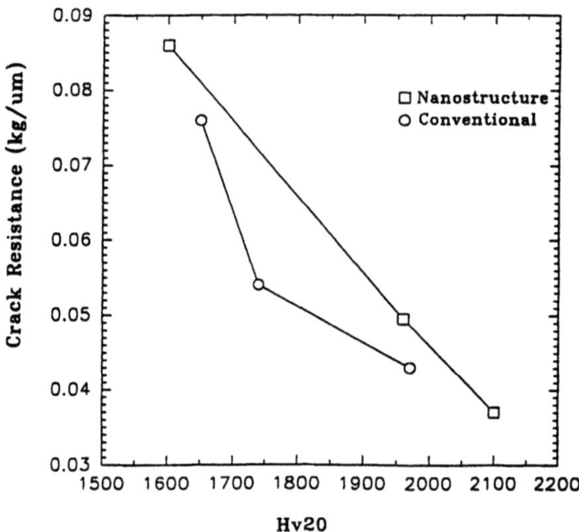

Figure 10. Improved crack resistance of nanostructured WC-7 wt. % Co, relative to conventional material (after Eason *et al.*[8])

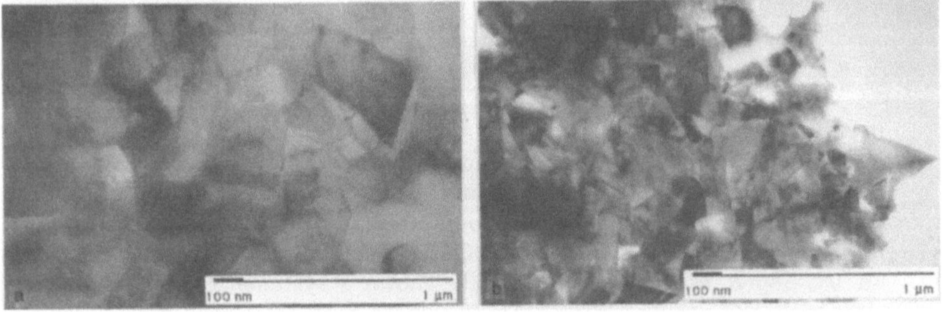

Figure 11. Transmission electron micrographs of liquid phase sintered WC-7 wt. % Co (a) without VC, and (b) with 0.4 wt. % VC. Note reduced WC grain size in (b).

to form a rigid skeletal structure. Typical micrographs showing significant difference in WC grain size due to VC doping are shown in Fig. 11.

So far, all sintering tests on VC-doped alloys have been performed on samples in which the VC powder had been blended with the nanostructured WC-Co powder by mechanical milling. Since it is believed that a more uniform distribution of VC in the mixed powder would be advantageous, we are now introducing vanadium into the starting solution as ammonium vanadate. Preliminary results look promising, and have raised questions concerning the optimal VC dopant concentration, which needs to be addressed. According to one current theory, this concentration should correspond to the maximum solubility of VC in liquid cobalt at the sintering temperature, which is about 6 wt. %. Larger concentrations apparently adversely affect mechanical properties.

IV. Summary

A new and versatile chemical processing method, called "Spray Conversion Processing" has been applied successfully to the synthesis of high surface area powders of nanocrystalline W and nanostructured WC-Co.

Nanocrystalline W

1. Reductive decomposition of high surface area WO_3 (prepared by pyrolysis of AMT) at $T > 625°C$ yields nanocrystalline α-W (*bcc* structure)

$$WO_3 \rightarrow WO_2 \rightarrow \alpha - W.$$

2. Reductive decomposition of WO_3 at $T \sim 500°C$ yields nanocrystalline β-W (A15 structure)

$$WO_3 \rightarrow WO_{2.9} \rightarrow W_3O \rightarrow \beta - W.$$

Elemental β-W has an A15 defect structure, from which all or nearly all of the oxygen atoms of the W_3O precursor have been removed.

3. The A15 defect structure of elemental β-W can be diffusively disordered (amorphized) by reaction with flowing ammonia at low temperatures ($\sim 300°C$). The amorphous state of W(N) decomposes at higher temperatures in the range $650 - 850°C$ to form nanocrystalline W_2N.

4. Nanocrystalline α-W is not susceptible to diffusional disordering with ammonia gas, but instead forms a thin passivation layer of W_2N.

Nanostructured WC-Co

1. Reductive decomposition of spray dried AMT + $Co(NO_3)_2$ precursor powder yields a high surface reactive intermediate of W-Co powder, which can be gas phase carburized to nanostructured WC-Co powder.

2. The fastest carburization rates at the lowest permissible temperatures, with control of uncombined carbon formation, occur with CO/H_2 gas mixtures. A two-stage carburization process gives first W_2C and then WC.

3. Nanostructured WC-Co powders have been consolidated by thermal spraying and by cold compaction and liquid phase sintering. Low pressure plasma spraying avoids degradation of the WC-Co particles by oxidative reactions during their short residence time in the plasma flame.

4. Additions of VC to the nanostructured WC-Co powder mitigate WC grain growth during liquid phase sintering at $\sim 1400°C$. The resulting theoretically dense nanostructured WC-Co materials display superior hardness, wear properties and cracking resistance.

Acknowledgements

The support of the Office of Naval Research under contract # N00014-91-J-1828 is gratefully acknowledged.

References

1. L.E. McCandlish, B.H. Kear, and B.K. Kim, *Materials Science and Technology,* **6**, 953 (1990).
2. V.H. Hartmann, F. Ebert, and O. Bretschneider, *Z. Anorg. Allg. Chem.*, **198**, 116 (1931).
3. N.C. Angastiniotis, B.H. Kear, L.E. McCandlish, K.V. Ramanujachary, and M. Greenblatt, *Powder Metallurgy World Congress*, San Francisco, CA, June 21-26, 1992.
4. Von J. Neugebauer, A.J. Hegedus, and T. Millner, *Z. Anorg. Allg. Chem.*, **293**, 241 (1958).
5. W.L. Johnson, *Progress in Materials Science,* **30**, 81 (1986).
6. L.E. McCandlish and R.S. Polizzotti, *Solid State Ionics,* **32/33**, 795 (1989).
7. L.E. McCandlish, B.H. Kear, B.K. Kim, and L.W. Wu, *Protective Coatings: Processing and Characterization*, edited by R.M. Yazici, (TMS, Warrendale, PA, 1990).
8. J.W. Eason, Z. Fang, B.H. Kear, and L.E. McCandlish, (to be published).

The State of the Art of
Materials Research in México

M. José-Yacamán

Instituto de Física
Universidad Nacional Autónoma de México
Apartado Postal 20-364
01000 México, D.F.
MEXICO

The application of scientific knowledge to the modification and characterization of materials in Mexico has a long story. In fact, since ancient times, native mexican cultures had a highly sofisticated technologyy for ceramics and construction materials. Some recent work has shown that some of the cultures of Mesoamerica had indeed metallurgical technology, producing even mechanical alloys and the extensive use of some metals. The main metals for the old mexicans were gold and silver. They did not master the steel.

1492 marks the beginning of a big tragedy for the indian cultures. The spanish "conquistadores" intended to destroy the indian culture in order to introduce the europan culture. This goal however was never achieved and a new culture, non european, non indian, emerged from the ashes of the great Tenochtitlan. The colony of New Spain was formed under the rule of Spain. Despite of this imposed economic role as producer of raw materials, the colony produced some important intelectual figures such as the great Sor Juana Inés de la Cruz or Juan Ruiz de Alarcón, who contributed to the golden century of the spanish literature. There were also scientists, such as Carlos Sigüenza y Góngora, who represented a true renaissance man: mathematician, astronomer, poet and philosopher.

However the development of materials technologies will not start again until late 18th century. During that century, the mining in New Spain was one of the main sources of income for the spanish crown. It was very clear for the spanish rulers that there was a great need for conducting research and development in metals refining methods. This research has to be conducted near the production place. In order to obtain more income or, in modern language, to "add value to the raw material", the crown created the biggest scientific proyect of colonial Mexico: "The Royal Seminary of Mines". This institution was founded in 1792 and was intended to perform research in all areas of metals and its alloys, and to train the mining engineers and materials

Advanced Topics in Materials Science and Engineering, Edited by
J.L. Morán-López and J.M. Sanchez, Plenum Press, New York, 1993

chemists that the mining industry required. In fact, the Royal Seminary of Mines was strongly oriented to materials science and was the oldest institution of this type in the Spanish america. The first director was Fausto Alohar who had discovered the element Wolframio back in Spain. With the Royal Seminary of Mines started the golden age of Mexican Colonial Science during which many new technologies were developed. The most important example was the discovery of a new element by Andrés del Río, which he called Eritronium. Unfortunately, the Baron Alexander Von Humboldt, which acted as referee of the discovery of Del Río, took the paper back to Germany and apprently did not pay any attention to it. Later on some swedish researchers rediscover it and name it Vanadium, in honor of the Viking goddess Vanadis. Although later on Von Humboldt made some recognition to Del Río, the credit of the discovery went to Sweden. This was a very bad thing for the development of science in Mexico. In those years the discovery of a new element was a big materials achievement. This history shows, however, the high level of science in Colonial Mexico.

In 1810 begins a struggle for independence that ends with the birth of a new country, Mexico, which, for many historical and political reasons, struggles for almost one and a half century to find peace and conditions for developing science. The Royal Seminary of Mines, which was the origen of all the mexican scientific institutions, fades with the independence.

The monumental building of the Royal Seminary of Mines, The Palacio de Minería (Mining palace), was given to the Pontifical University of Mexico, which later became the National University of Mexico (UNAM), and was the house of the Engineering School. In 1938, in a room on the top floor of the Palacio de Minería, was founded the Institute of Physics, which was a key institution for the development of modern materials science.

In the early seventies, materials science stars to develop in Mexico. In the UNAM it started with the Solid State Physics group of the Institute of Physics. That group, headed by Alonso Fernández, made a seminal contribution to the science in Mexico. The members of this group developed work on crystal growth, electron microscopy, dislocation studies, EPR, electrical properties, metals and instrument development. From that group emerged many of the leaders of the development of materials science in Mexico.

In 1970 the Center for Materials Research of UNAM (CIM) is born and his first director, J.A. Nieto, starts to recruit scientists. Most of the founders of the CIM were physicists interested in materials and that gave its character to this center that also became an important institution in Mexico.

Inside of UNAM, there were also developments in the School of Chemistry basically in the area of metallurgy. Another institution which strongly developed materials science was the Polytechnic Institute of Mexico (IPN). As a part of the School of Physics a lab for materials research, which had state of the art equipment such as SEM, microanalyzer, and mechanicals testing, was founded in the IPN in 1967. The big contribution from the IPN came from the solid state group headed by F. Sánchez, who latter on moved to CINVESTAV and produced a good number of scientists working on materials science in Mexico. Nowdays, materials science is strong and developing at full speed. A list of institutions working in Materials Science is shown in Table I.

Materials science in Mexico has reached an interesting level of development. Mexico has roughly 6000 researchers members of the Sistema Nacional de Investigadores. This represents the most active part of the scientific mexican community. According to the 1991 directory of the SNI, there are about 200 investigators conducting research at the Ph.D level which can be clasified as materials research. That is, only 4%

Table I

List of Institutions That Participate in Research
on Materials Science in Mexico

UNAM (National University of Mexico)
Faculty of Engineering
Institute of Materials
Faculty of Chemistry
Institute of Physics
Faculty of Sciences
Lab. Cuernavaca
Lab. Ensenada
Lab. Temixco
CINVESTAV (Polytechnical Institute Graduate School)
Department of Electrical Engineering
Department of Physics
Campus Merida
Campus Saltillo
UAM (Metropolitan University)
Campus Azcapotzalco
Campus Iztapalapa
UASLP (University of San Luis Potosí)
Institute of Physics
UAP (University of Puebla)
Institute of Sciences
CICESE (Ensenada Research Center)
Lab. of Applied Physics
UNISON (University of Sonora)
Institute of Physics
IIE (Electrical Research Institute)
Cuernavaca
IMP (Mexican Petroleum Institute)
Division of Basic Sciences
ITESM (Monterrey Institute of Technology)
Campus Mexico State
Campus Monterrey
UMSNH (University of Michoacan)
Metallurgical Research Institute
UANL (University of Nuevo Leon)
Engineering Department
TECHNOLOGICAL OF SALTILLO
Metallurgy Department
IPN (National Polytechnical Institute)
School of Chemical Engineering
School of Physics

Table II

Areas of Strength in Material Research in México

Characterization of materials, TEM, AUGER, SIMS, X-Ray, etc.
Superconductors (High Tc)
Nanostructured materials
Electronic properties
Magnetic properties
Metals and alloys
Opto-Electronics
Molecular sieves
Semiconductors
Surface physics
Crystal growth
Corrosion
Polymers
Ceramics

of the Mexican science community is in materials. However, most of this group are physicists. A list of the areas in which Mexico has some research strength is shown in Table II.

This also establishes one of the problems of mexican material sciences: the lack of chemists, engineers and related disciplines working on them. The first program in materials science outside Mexico City was established in Ensenada in 1985 by an agreement UNAM-CICESE. The first Ph.D on materials science outside Mexico City was graduated in Mexico in 1988. Later on, programs were established in the school of science and chemistry of UNAM.

We need to strenghten our current programs in materials science. This area has been strongly hit by the so called "Brain Drain". We have now identified at least 65 mexican materials scientists working in the USA: *i.e.*, about 1/4 of the mexican community. That means that we have to replace the lost people and generate new Ph.D's. We need to form new human resources for the industrial development. A recent study performed by CONACYT shows that, if the present rate of growth of the mexican economy continues by the year 2000, the main cost factor in production will be engineers and applied scientists. The engineering labor will be so costly and scarce that it will seriously limit economic growth. If this problem is not addressed, Mexico will have serious problems in the future.

An important point that should be mentioned is that, in November 1990, the "Academia Mexicana de Ciencia de Materials" was officially formed. This society has a membership of $\sim$ 300 members including graduate students. The Mexican MRS is now part of the international union of materials research societies, IUMRS, and it is the only mexican adherent body. We had our first national congress on April 1991 in Mexico City, and the second one took place in May 1992 in Guadalajara, Mexico. We are also co-organizers of the First International Meeting on Nanostructured Materials in Cancun in September 1992. This society in also now an affiliated society to Acta Metallurgica Inc., and it is part of the International Federation of Societies of Electron Microscopy.

We have now some suggestions that can improve the US-Mexico-Canada cooperation programs in materials science:

- Mexicans Scientists working in Mexico should establish close ties with the Mexicans Scientist in US and Canada. That might help to reinforce the mexican graduate programs.
- We should look into binational research founds.
- In view of the free trade agreement, we should look for new ways to cooperate in science and in R&D.

The Role of U.S. Federal Laboratories in Research and Development of Materials

Lyle H. Schwartz

Materials Science and Engineering Laboratory
National Institute of Standards and Technology
Gaithersburg, Maryland 20899
U.S.A.

It is unquestioned that technology is the cornerstone of international competitiveness in manufacturing, for its advancement provides the basis for new industries and the maintenance or resurgence of existing ones. Equally well recognized but sometimes less well publicized, is the role of materials technologies and their technical infrastructure, materials science and engineering (MSE), in influencing the competitive posture of the whole of industrial technology. Materials technologies are enabling, they are essential for the health of every manufacturing industry and must flourish if the technological base is to do so. The competitive demands confronting modern society cannot be satisfied without skillful exploitation of this field to yield quality manufactured products that do totally new things or old things in much better ways. Most technically advanced nations have recognized the importance of MSE and have targeted materials technologies along with biotechnology and computer and information technology as major growth areas in the next decades. In this highly competitive arena the once premier position of U.S. materials technology has eroded over the past decade or so. Compared with our major trading partners, we are only at parity in many materials areas and are being overtaken or surpassed in many others.

This erosion of leadership occurred in the U.S. even though national leaders have long been aware of the criticality of materials technologies to national economic and strategic security. Many studies conducted over the years reflect this awareness and among these, the most comprehensive in recent decades are two landmark National Research Council (NRC) assessments, "Materials and Man's Needs" (1975) and "Materials Science and Engineering for the 1990's: Maintaining Competitiveness in the Age of Materials" (1989). The earlier study considered national MSE issues in a climate in which the U.S. science and industrial technology base still largely dominated and was thought unchallengeable. In the 1980's the tide had turned and the latest NRC assessment addressed the issue of the viability of U.S. technology itself, in the context of today's world of technological equals. To this end the study listed 19 separate recommendations, one of which, repeated below, presents in a few words the foremost strategic approach used by most nations to foster international competitiveness, namely cooperative research and development:

> "Universities, industry, and government laboratories should develop joint programs in areas of national importance. Government laboratories should play a central role in this effort."

Cooperative research, of course, is nothing new, but recent times have seen a buildup in the U.S. and abroad of a plethora of these technical linkages, due in part to the complexity of modern manufacturing and the interdependence amongst industries. In this environment R&D capital costs are substantial and return on research investments often do not justify separate technology development efforts by individual companies. Cooperative research at the pre-competitive level is viewed as one way to mitigate these problems as it entails the joining of resources, technical and financial, to pursue areas of collective interest in furtherance of specific individual needs. In effect, a form of vertical integration of requisite technical capabilities is achieved along with a significant leverage factor on individual investment.

However, our major competitor nations have generally been more active on this front and perhaps more creative than the U.S. in developing a full slate of cooperative efforts between government and industry. They have installed cooperative research into the basic framework of their economic strategies with Japan and West Germany being outstanding examples. Several general characteristics typify the cooperative venue of these nations. These include direct industrial participation along with a government-industry consensus building process, overt government orchestration in tandem with various forms of funding enticements to industry, targeted technologies like materials, and wide ranging research thrusts with pointed activities focused on applied problems and the engineering end of the innovation to commercialization sequence, *i.e.*, manufacturing.

Another, and perhaps the key tactical approach common to Germany and Japan, is that they have placed many of their government laboratories and national research centers in a pivotal position in the technology development process. The designated role of the selected national laboratories is to fill the gap between innovation and product development through organized pre-competitive cooperative programs with industry. Assisted by industrial advisory boards, technological efforts undertaken by the laboratories are industry driven and have a commercial orientation generally with a finite lifetime. Mid-to-long term funding guarantees, often on a 50% government – 50% industry basis, assure program stability and industry involvement at a reduced cost for individual participating companies as compared to similar go-it-alone efforts.

In West Germany, the Fraunhofer Gesellschaft, currently comprised of some 34 separate institutes, has been established as the government's central applied R&D organization with the primary charter to develop technologies directly with, and for industry. Collectively, the Fraunhofer Institutes encompass nine generic technical thrust areas, three of which cover materials and components, production technology and process engineering.

In Japan, the government laboratory infrastructure differs from Germany in organizational detail, but is directed similarly towards co-development with industry of commercially oriented technologies. Cooperative arrangements combining the technical resources of both government and industry generally are effected through government authorized industrial associations in specific technology areas such as advanced materials. Factored into this process in an overt way are the six major laboratories (two materials oriented) of Japan's Science and Technology Agency (STA) and the sixteen national research institutes (eight designated industrial) of the Ministry of International Trade and Industry (MITI). The STA laboratories cover the basic end of the research spectrum with the MITI labs handling the more applied problems. Both lab systems work interactively with each other and directly with their industrial partners.

In comparison, the resource base for materials research at U.S. government laboratories is also enormous. Overall, there are more than 700 small and large Federal laboratories supported by 14 Federal agencies. About 2000 people, divided about equally between defense and civilian type laboratories, are engaged in some facet of materials research. All of these laboratories are mission oriented with the Department of Commerce's National Institute of Standards and Technology (NIST) and its Materials Science and Engineering Laboratory being the only organization with the designated charter to support the industrial materials sector. Other Federal laboratories, though charged with different missions, also have extensive research capabilities relevant to materials technologies. These are limited in number and include the National Laboratories funded by the Department of Energy, and various laboratories operated by the Department of Defense, National Aeronautics and Space Administration, and the Department of Interior's Bureau of Mines. However, it is clear that the U.S. currently has no direct counterparts to the MITI labs in Japan and the Fraunhofer Institutes in Germany in its national laboratory complex.

Even so, the role of U.S. government laboratories has evolved towards more effective support of industry. Cognizance about the need to better utilize the national lab system began in the late 1979–early 1980 period. The Packard Commission in its 1983 report emphasized closer interactions between the laboratories and industry while the 1985 report of the Young Commission on Industrial Competitiveness brought attention on technology transfer from the national laboratories. In concert, a series of legislative acts have been approved that set the current basis and operational framework for better ties between the private sector and the national laboratories. The Stevenson-Wydler Technology Innovation Act of 1980, the primary law facilitating access to the Federal lab system, mandated that Federal organizations and affiliated laboratories pursue technology transfer activities. Later legislation, the Federal Technology Transfer Act (FTTA) of 1986, amended the Stevenson-Wydler Act to create a legal and management framework and a decentralized approach for the commercialization of technology developed by government-owned, government-operated laboratories. The National Competitiveness Technology Transfer of 1989, part the National Defense Authorization Act for Fiscal Years 1990–91, extended coverage of the FTTA to include the government-owned, contractor-operated laboratories of DoE. Executive Order 12591 issued in 1987 by President Reagan gave implementation guidelines to the Federal laboratories and importantly directed that cooperative R&D efforts be promoted between these laboratories and the private sector.

The stated purpose of the FTTA was to foster technology transfer nationwide through the implementation of cooperative research and development agreements (CRADA's) between Federal laboratories and industrial companies. Its intent was to help American industry take advantage of the $ 75 billion ($ 1.5+ billion on materials) the Federal government spends annually on R&D. Federal participants include all major government laboratories and centers and their parent agencies.

The U.S. approach to utilization of its national laboratory resources has focussed on technology transfer. This tactical methodology has its drawbacks as it is based on the premise that government laboratories frequently produce something that industry wants already packaged in a way that industry can use. Thus it can be only as effective as the extent to which there is a one-on-one match between industrial needs and the research outputs of mission oriented government laboratories. Such a match does not usually occur in applied research without planning and coordination between the researcher and the intended user. What is required to achieve the desired match is a process of joint government-industry technology development activities where the government laboratory and appropriate industrial partners identify a scientific and applied program in advance, match complementing technical strengths and work the problem together from start to finish to achieve a new or improved technology. In these partnership Government activities focus on the pre-competitive, more generic basic and applied research to set the stage for product development and commercialization by industrial participants. The newly developed CRADA strategy provides the mechanism for such partnerships.

Programs having these characteristics were pioneered by the National Bureau of Standards with cooperative research which started more than 60 years before its name was changed to NIST and its responsibilities were broadened by the 1988 Omnibus Trade and Competitiveness Act. NIST works collaboratively with a single company, multiple companies, trade associations, universities and other government agencies. Cooperative ventures take several forms, from industrial consortia, to the popular Industrial Research Associate Program where companies pay their staff to work in NIST laboratories. Currently more than 200 industrial associates work each year side-by-side with NIST researchers. Projects last in duration from a few weeks to a few years, or more, until the job is accomplished and the newly developed technology is taken back to the home organization to be transferred into new ideas, new processes, and new products. NIST in turn benefits from the joint programs, learning first hand the needs and views of its industrial partners and thus is better able to adjust its research directions for continuing service to industry.

NIST's new role in advanced materials technologies is carried out through joint technical collaborations with industry that begin as early as possible in the R&D process. That is, industry and NIST researchers work together from the inception of the ideas up to the stage at which a technology is ripe for commercial applications. The research goal is generally not to create a new material or process, but to enable industry to do so more effectively. The following three programs are current examples of such cooperative ventures:

1. NIST and the Automotive Composites Consortium (ACC) are working on a co-operative program to improve the processing of structural polymer composite materials. The ACC partnership of Chrysler Corporation, Ford Motor Company, and General Motors Corporation is aimed at developing the technology industry needs for processing reliable, cost-effective structural polymer composite materials. The ACC was established in 1985 by the three auto producers to jointly conduct pre-competitive research. The companies selected structural composites as the topic for their first cooperative effort.

   The cooperative project focuses on composites made by resin transfer molding (RTM) and structural reaction injection molding (SRIM). Leading industry representatives at two NIST workshops held in recent years indicated RTM and SRIM molding as among the most important processing methods for the next 5 to 15 years. Consequently, new NIST programs on polymer composite processing were focussed on these methodologies. These new programs have allowed NIST to develop the expertise that enabled the establishment of a CRADA with the ACC. In cooperation with the ACC, NIST researchers will develop and use computer

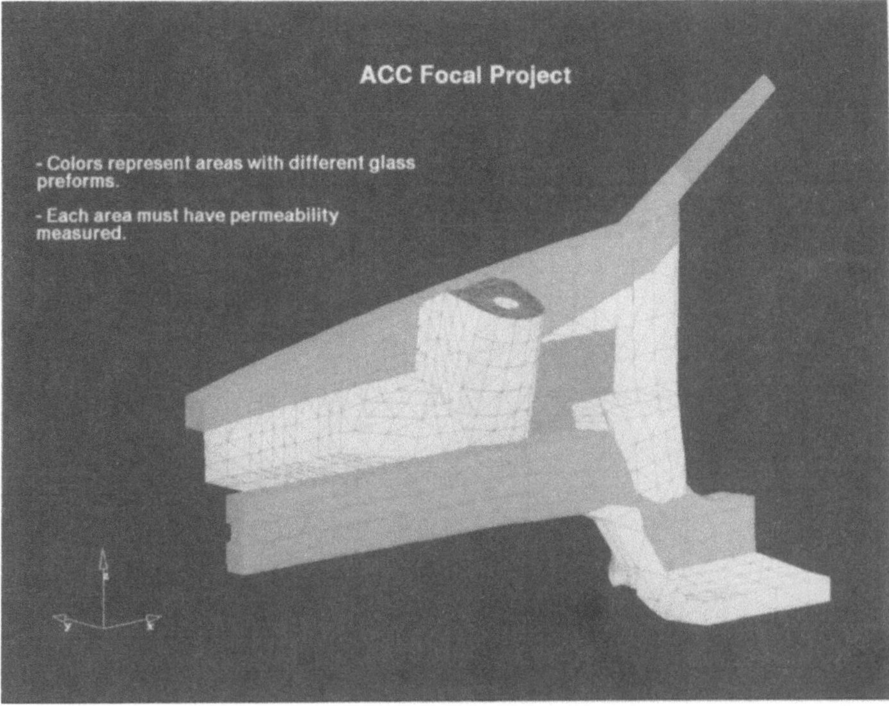

Figure 1. Simulation description. The finite element model. The mesh consists of 3310 triangular elements, and 1624 nodal points. The model is shaded to indicate elements (and hence, regions) of different thickness. In terms of permeability, the different shaded regions consist of unidirectional braid, random mat on top of unidirectional braid, and random mat.

models (Figs. 1 and 2) for both processing methods to simulate the fabrication of a complex demonstration part made by the ACC using the latest processing and performance technologies. The demonstration part is the front end structural member of a discontinued model of the Ford Escort to be constructed primarily of composite materials.

Under the program, the partnership agreed to develop technologies in three areas: processing, materials properties, and crash energy management. Advances in these areas will be demonstrated in the fabrication of the sample part using RTM and SRIM processing.

The RTM and SRIM processes involve injecting resins into a mold that has been filled with fiber reinforcements. By combining plastic resins with glass fibers, very versatile, lightweight, and high-strength materials can be made. The challenge is to achieve the optimum flow of resin so that the mold is completely filled in minimum time and with minimum pressure while avoiding bubbles and voids. The NIST computer models are designed to predict flow patterns and pressures during RTM and SRIM processing. Information from the simulations will be used by the ACC to optimize processing conditions and to improve mold design.

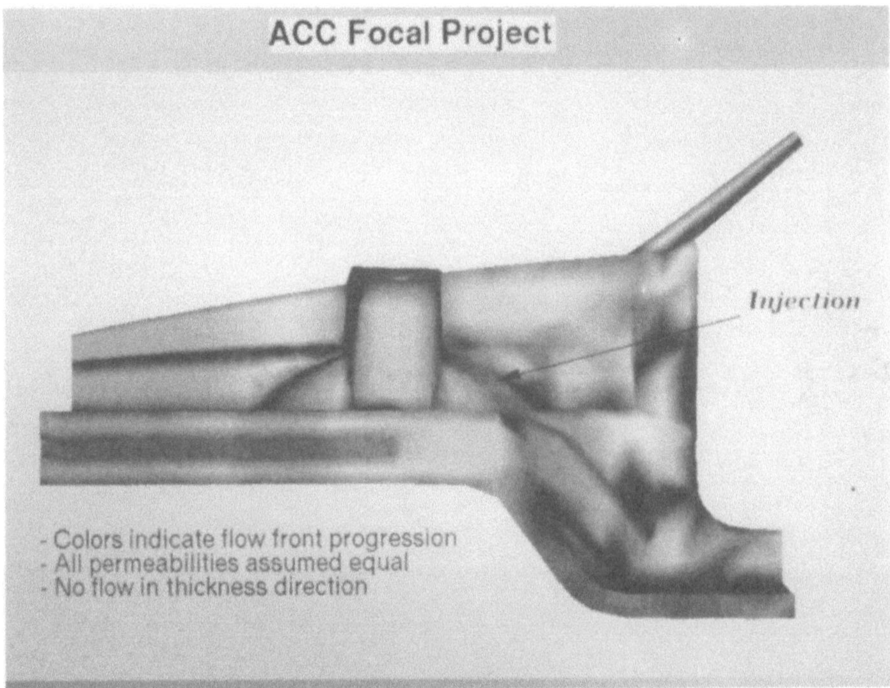

Figure 2. Results for flow front *vs.* time from a test simulation using the 2.5D Darcy's law simulation. The shading indicates the flow front progression. For this test case, the preform was assumed to be isotropic with a permeability of 10^{-5} cm^2 and a viscosity of 1 Poise. The picture was generated using the package MovieStar.BYU.

This collaborative effort with the ACC offers NIST researchers an opportunity to compare model predictions with data from the fabrication of a real, complex part. The information also will be used to develop generic tools and procedures needed to establish RTM and SRIM processing as efficient routes for producing a variety of polymer composite parts. We anticipate that this generic information will be broadly applicable to industries other than automotive and are exploring further industrial interaction in the aerospace arena in particular.

2. In the late 1970's and early 1980's major advances in several technologies resulted in their convergence to form a new systems approach for materials processing. This approach known as intelligent processing of materials (IPM) can be used to control the evolution of microstructure in real-time in contrast to conventional materials processing where process variables such as temperature and pressure are automatically controlled to preselected values, but the microstructure may not be directly controlled. These advancing technologies include (1) nondestructive evaluation (NDE) sensors for determination of materials properties and microstructure; (2) physical and mathematical process models based on the maturing materials science concepts of relating process variables to microstructure, and microstructure to performance; and (3) fast, inexpensive computer hardware and appropriate software, *e.g.*, expert systems and artificial intelligence.

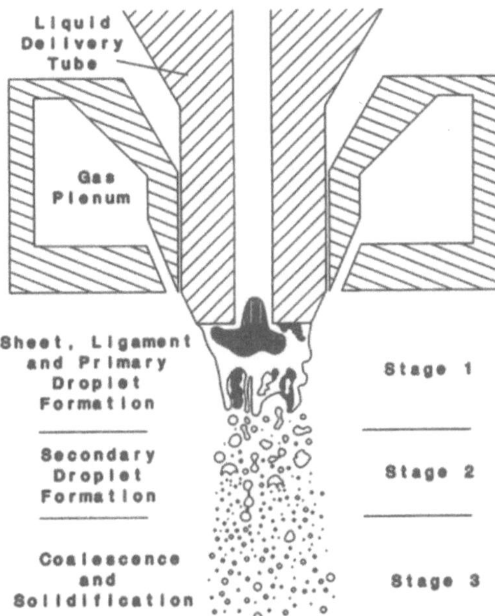

Figure 3. Schematic representation of the break-up of a molten metal stream by gas during rapid solidification.

Rapidly solidified metal powders are very important to systems requiring advanced materials, since they provide the basis for alloys which exhibit, after consolidation, high strength-to-weight ratios, improved fracture to compare model predictions with data from the fabrication of a real, complex part. The information also will be used to develop generic tools and procedures needed to establish RTM and SRIM processing as efficient routes for producing a variety of polymer composite parts. We anticipate that this generic information will be broadly applicable to industries other than automotive and are exploring further industrial interaction in the aerospace arena in particular.

When metal powders are produced by atomization of liquid metals (Fig. 3), the microstructural characteristics of the solidified particles are directly related to the size of the particles. The unique benefits of rapid solidification may be attained only if the particles are solidified under extremely rapid cooling rates in the range of 103 to 106 K/s. Since cooling is usually proportional to particle size, it is necessary that the droplets be within a certain size range to obtain the desired refined microstructure. Currently, atomization is carried out without any provision for real-time sensing of particle sizes and size distributions or controls based on these quantities. Therefore, the optimum values for the processing parameters such as melt flow rate, temperature, atomizing gas pressure, nozzle geometry and, perhaps, ultrasonic fields are selected by trial-and-error using several batches of material. It is then necessary to manually sieve the resultant powder to obtain the required particles size distribution. This is undesirable because as much as 50 percent of the powder may be of the wrong size and must be recycled. In addition, this is undesirable because sieving and recycling introduce impurities into the metal powder that may degrade the final product.

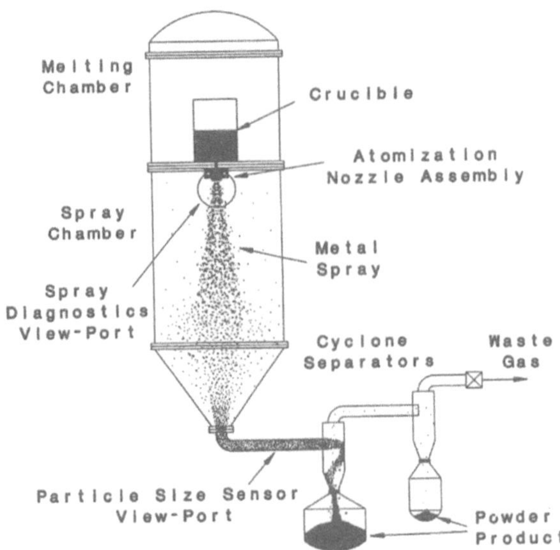

Figure 4. Schematic representation of NIST's supersonic inert gas metal atomizer.

A research program on the intelligent processing of rapidly solidified metal powders was begun in 1987. The program, developed through workshops held at NIST, and involving a number of producer and user companies, focussed on the automated, real-time, control of the atomization and solidification of liquid metals to form powders. The industrial consortium working with NIST on the program included Crucible Materials Corporation, General Electric Company, and Hoeganoes. Subsequent to the successful conclusion of this program in 1990, a second consortium was formed which includes the Department of Energy (DoE), G.E., Crucible Materials, and Pratt and Whitney, the goal of which is to develop an integrated, rapidly-solidified-powder system. It is significant that because of the generic nature of the research program, normally competitive companies such as G.E. and Pratt and Whitney are able to work together to develop new technology.

The research at NIST is aimed at developing an integrated system of particle size sensors, a process model and fundamental understanding of the fluid dynamics of the liquid jet break-up processes that lead to droplet formation, and on-line computer decision making using expert systems and artificial intelligence (Fig. 4).

A process model and a clearer understanding of the atomization process are being developed that relate process variables such as atomization dye gas pressure, atomization chamber gas pressure, metal viscosity, and die geometry to the liquid metal stream disruption process and metal droplet size. The modeling effort is based on a broad range of laboratory studies utilizing high speed photography, Schlieren interferometry, laser holography, and other techniques. This experimental effort is being coupled with computational modeling of the fluid dynamics and is leading to a strategy for control of the process. In order to produce the desired understanding of these flow processes, the researchers study gas-only flow, gas-water flow, and gas-molten metal flow.

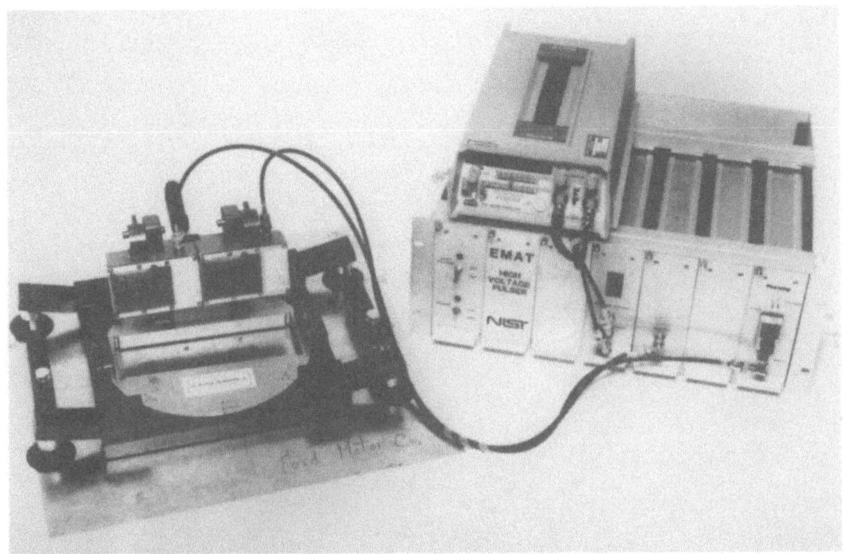

Figure 5. EMAT system developed in collaboration with the Ford Motor Company.

The computer control system, based on artificial intelligence and collective learning, is currently being developed. The computer control system can also provide simulations of the atomization process. The operator, via the user interface, can plan and execute atomization runs, monitor process conditions during runs, and analyze process data subsequent to the runs. The intelligent control system is responsible for reconciling user requests with current process conditions in order to meet production objectives.

3. Since the early 1980's, NIST has been collaborating with the steel industry to develop process control sensors to improve productivity and product quality. In 1983, NIST and the American Iron and Steel Institute (AISI) established a joint research program which led to the successful development of two of these sensors, one for porosity detection and the other for measurement of temperature distribution. The research, conducted mainly in NIST laboratories, involved close day-to-day collaboration between NIST scientists and staff from the steel industry assigned to NIST and supported by the AISI.

Looking to the future, NIST continues to interact with the metals industry through research planning activities and updates the technical content of the NIST program to reflect changes in research needs. Advanced manufacturing techniques increasingly demand materials of greater reliability and uniformity at competitive cost. Specific demands for properties and performance can be met by control of the processing of the materials from synthesis or raw material production to forming/finishing, *i.e.*, by use of intelligent processing.

This NIST experience in intelligent processing of advanced materials is now being applied to advanced processing in the steel and aluminum industries. In 1989, NIST was invited to participate in an AISI-organized forum to identify the long-range research opportunities for the steel industry. The industry participants concluded

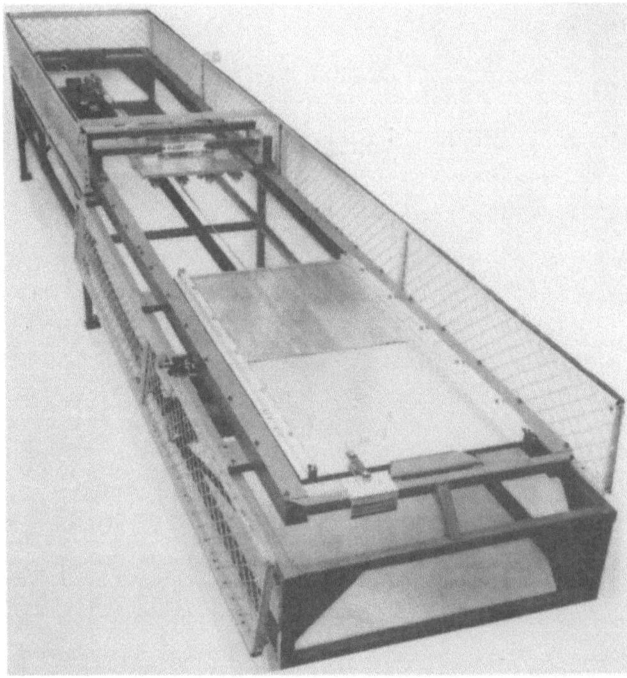

Figure 6. Ultrasonic formability sensor mounted on test bed.

that research opportunities could best be addressed in the context of three long-range development projects: direct production of liquid steel; near-net shape casting; and finishing and coating operations. Successful integration of these three contiguous unit operations would yield important benefits by responding to the increasing restraints on traditional batch processing, by developing new product properties to match customer needs, and by taking advantage of this era of computational plenty to control processes for increased efficiency and product quality. A decision was made to organize an industry-led workshop to highlight the recent advances in sensing, modeling, and process control, to identify areas of need in the primary metals industries and the role of intelligent processing of materials concepts in solving these needs, and to develop a strategy for implementation of research results.

Planning for the workshop was initiated through a steering committee chaired by the author with representatives from steel, aluminum, and copper trade associations, individual companies, the academic community, and Federal agencies. The workshop, sponsored by NIST, DoE, and the AISI, was held in August 1989. Industry, university, and government participants reviewed information provided by researchers and operating staff from industry and developed a plan for coupling the advancing state of materials processing in the primary metals industries. The report of the workshop summarized a research agenda for the intelligent processing of steel by defining the key elements producers need for the intelligent processing of steel, assessing the status of available technologies, and outlining strategies for implementing their use.

The steel industry has already acted on the results from the workshop. An AISI advisory committee has reviewed the findings and has recommended to the AISI that priorities be determined from the research agenda and a long-term national research program be established with an industry consensus. NIST has taken an active role in this planning process and, at the request of the AISI, has submitted proposals for inclusion in the national research program. Of these proposed projects, one is concerned with the intelligent control of steel sheet texture during processing. Sheet texture (or metal grain orientation) is vitally important in controlling the quality of the subsequent formability of sheet steel by automotive companies and other industries utilizing stamped sheet steel. A system has been developed at NIST to do this. The system (Fig. 5) consists of electromagnetic acoustic transducers (EMAT's) to generate and receive guided waves which propagate in the plane of the sheet, electronics to send and receive the ultrasonic signals and to measure wave speeds, and a fixture to position the transducers. NIST then collaborated with the Ford Motor Company to develop a test bed to demonstrate the feasibility of measuring the formability of rapidly moving steel sheets. The device propels sheet specimens along a track at controlled, repeatable velocities that can be increased up to 150m/min. The ultrasonic formability sensor will be mounted on the bridge (Fig. 6) of the test bed. The ability to make non-destructive measurements on a moving sheet for on-line quality is a significant improvement over the current practice of specimen removal, preparation, and testing.

Summary

The roles of Federal laboratories in technology transfer and in joining with industry in generic precompetitive research is still evolving. For NIST, with its unique role to provide appropriate support to industry, the experiences gained in this arena underscore the importance of two complementary factors in assuring NIST's successful interaction with industry. They are, first, the involvement of industry from the earliest planning stages of a program, through planning workshops and by having industry-sponsored research associates work hand-in-hand with NIST scientists and engineers; and second, long-range generic research laboratory programs to ensure cutting edge/capabilities at NIST must be maintained to support associated industrial research efforts.

Acknowledgements

The assistance of Mr. Sam Schneider in the preparation of this manuscript is gratefully acknowledged.

Background Reading

- *Materials and Man's Needs*, Supplementary Report of the Committee on the Survey of Materials Science and Engineering, National Academy of Sciences, Washington, D.C. (1975).
- *Advanced Materials by Design: New Structural Materials Technologies*, U.S. Congress, Office of Technology Assessment (Washington, D.C.: U.S. Government Printing Office, 1988).

– *Pay the Bill: Manufacturing & America's Trade Deficit*, U.S. Congress, Office of Technology Assessment, OTA-ITE-390 (Washington, D.C.: U.S. Government Printing Office, June 1988).

– *The Federal Technology Transfer Act of 1986: The First Two Years*, Report to the President and the Congress from the Secretary of Commerce, DOC (July 1989).

– *Materials Science and Engineering for the 1990's, Maintaining Competitiveness in the Age of Materials*, National Academy Press, Washington, D.C. (1989).

– *International Cooperation and Competition in Materials Science and Engineering*, National Institute of Standards and Technology, NISTIR 89-4041. (1989)

– *Picking Up the Pace: The Commercial Challenge to American Innovation*, Council on Competitiveness, Washington, D.C. (1989).

INDEX